4D Printing 1

Series Editor
Jean-Charles Pomerol

4D Printing 1

Between Disruptive Research and Industrial Applications

Frédéric Demoly
Jean-Claude André

ISTE

WILEY

First published 2022 in Great Britain and the United States by ISTE Ltd and John Wiley & Sons, Inc.

ISTE Ltd
27-37 St George's Road
London SW19 4EU
UK

www.iste.co.uk

John Wiley & Sons, Inc.
111 River Street
Hoboken, NJ 07030
USA

www.wiley.com

Library of Congress Control Number: 2022935187

British Library Cataloguing-in-Publication Data
A CIP record for this book is available from the British Library
ISBN 978-1-78630-731-6

Contents

Foreword

Jean-Claude André, who has already provided us with three very insightful books on additive manufacturing, is taking it up a notch! Indeed, with the help of Frédéric Demoly, he now offers us an "encyclopedic" book on 4D. Let us recall that Jean-Claude André is the first Frenchman to have filed patents and to have believed in what we now call "additive manufacturing", which was just called "3D printing" at the time.

The fourth dimension is that of time, and the question we face is that of producing, in additive manufacturing, objects capable of changing shape or function over time. In old and simplified terms, we could say that it is a question of using shape memory materials or materials that can be deformed under the effect of external, thermal, photochemical, mechanical or electromagnetic disturbances in 3D printing. Biomaterials are also included in 4D, whether they are biocompatible or living materials, the latter being referred to as bio-printing. This important theme is hardly if at all developed considering the 4D field seems to be so important in its future achievements.

This book covers several aspects. First, it can be read as a classic science book that explains the state of the art and tells us everything there is to know and everything that is being done in 4D printing. Then, it goes through all the problems that arise for the designers of artifacts in the field: mechanical problems, material problems, choice of applications and industrial problems. Finally, this book can be read like an adventure novel. Indeed, it follows the research adventure step by step; the difficulties, the obstacles and the facilitators (or even the impediments to straight thinking) that researchers encounter on their way. It encourages reflection on creativity in research and the role of breakthrough innovations. The historical overview of inventions sheds light on the (sometimes impenetrable!) paths and methods of research, which lead to a profound epistemological reflection on research and innovation in the context of successive technological revolutions over several centuries.

In particular, our authors shed light, in a very original way, on the notions of rupture or "disruption" to use a fashionable term. Are 3D printing and now 4D printing disruptions or the intelligent use of known techniques? Will 3D end up imposing itself in manufacturing, thus opening the way to frequent use of 4D? These are some of the questions raised by the industrialization of additive manufacturing and its multiple applications.

Moreover, the authors do not avoid any question concerning the impact and social acceptance of new technologies and their applications. All the industrial applications that can already be envisaged are described, while taking care not to act as if the story is over (in fact, it has only just begun) and as if, consequently, the creative process at work in 4D research will hold no more surprises for us.

Those who have read Jean-Claude André's books on 3D know of his enormous erudition, which he once again makes available to us by peppering the book with numerous quotations, often delightful, always very instructive and inspiring. All of this is accompanied by an extensive bibliography, very useful for students and researchers alike.

From this book, the reader will not only learn everything there is to know about 4D, the state of the art, materials, applications, developments and problems, but will also understand how research progresses between incremental development, creativity and disruption. It is a whole world, full of reflections on science and its epistemology in relation to creativity and innovation, which is revealed to readers by two great researchers, pioneers of additive manufacturing in all its dimensions.

Jean-Charles POMEROL
President of the AGORANOV incubator and
of the scientific council of ISTE Group
April 2022

Preface

[...] no lasting accord exists between those who seek, ponder, and dissect and pride themselves on being capable of thinking tomorrow other than they do today, and those who accept the Faith, or declare that they do, and oblige their fellow men to do the same, on pain of death. (Yourcenar 1968)

Culture [...] is made up of all the values, representations, ideas, knowledge and myths that enable us to think and act. In order to think and act, we must establish landmarks, build a system of reference, in short, create a small universe that allows us to situate ourselves in time and space. (Lesgards 1994)

There is another character of permanence in time than that of character. It is that of the word held in the fidelity of the given word. I see in this behavior the emblematic figure of an identity polarized from that of character. The kept word speaks of a maintenance of oneself that does not allow itself to be inscribed, like character, in the dimension of something in general, but only in that of *Who*? Here too the use of words is a good guide. One thing is the perseverance of character; another is the perseverance of the fidelity of the given word. One thing is the continuation of character; another, constancy in friendship. (Ricoeur 1990)

What we usually call "friends" and "friendships" are merely relationships and associations formed by circumstance or interest, which keep our hearts together. But in the friendship of which I speak, they mingle and merge with each other in such perfect unity that they erase and no longer find the link that united them [...]. I feel that this cannot be expressed, except by replying: "Because it was him, because it was me". (Montaigne 1995)

Warning

3D printing, through digital means, permits the creation of objects with very complex shapes from passive materials (polymers, ceramics, glasses, metals, alloys, etc.) by adding material elements or materials to make the desired object, with the desired precision. The new idea (reportedly coming from the University of Bath in the UK in 2012 or MIT in 2013) is to add functionality to materials (referred to as active, stimulated, informed, smart) in the right place, which can thus have their form or functionality change over time under external stimulation. It took about 25 years for 3D additive manufacturing to develop an industrial market (currently about 30 billion €/year). Today, 4D printing represents a modest market of about €100 million/year, but with an annual rate of publications/market increasing by 40% per year. However, despite many promises, we are still far from Terminator 2 with global stimulation (and not always easily localized) and materials that are probably still too soft and response times that are too long, but it works and it is spectacular. What we are looking for today (with many other colleagues) is to break the locks presented in this book (perhaps to find others?).

But, even if they answer the subject of this four-handed book, there are (too) many problems to solve (at least for both authors). So, if you have already bought this book, this warning may be enough for you. By the way, Pierre Bayard (2007) will be able to help you have an exciting exchange about an unread book with someone who, sometimes, might not have read it either! If you have a little more time to spare, the introduction will already give you a better idea of the different locks to be broken, and there are many of them, because saying that we add a parameter, time, to the three dimensions of space, hides many criteria or parameters with which we can play to fail or succeed in a difficult process requiring a few divergent ideas followed by many interdisciplinary mutualizations, the difficulty of which is well known.

Furthermore, we have polluted this work with so-called "scholars" who reject the arbitrary division between the "cultivated ignoramuses" and the "learned uncultured" (Serres 1991), because, in agreement with Darbelay (2020), we do not want to keep the so-called hard sciences and the humanities in watertight silos (besides, perhaps it is easier to practice interdisciplinarity between truly disjointed disciplines?). We have therefore introduced a few quotations into the text, quotations that are often presented as elements of our reflections and which, as Darbelay writes, "blithely cross the borders that are still often posed today between literary and scientific cultures".

For our part, if we wanted to move forward on this spectacular subject of 4D printing with some indispensable epistemological insights, we had to start with an agreed, positive scientific literature, keeping solid scientific roots from "good" authors who are recent but already considered to be classics or references of the field (which is not yet a discipline). Having known since 1984 the history of the source technology, additive manufacturing, it was also possible to relate it to this emerging field (2013), following the progress of 4D science and technology, conducted in a relatively stable organizational framework (before the Covid-19 pandemic).

From an epistemological point of view, as proposed by Némo (2020), the need arises to estimate the importance of the accumulation of scientific knowledge from different horizons in a technological adventure where the stabilization of certain concepts remains to be done, before arriving at their convergence to produce technological devices of current use. How will reading several articles and drawing from them singular knowledge, generally disciplinary and not integrated in a robust paradigmatic logic, help the reader in their quest for a coherent and possibly practical narrative? Némo (2020) writes: "if the philosopher's ambition is 'synoptic', he must be able to extend his gaze over the whole of reality, as humanity knows it". Knowledge is dispersed, has multiple disciplinary origins and is progressing rapidly. There is therefore a need to have current knowledge, as unfiltered as possible, to "construct the notions and abstractions relevant to the present time". But, as we have understood, this is the operation for a conceptual object still under construction. It aims to reduce the fragmentation of a specialized world of digital fabrication by creating, as far as possible, a bit of common sense that can be used for collective action. However, as long as certain preconceived ideas and habits that have emerged since 2013 persist, it will be difficult for 4D printing to be able to question, clarify and better take into account the socio-economic determinants associated with its objects, its practices, its potentialities as well as its present and future purposes. For the authors, this book is not a refusal to grant self-assigned monopolies to certain disciplines and ways of thinking, quite the opposite so long as the game is worth playing. They just want to exploit all the richness possible for a field they believe in.

Finally, with all necessary humility, because we are only at the beginning of a possible great scientific and technological adventure, we will try to show you based on the existing research which avenues seem to us playable, those that are possible, but not explored, or those that probably have only a modest future. By posing as censors, we know that we are taking the risk of making mistakes, of not satisfying many of our dear colleagues who take their full measure in a more easily publishable conformism, etc. It is a question of engaging in a form of collective learning with all the disciplinary or interdisciplinary components, with new knowledge of understanding and action, capable of making sense of decisions that will lead to the future. We are taking this risk because we believe in a very positive future for this open technology which may not be generic like its parent discipline, additive manufacturing. But, with this exponential growth, you never know, there may be publications that break out of the conformisms elevated to the rank of ethics by disciplinary peers, publications that we might have missed in our bibliographic analyses (which, let us remember, double in number every 17 months). So, you will have the right to criticize this book.

So, if you think we are wrong, if you have great ideas, do not hesitate to contact us (frederic.demoly@utbm.fr and jean-claude.andre@univ-lorraine.fr). And, we promise you, if we have to protect your idea or make a publication, you will *at least* be associated with the validation work (proof of concept), but you can do it without us. To come back to Pierre Bayard, we endorse what he wrote (from his book, which has been read!): "What better gift can you give to a student than to make him aware of the art of invention, that is to say, of self-invention?"

A two-volume work

Humans, a long time ago, tried to use the materials they had at hand to make tools, weapons, etc. To do so, apart from the opportunities associated with these materials, they had to have information on their mechanical properties, their availability, their fatigue and of course on their shaping to achieve the desired functionality. This multi-millennial situation is still relevant today with progress in all of these areas.

In 1984, 3D printing or additive manufacturing emerged as a shaping technology that uses different types of materials (and several technologies) to produce complex objects in a deterministic way, using digital sciences. But it was around 2013 that it was proposed to use one or more materials that can be activated by energy stimuli that allow, at the same time, not only to give shape to this object but also to evolve its functionality: movement, change of surface properties, color, etc. In fact, to the exploration of the three parameters of space covered by additive manufacturing, we add a new degree of freedom, time.

Since then, the field has been developing very rapidly because of the obvious promises allowed by these freedoms of action. The analysis made in this book indicates a growth rate of 44% per year which is much higher than that of 3D printing alone which is only a little less than 20%!

For this reason, while we had committed ourselves to propose to the Editor only one volume at the beginning, in front of the avalanche of new results, we were led to propose a much more vast piece of work, taking into account (provisionally) the scientific situation of the moment. Thus, two volumes on the theme are proposed:

– Volume 1: Origins and application targets of 3D printing.

– Volume 2: 4D printing between science and technology.

The first volume deals with the foundations of this recent technology and, as it is a technology that must find practical applications, an analysis of this potential has been carried out in an attempt to identify possible prioritization possibilities for scientific and technological actions. The second volume, more classical, concerns the current state of the art with openings in the design of complex objects.

We hope you enjoy reading these two volumes,

Frédéric DEMOLY
Jean-Claude ANDRÉ
April 2022

The question of how a new idea can arise in the mind of a man [...] may be of great interest to empirical psychology, but it does not reveal a logical analysis of scientific knowledge. The latter is concerned not with questions of fact, but only with questions of justification or validity. (Popper 1978)

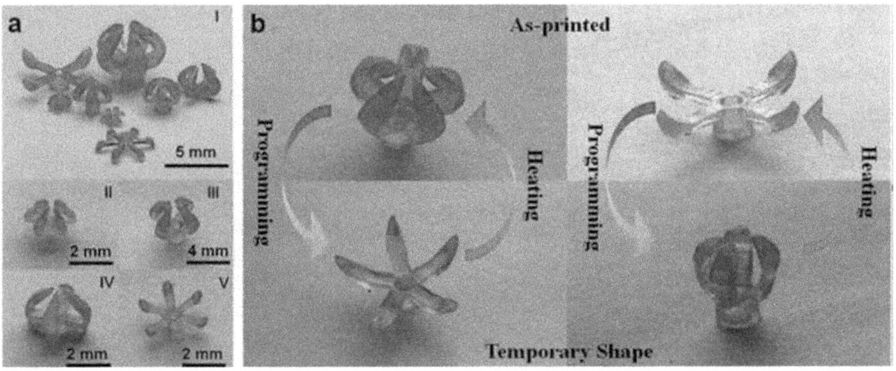

Evidence of a 4D effect (reversible closure by thermal effect according to Ge et al. 2016). For a color version of this figure, see www.iste.co.uk/demoly/4Dprinting1.zip

References

André, J.C. (2017a). *From Additive Manufacturing to 3D/4D Printing 1: From Concepts to Achievements*. ISTE Ltd, London, and John Wiley & Sons, New York.

André, J.C. (2017b). *From Additive Manufacturing to 3D/4D Printing 2: Current Techniques, Improvements and their Limitations*. ISTE Ltd, London, and John Wiley & Sons, New York.

André, J.C. (2017c). *From Additive Manufacturing to 3D/4D Printing 3: Breakthrough Innovations: Programmable Material, 4D Printing and Bio-printing*. ISTE Ltd, London, and John Wiley & Sons, New York.

Bayard, P. (2007). *Comment parler des livres que l'on n'a pas lus ?* Editions de Minuit, Paris.

Darbelay, F. (2020). L'érudition : faire du neuf avec du vieux ? Interdisciplinarité, polymathie, indiscipline. *Hermès – La Revue*, 87, 205–212.

Ge, Q., Sakhaei, A.H., Lee, H., Dunn, C.K., Fang, N.X., Dunn, M.L. (2016). Multimaterial 4D printing with tailorable shape memory polymers. *Scientific Reports*, 6, 31110.

Lesgards, R. (1994). Savoir et vision dialogués. In *Sciences et imaginaires*, Maréchal, I.A. (ed.). Albin Michel, Paris.

de Montaigne, M. (1995). *Essais*. PUF, Paris.

Némo, P. (2020). Philosophie et connaissances "érudites". *Hermès – La Revue*, 87, 68–75.

Popper, K. (1978). *La logique de la découverte scientifique*. Payot, Paris.

Ricoeur, P. (1990). *L'ordre philosophique*. Le Seuil, Paris.

Serres, M. (1991). *Le Tiers-Instruit*. François Bourin, Paris.

Yourcenar, M. (1968). *L'œuvre au noir*. NRF, Paris.

Preamble

4D Printing, Between the Why(s) and the How(s)

> All techniques form industries and trades. On the whole: techniques, industries and trades, form the technical system of a society. (Mauss 1947)

> The human brain does not like complexity […]. Innovation, like any change of habit is a new complexity to manage. (Leblanc 2017)

> The universal triumph of science will ensure the maximum happiness and morality for mankind. (Berthelot 1897)

P.1. Introduction

> If the human condition consists of the fact that Man is a conditioned being for whom everything, whether given or manufactured, immediately becomes a condition of his subsequent existence, Man "adapts" to an environment of [tools and then] machines from the moment he invents them. [The most refined tool remains at the service of the hand, which it can neither guide nor replace. The most primitive machine guides bodily labor and eventually replaces it altogether. (Arendt 1968)

Indeed, the line between the agreement and rejection of a technology depends on whether it is seen as neutral or value-laden, or simply symbolic. What narrative does it tell, because "technological development transforms the meaning of what is human" (Heidegger 2003)?

> To innovate the ship is to innovate the shipwreck; to invent the steam engine, the locomotive, is to invent the railway catastrophe [...] each period of technical evolution bringing, with its batch of instruments, of machines, the appearance of specific accidents, revealing the 'negative' rise of scientific thought. (Virilio 1996)

For the moment, 3D printing, which is still a modest phenomenon from an economic point of view, enjoys a good brand image. Contrary to the negative considerations linked, for example, to nanotechnologies (Frochot et al. 2020), there are practically no rejections from society for technologies such as bio-printing or 4D printing, and, in the gray literature, there are positive opinions with spectacular images.

"*Homo Sapiens*, *Homo Faber*, our species has two characteristics, both fundamental: wanting to understand the world and wanting to act on the world to transform it, normally for our benefit" (Lazar 2014). There is, to the authors' knowledge, no angel Gabriel, who suddenly appeared in the sky of science and technology in 1984, who could have imagined and promoted a mythical and disruptive history of consumerist humanity. 3D printing is in fact only the continuator of an intellectual tradition, that of technology inaugurated by Archimedes, Galileo and many other engineers over several thousand years; because everything "was on the shelf", it is the heir to a knot of problems that originated in mechanization, especially since autonomous transport, clock-making, weapons, etc., and the increasingly intimate marriage between science and technology due to a recent submission to the digital.

Before year zero of our era, oil presses, semicircular arches, concrete, windmills, cranes, and scaffolding of all kinds, siege engines, etc., had been invented. Fabry (2015) recalls that Ctesibios of Alexandria, in the third century BCE, invented a hydraulic system as well as the piston, the keyboard and the very first working automata. Similarly, the machine of Antikythera illustrated the ability to develop complex metallic gears. Heron of Alexandria left us the principle of a steam engine with several applications, etc. If in a world subject to irrepressible technical progress, disciplinary sciences can only coexist by sharing, by interfacing with other specialties to achieve cohesion or even some form of technical and/or social utility, Halleux (2009) reminds us that invention was already based, two millennia before our century, on two mechanisms: chance and/or serendipity (mines discovered as a result of the partial fusion of ores after a forest fire) (Van Andel and Bourcier 2009) and the imitation of Nature (mimesis) (e.g. the boat imitating the shape of the fish).

It was around the time of the Renaissance that machinery developed (industrial capitalism), even though slaves continued to be exploited with the increasingly fruitful mastery of energy and its transformation (and use) of time. There was thus a strengthening of reflection on the unexpected and the scientific rationalization of facts leading to innovation. This change was then authorized, according to Feyerabend (1988), by a provisional form of epistemological anarchism induced by disturbances of various origins that broke with the course of events. Then, through continuity, what had been accepted and integrated into the industries was incrementally improved. Thus, over several centuries, several ruptures have taken place.

For a long time, since the beginning of industrialization, we have been content with a modest degree of automatism, but without any particular margin of indeterminacy; this is what could be perceived in the concepts of Industry 1.0 up to 3.0, with each of these technological eras having their coherence and increasingly higher degrees in terms of automation. Simondon (2012) reminds us of the interest: "The machine, a work of the organization, of information, is like life and with life, what opposes disorder, the leveling of all things tending to deprive the universe of powers of change." Before the 19th century, the craftsman, trained "on the job" for many years, was the owner of the technique (see *Encyclopédie de Diderot et D'alembert*), then the tools were entrusted to the machines and in assembly line work, the operator lost their prominent role. In Industry 4.0 (André 2019), to put it simply, they would at best be the servant of the robot if they are not capable of being a creative participant in the production system.

By making the production system sensitive to external information, production – thanks to a concrete invention, real operative anticipation – becomes more agile, flexible, reliable, and probably more entropic than the classic predictable automaton, but is likely to progress by learning, which implies a much higher degree of technicality, approaching almost human behaviors via artificial intelligence (AI). So, now, a machine can have double the information: external via digital communication modes and AI on the one hand, internal from measurements and various sensors (IoT for *Internet of Things* or, in French, *Internet des Objets*) present on site. The enterprise is not only the sum of the machines that define its hardware, its backbone, it is now connected to the world ("software"). The production system, through its networking on a global scale (reticulation), is therefore characterized by its globality and above all its radicality, excluding, or almost excluding, anything that may be foreign to it (if only for financial reasons). In this context, it can become a robust and confident rival of the isolated Man, a rival that has every chance of being victorious.

> For the most part, creativity stems from the desire to disrupt the present, in order to create a future that is at once connected to that present and yet at odds with it. (Du Sautoy 2020)

David Gunning (2018) of the Defense Advanced Research Projects Agency (DARPA) in the USA explains the emergence of AI, one of the key components of Industry 4.0, from three waves: the first began in the 1950s and concerned symbolic reasoning (of a qualitative order), such as the generation of proofs of all theorems in the "*Principia Mathematica*"; the second concerns statistical learning (of a quantitative order), which involves the development of models to learn from real data (see Figure P.1); and the third is called the wave of explicable or contextual adaptation, in which an attempt is made to combine perception and sophisticated reasoning with abstraction, in order to get closer to what we understand about human intelligence. It is this decisive turning point that changes the game (e.g. Moyer (2016) and Cain (2017)). Classical publicly known examples in this area involve games (of Go, chess, etc.) (e.g. Brown and Sandholm (2018)).

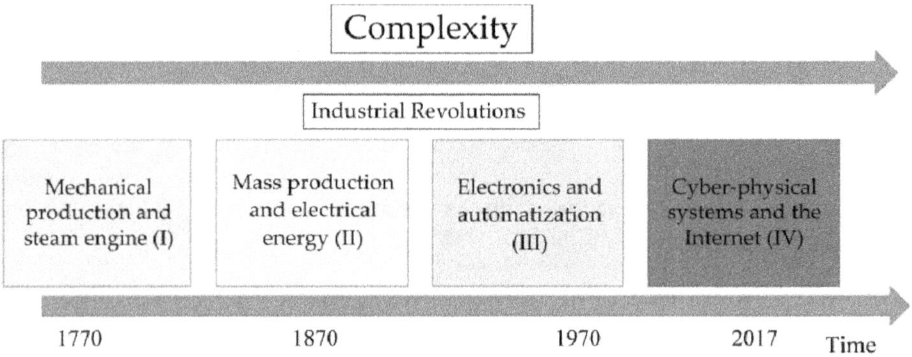

Figure P.1. *Technological change: from Industry 1.0 to 4.0. For a color version of this figure, see www.iste.co.uk/demoly/4Dprinting1.zip*

Cédric O, Secretary of State for Digital Transition and Electronic Communications in France, said:

> The policy led by the government has made French Tech the 1st digital ecosystem in the European Union. This dynamic contributes to strengthening French technological sovereignty, creates jobs throughout the country, and accelerates the environmental transition. To ensure the pivot towards Industry 4.0 and face the challenges of attractiveness, reindustrialization, and decarbonization of the economy, it is necessary for French Tech and French Manufacture to pool their strengths. This is why we welcome this initiative by Bpifrance, which will benefit the entire French economic fabric.

It is indeed a question of exploring an integrating concept for different applications, but above all with the desire to make better use of the national workforce in the manufacturing competition.

The concept of Industry 4.0 corresponds to a new way of organizing the means of production: the objective is the implementation of so-called "smart factories" capable of greater adaptability in production and more efficient allocation of resources, thus paving the way to a new industrial revolution (Valero 2016). After automation, this new revolution would be based on the smart factory, characterized by an interconnection of machines and systems within production sites, and also between them and the outside (customers, partners and other production sites). Industry 4.0 corresponds, in a way, to the digitization of the factory (see Table P.1 inspired by Schwab (2015), representing the temporal evolution of ruptures (Kondratieff cycle) of the industry, and Figure P.2 by Pau (2014), explaining the forms of activities). Through the use of the IoT, cyber-physical systems and additive manufacturing, that is virtual networks used to control physical objects and then produce them, the smart factory is characterized by continuous and instantaneous communication between the various tools and workstations integrated into the production and supply chains. The use of communicating sensors provides the production tool with a self-diagnostic capacity and thus allows its remote control, as well as its better integration into the overall productive system... (Hinks 2015; Schaer and André 2020a).

Typicity	**Period**	**Technology areas**
1.0	>1780	Steam engine, coal, early industrialization
2.0	>1870	Division of labor, electricity, automobiles, mass production
3.0	>1970	Electronics, automation, information and communication technologies, Internet
4.0	>2000?	Internet of Things, cyber-physical systems, etc.

Table P.1. *Kondratieff cycle*

> All learning consists of crossbreeding. Strange and original, already mixed with the genes of his father and mother, in thirds between them, every child only evolves by new crossings, as all pedagogy takes up the begetting and birth of a child: born left-handed, he learns to use the right hand, remains left-handed, is reborn right-handed, at the confluence of the two senses [...]. His mind resembles Harlequin's coat. This is as true for instruction as it is for raising the body.

> The half-breed, here, is called the third-breed. A scientist, rather, by nature, he enters into culture because science today embraces the questions, by itself unpredictable, of pain and evil. It is enough to learn two things: the exact reason and the unjust evils; freedom of invention, and therefore of thought, follows. Finally, this applies to conduct and wisdom, to education. It consists of and requires marrying the most foreign otherness, being reborn as a half-breed. Love the other who engenders in you a third person, the spirit. (Serres 1991)

As an example of the current dynamics, Leonhardt (2018) makes a number of statements such as: "Technology is exponential, humans are linear"; "Data is the new oil"; "Artificial intelligence is the new electricity"; "Internet of Things is the new nervous system". Listening to him in Brussels, in front of several hundred European principals, everything seemed possible, the audience was spellbound, there was nothing to do but to pick up the fruits of knowledge to embark on a largely digital future… easy to implement.

In the same vein, McNamara of IBM, quoted by Knapton (2017), is reported to have written:

> We may see nano-machines with artificial intelligence implanted in our bodies. These will provide significant medical benefits, such as the ability to repair damage in cells, muscles, and bones, and perhaps even augment them. Beyond the use of this technology that is already being explored today, we see the creation of a technology that can merge the biological with the technological, and thus be able to directly augment human cognitive ability, potentially offering considerable mental enhancement, while being able to use vast amounts of computing power to augment our own thought processes. Using this technology, incorporated into us and our surroundings, we will begin to be able to control our environment simply by thought and gesture […]

Major trends and their assimilation by the public, such as those gathered below, are to be taken into consideration (for a part, André, 2017b):

– internationalization/globalization of production processes (and dematerialization of the same);

– permanent public demand (spontaneous or stimulated) for the new, despite the depletion of reserves; health and well-being are plebiscited;

– "To scientifically organize humanity, such is the last word of modern science, such is its audacious, but legitimate claim" (Renan, 2014);

– increasingly rapid assimilation of new technologies by the public (see Figure P2, inspired by Kurzweil (2005) and Figure P3, extracted from André (2017b), on which the arrows indicate assimilation speeds);

– social demand increasingly demanding "zero" risk and NIMBY ("Not In My Backyard"); individualization;

– effects of technological progress on employment; etc.

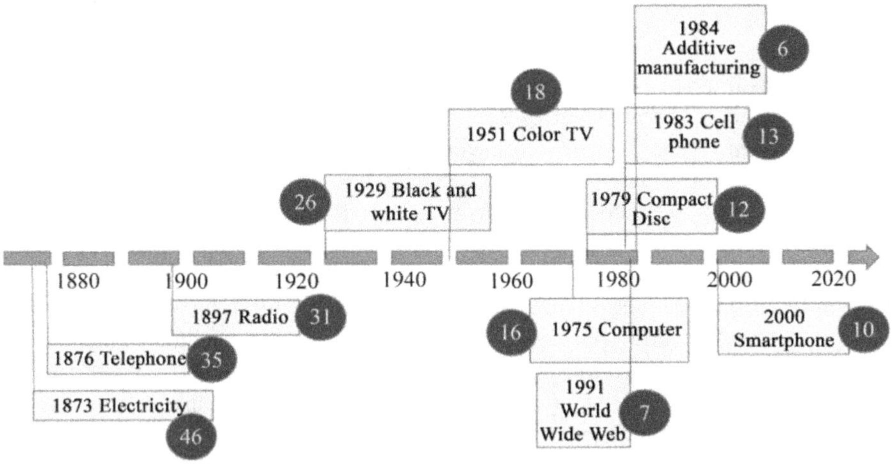

Figure P.2. *Reduction of technological eras (first date: origin; in yellow on blue background: time gap between origin and industrialization) – The figures indicated correspond to the acceptance time of products and processes*

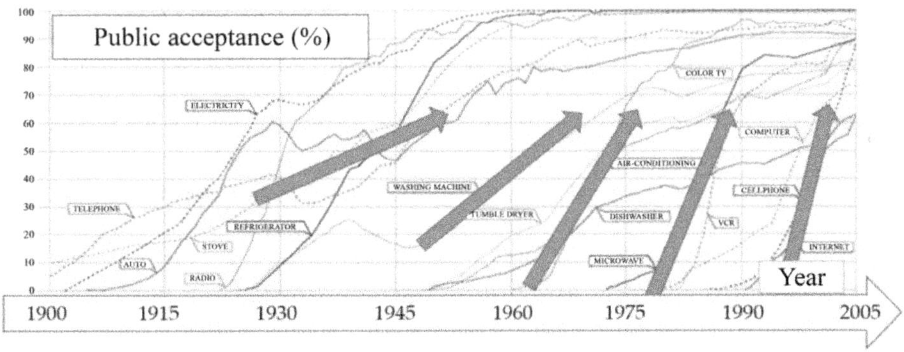

Figure P.3. *Increasingly rapid appropriation of technologies. For a color version of this figure, see www.iste.co.uk/demoly/4Dprinting1.zip*

There are relationships between techno-economic phenomena and socio-cultural manifestations because the gestural handling of a tool, the arrangement of an operating chain, etc., influence the social and individual behaviors of the members of an organized human group (Cresswell 2010). There are thus dialectical relationships between techniques and culture(s). This is a vast problem since it involves a reciprocal relationship between technical processes and the standard of living of the social body. What interests us here is the notion of a chain of operations that corresponds to the operations that transform a refined raw material into a technical product, in the knowledge that the introduction of new techniques into society brings out a set of interdependent factors (whose "rheological" manifestations must be taken into account). Historically, technical progress occurs in stages. "Beyond a certain level of sophistication, technical progress is qualitative rather than merely quantitative" (Cresswell 2010).

If a long time ago, industrial activity was centered on centralized energy, steam, the explosion engine and especially electricity have made it possible to include or distribute motors and actuators in any object of power (for example, a windmill, water mill, etc.), whatever its size, as a way of transforming, at will, energy flows that can be used in any space. In classical technologies, discrete elements were integrated into each other to form a system that operated a temporal change, and this is what 3D printing achieved in this previous world. So, technical progress was achieved because new elements must be "compatible" with other techniques. But Gille (1978) speaks of the "equilibrium" of a technical system or the "saturation" of the techniques that make it up. For Gille, the "correspondence" of a social system with a technical system must be understood through "ways of life", "social habits", the "nature of communities", "ways of thinking" (Lemonnier 2010). If for a given society, a technical ensemble forms a system, it is because it is necessary to take into consideration the complexity of the interactions between the elements it uses (hence the procedures, norms, etc.). It may be a question of the combination of operating chains, the gesture-tool pair, the mobilization of specific knowledge, etc.

The crisis caused by the Covid-19 pandemic showed that spatial mobility was important for citizens, but that other forms, other capacities of movement could be in line with favorable individual imaginaries. They are not exclusively, or even primarily, based on the implementation of socio-technical rationalities ranging from identified problems to the search for the "best" solutions to solve them, but in recent history that distances humans from their environment through their physical actions. Our relationship with time is marked by a kinetic which accelerated considerably with the appearance, then the generalization of the use of digital technologies, a little before the 1990s. It is hard to imagine what the remote control of a radio or television set has done to the lifestyles of citizens (at least in their minds), who have become increasingly passive because technology allows it. With stakeholders in isolation, thanks to seductive and effective storytelling, this trend toward ease continues, particularly with Human-Systems interfaces and informed, intelligent or active matter.

In this field, disruptive innovations remain rare events because they use specific methods with an anticipatory and divergent vision of the market, by going beyond the obvious. The intuition that wins, and is supported as such, is associated with calculated risk-taking. While the overall process relies heavily on intuition and creativity, there are more structured steps that allow for proactive research (trial and error methods, for example). The success of well-anticipated "tests", with premature products, quickly puts the competition "offside" by destabilizing the competitors (strengths transformed into weaknesses). "Not only must they apprehend the new rules but also – and this is more difficult – unlearn the old ones" (Sarazin 2019).

While it is possible to shed light on disruptive digital technologies (Foster and Graham 2017; Yeung 2018) in the organization of globalized production networks, the same is not quite true for the active matter and its somewhat enclosing, but 4D printing component. For Paché et al. (2018), "digital integration from different sources, therefore, allows real-time access to this massive data whose 'intelligent' analysis facilitates the continuous reconfiguration of industrial and logistical systems leading to the design, production, and provision of tailor-made products". The path in this respect is thus (almost) clear. The need for novelty has invaded the economic sphere and thus professional and personal activities. The advent of this way of living "now punctuates the rhythm of contemporary society, both at the level of work and of individual experience" (Aubert 2018).

The integration of 3D additive manufacturing technologies into the innovation ecosystem was recently studied by PricewaterhouseCoopers (2016); it is the subject of Table P.2 which indeed places 3D printing at the center of current and future devices as a recognized method of manufacturing (see André (2017a)).

> Our primary and secondary education system is far removed from the business world. Most teachers are not familiar with this world and their teaching does not value entrepreneurial success and innovation among students. This is an ideological problem specific to France. Under these conditions, how can we be surprised that there is no culture of innovation among young French people? (Fink 2018)

Period	Frame	Comments
>1990	Emergence of digitalization	Ubiquitous digital products: computers, digital phones Conceptualization and proof of concept of additive manufacturing; first companies Web/Internet platforms Process automation Digital industrial standardization Putting business models online
2000–2015	"Disruptive" digital technologies	Mobile phones Digitalized and automated industrial processes (including additive manufacturing) Digital functions User interfaces; shared access Cloud computing Sensors, Internet of Things, digital video
2016–2020	Industry 4.0 and Society 5.0	Horizontal and digital integrations along the value chain New business-to-business companies Real-time data analysis in relation to users Empowerment of computers in the process Digital manufacturing (including industrial scale additive manufacturing)
>2020	Digital ecosystem	Integrated and flexible networks along the value chain Digital manufacturing (including bio-printing) Virtualized processes Digitalized procurement Virtual user interfaces "Complete" dissemination of the Internet of Things Intelligent algorithms (decision support, adaptive processes, robotics, etc.)

Table P.2. *Positioning of additive manufacturing in industrial devices*

The term "Society 5.0" appeared in Japan in 2016 (France Diplomatie 2019) and complements the concept of Industry 4.0 (proposed about 10 years ago in Germany). Thus, innovation "is not defined by waves of innovation, but by the way, innovations have changed society", according to Salgues (2018). This 5.0 society is digital, it is based on several other pillars: decentralization of decisions, interoperability and automation, transparency of information, and its virtualization and robotic assistance. The major societal issues require progress in global governance and reasoned use of the potential of new technologies (Moulonguet 2017).

Box P.1. *Company 5.0*

Regardless of this aspect, does 3D printing ultimately doom giant factories and their production lines? The managers of Local Motors are convinced of this (Géneau and Caulier 2016). This car manufacturer is betting its development on micro-factories of less than 4,000 m^2, installed near consumers. In September 2014, Local Motors printed a prototype car named Strati in 44 hours for the body and chassis. A traditional car is made up of 20,000 different parts, but for all that, traditional manufacturing is holding its own. Both the competition and the production methods remain open. Nevertheless, over the next few decades, the pace and scope of technological developments (which are politically well accepted) are expected to accelerate further, transforming a range of human experiences and capabilities, while creating new tensions and disruptions within and between societies, industries and states. State and non-state rivals are likely to compete for leadership and dominance in this accepted field of science and technology (and their practical uses), leading to cascading risks and implications for economic, military and societal security.

From a historical point of view, what we see is that each time a technological mode supplants another, older one, it spreads like a natural water network in most of the possible niches, as shown in Figure P.4 (André 2006), to such an extent that it becomes increasingly difficult to find (or maintain) devices based on the older models (see examples of recordings, all of which are now digital).

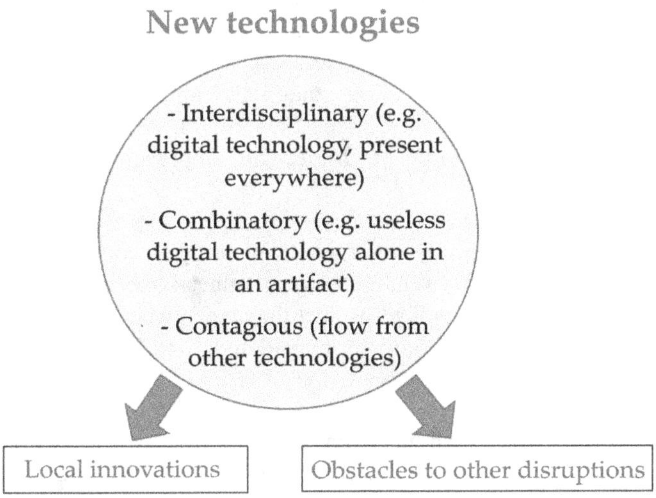

Figure P.4. *Wide-spread diffusion of "winning" technologies*

A principle of belonging to a community, associated with economic success or social recognition, allows, thanks to the reduction allowed by the paradigm (or the refounding concept), the scientific problems considered as acute (and by continuity, industrial artifacts?), to be solved efficiently:

> by concentrating attention on a limited sector of relatively esoteric problems, the paradigm forces scientists to study certain domains of Nature with a precision and depth that would otherwise be unimaginable. [...] The establishment of a paradigm provides the scientific community, among other things, with the means to choose problems that can be assumed to have a solution, as long as the paradigm is taken for granted. (Kuhn 2008)

P.2. Toward a more "total" integration of autonomy and matter

> Industry knows how to solve problems, but does it know how to pose them? The engineer himself is produced by the school in order to reproduce its idea of industrial needs. He is seen as a factor of production, not as an innovative and disruptive germ posing the real problems (relevance) in an unsparing way (impertinence). Continuing their allegiance, the Schools follow industry instead of preceding it, and contribute to reproduction rather than innovation. Rationalizing and creating are, in fact, complementary but opposite approaches: on the one hand, squaring, a laborious decomposition of elementary difficulties, referencing solutions that have already been tried and tested; on the other, the emergence of a form, an artistic composition, escaping the framework of logic, an approach of creativity. (Gaudin 1984)

If we follow Gaudin (1984), any rupture would not come from the (French) industrial domain. Indeed, this sentence from Brunet-Labbez (2013) is enough to haunt you: "The French reaction is to protect the acquis when it is already obsolete." We should, and this is what should be considered for 4D printing, rather follow Bachelard (1966) when he writes:

> There always comes an hour when it is no longer worthwhile to look for the new in the footsteps of the old when the scientific mind can only progress by creating new methods [...]. Scientific concepts themselves can lose their universality [...]. Concepts and methods, everything depends on the field of experience; all scientific thought must change in the face of new experience; a discourse on the

> scientific method will always be a discourse of circumstance, it will not describe a definitive constitution of the scientific mind.

The emergence of the 4D concept, which must be pushed to its limits, is an extraordinary opportunity to change the deal, not by seeking to add to what already exists (3D printing) as new performances (this is the objective of this book!) in an incremental way, but to take up the question at the bottom of the matter and intelligence side. We will therefore respect, at least in part, what Foray (1992) wrote: "Old technology imposes its own standards of economic evaluation on new technology, elaborated by reference to its natural qualities, thereby introducing a kind of bias at the time of the economic calculation exercise."

> Values must be related to the sphere in which they are valid, they must not be reified into dogmas, principles, truths, "objective" good or evil that would claim to be universally imposed. (Conche 2007)

Indeed, many writings speak of a need for reactivity, adaptability, to the detriment of competence and long-accumulated know-how, of the culture of the profession, values that are being devalued in a world where digital technology is constantly creating new needs, relational and/or for ourselves (communicating objects, IoT, ubiquitous computing; technologies of intelligent environments), making life easier and easier (for those who can access them). But, Gendron and Pierssens (2019) believe that with technological change, we need the tools to grasp the ruptures and their consequences in our lifestyles and social organization. They write:

> Because it sets itself apart from these realities from the outset, with which, therefore, it never enters into competition, fiction proposes another order of truth that has no claim to bring order to the social world. It is much more often a question of telling it differently, through a narrative thread that aims for the springs of emotion rather than an explanation, justification, and ordering with a view to action and social coordination. To the knowledge of the authors, there are only two main sources of active or intelligent matter, one of which is the science-fiction film "Terminator 2" and the other, more stabilized, is 4D printing, a new expression of temporal objects. Between total fantasy and verifiable reality, the space is therefore immense.

("The difference between fiction and reality is that fiction must be believable" (Twain 2018)).

Who has not dreamed of driving a car capable of transforming itself to fly like a plane! The "Back to the Future" trilogy did it through its mad scientist "Doc" and his doped DeLorean, a vehicle with the ability to reconfigure its wheels like an airplane's landing gear in order to fly. In the same way, the "Transformers" films have highlighted humanoid robots with the ability to transform themselves into technical systems (airplane, car, helicopter, toaster, etc.). It is important to note the ingenuity of the special effects teams that allow these intelligent robots to change from one configuration to another. In this way, we can note the concept car "Vision Next 100" and the idea of "alive geometry" of the German manufacturer BMW, which present a transformable body and a reconfigurable cockpit according to the driving mode. Beyond the extremely spectacular aspects, it is the development process of these reconfigurable or deployable structures and the mechanisms of transformation that should be highlighted, as they are part of a certain form of "opaque" art that is difficult to formalize. The concern for realization is, as a rule, hardly developed and detailed, but it is possible to glimpse visions approaching 4D printing and programmable matter. For example, in "Mission Impossible III", a portable hybrid (subtractive and additive) manufacturing machine is used to replicate the face of one person – based on a 3D photo – to be worn by another. The mask, thus created, presents different colors, different textures and both rigid and flexible regions to mimic the features of the face to be set in motion.

In an even more futuristic register, the programmable matter frees itself from the phases of manufacture and seems to present dispositions to change form/structure/texture/color via external events or intentions of change. Thus, in "Terminator 2" we can recall the T-1000 prototype made of a programmable polymimetic alloy (liquid metal) from "Skynet", capable of reconstituting itself indefinitely under the effect of bullets. Similarly, the movie "Transformers Age of Extinction" introduces the programmable material called "Transformium", which is composed of multiple voxels capable of changing properties and shapes and also of moving in space, allowing its owner to obtain any object or structure. These voxels – spatial, mobile and deformable elements – can also be imagined as micro-robots capable of self-organizing collectively to create an object or a structure. This is notably the case in the Walt Disney Studios animated film "Big Hero", where the young inventor presents a set of magnetic micro-robots that can interface collectively to form an imagined shape. The network of micro-robots is then in direct connection with the brain of the person who controls them.

Box P.2. *Other examples*

The notion of an "all-connected" world can be linked to a number and a date: since 2014, there have been 12 billion machines and communicating objects in the world, compared to the 6 billion humans connected to mobile networks and 2 billion to the Internet, according to Allard (2016). We can already understand why the so-called Big-Data era is beginning. But, if this "all-connected" world still presented itself in recent imagination under the appearance of a world connected via mobile phones or devices attached to a bracelet, some have thought of integrating the digital directly into a chip placed inside the body. The disappearance of computers has already begun, it will go beyond the old dream of technology as a service, to technology made both domestic in our lives and also totally invisible. In any case, digital discretization increasingly allows us to add, process, calculate and model, in short, to control (everything) in a very short timeframe and with a globalized connection.

With 4D, this relationship with information and communication technologies could be more passive and peaceful for people, with the interaction capabilities of a robot but placed directly in empowered matter, becoming functional in an adequate way. Indeed, with 4D printing, the information transfers between energy stimulation sources and 4D adaptive structures only rarely use electrons (for example, those that process the algorithms of our computers). The current time bases are for the moment superior to the second and the stimulations/interactions with the 4D objects must be proximal. Unlike digital technologies, interactions occur in a confined space. "We are in a treadmill society: we sit down, we let ourselves be carried along and we do nothing more" writes Zimmermann (2021). So it is not a question of moving toward an augmented human, but rather toward an augmented human environment. A vast program, but profoundly more ethical! Moreover, in a no doubt reductive way, we can compare the two systems, that of digital and that of 4D printing, by remembering this sentence by Stiegler (2004): "Power is always at its core a power to calculate, whereas knowledge is essentially a knowledge of the incompleteness of knowledge: it is always the knowledge of a non-knowledge." 4D printing fits into this class which requires us to go further to increase current knowledge.

> The characteristic of modern societies is that they can only stabilize themselves dynamically, which means that they constantly need growth, acceleration and condensation of innovation to maintain, depending on the case, their structure or the status quo. (Rosa 2018)

Thus, with 4D printing, it is not a question, as illustrated in Figure P.3, of necessarily going faster, but of engaging in a radical change in our daily modes of relationship between a fully informed material, providing the required services, and the working or leisure citizen. After proofs of concept that are out of the ordinary, real possibilities of technological breakthroughs, it is a question of playing on forms of participative design that envisage collaborative work with the actors who can contribute to the development of a new product/service, and thus break down barriers between users, designers and developers, and even raise awareness upstream with new ideas. In this framework, knowledge explores the possibilities of becoming 4D "temporal objects", of their consistency, leading in a way to rational anticipations. If it is a question of building bridges between users and designers (which will be claimed several times in this work), it is above all the research component that will be presented. With digital technology, it is indeed possible to integrate the personalization of tools or artifacts by going beyond the mere aspects of technical parameterization to engage in the consideration of the concept of appropriation in everyday practices (Dourish 2003; Lewkowicz 2017).

> Participatory design offers a set of techniques and tools to consider users in an iterative way, until the system meets the requirements defined at the beginning (Lewkowicz 2017).

Research, when it succeeds, produces an invention, which is the subject of this book. It should not be confused with innovation, because functional and organizational factors must be brought together in terms of marketing, design and creativity. Marketing has, according to Manceau and Morand (2014), evolved "to address markets driven by supply and innovation, rather than by unmet needs". In this recent context, innovation is interested in how to create emotions and experiences rather than just providing features and benefits. Therefore, demand can be driven by innovations, not the other way around.

The 4D disruption which, for the authors, should be brutal, would allow us to move from "quantum" uses or discrete systems with their stocks, instructions for use, historical procedures, and norms, to something innovative, which would be a return to the continuous, to the local, leaving aside an "ancestral" (but robust and proven) technical know-how to go toward much more culturally divergent concepts of a new mastery by Man of matter (at least in appearance). We would come closer to this sentence of Clarke (1962): "All advanced technology is indistinguishable from magic"...

Then, since the possible repair in this new field imposes the maintenance of different levels integrated in the process itself of creating an object, we would no longer replace elements: no more need for skills, standardized modes of communication, disassembly, storage, maintenance... Here is a future that could, independently of the other performances associated with 4D (including positive environmental performances), constitute a real disruptive leap in technological functioning: we no longer repair, but we throw away (unless we know how to make this informed matter hidden in the 4D object autonomous or reprogrammable)! It would therefore be the "intelligent" use of the object that would paradoxically create value and not the object itself, which would become more or less invisible! Moreover, 4D objects must be connected objects: "[They] do not pose any problems as such, what challenges us is the extent to which we have become, willingly or unwillingly, connected subjects, and therefore disconnected" (De Brabandère 2017). But this is undoubtedly another story to be written, perhaps for later, unless we know how to integrate it into the concept of Industry 4.0.

Technology is the answer, but what was the question (Price 1966)?

Technical and formal innovation has involved and combined, in one way or another, a theoretical reflection that is now mature in principle, with a creative practice that should link together the different energies involved in the 4D context. It could be interesting to try to reveal the various stages, methods and stakes which led the designers to push back the limits of the transfers of energy(s), via various types of actuators, to develop, as well as to innovate, renew and create new materials, tools or uses through the production of these machines of transfer and transformation of energy(s) in the world before. But it is more desirable to produce a situation which will be used to engage reflections on this setting in movement or on this change of functionality in integrations in 4D objects.

Indeed, this prior knowledge can reveal how traditional systems of action and their cultural practices can inform and strengthen research into 4D printing. Like the tendency to focus on spectacular cases, anticipating the possible exclusion of cultural knowledge and traditional practices for a pre-foundational dogma can impoverish 4D development. It is therefore a question here of re-examining the question of the links between external stabilized knowledge concerning what "moves" or "produces movement", or more simply "causes interaction", to consider possible ramifications. It is not a question of reconstructing the modes of thought, the socio-cultural and scientific markers, the impact on the social recognition of the designers and manufacturers of actuators, as well as the types of production and transmission that have led to the establishment of this approach within developed societies, but of examining whether this immense field of knowledge can serve as a support for the applicative development of 4D printing.

An invention begins with the association of at least two existing techniques (Kaplan 2012) to lead, in Simondon's sense (2012), to what he calls a "concretization": here, 3D printing (André 2017a,b,c) and programmable matter. Initially, it is classically a juxtaposition of knowledge, but in time, one can imagine a mutualization, a more complete integration (which is still only slightly envisaged). "Converging technologies (CTs) refer to the convergence towards a common goal of discoveries and techniques of fundamental science and technology: CTs are technologies and knowledge systems that complement each other in order to achieve a shared goal. Together or in isolation, [4D] technologies can contribute to this convergence" (Nordmann 2004).

> Christensen (2011) classifies innovation into three categories: disruptive, sustainable and efficient. In its convergence brief, National Academy Press (2021) presents disruptive innovations as coming from the complex and costly to arrive at readily available products. They would be likely to create jobs, in large numbers. On the other hand, others do not create jobs, and at best preserve them (replacement of old products by better performing

components). Others streamline processes, reduce costs and cut jobs. Innovation can be stimulated by the convergence of scientific disciplines through greater inclusiveness of disciplines (including social sciences, humanities and performing arts). Indeed, it can be useful to understand not only the engineering, but also the local culture, language, history and politics associated with a given manufacture (OECD 2021).

Box P.3. *Convergence and innovation*

> Convergence [...] refers to the increasing and transformative interaction between seemingly distinct scientific disciplines, technologies, communities and areas of human activity, in order to achieve mutual compatibility, synergy and integration and, through this process, create added value and expand into emerging areas to achieve common goals. (Roco et al. 2013)

> Convergence: "the merging of ideas, approaches and technologies from a wide range of knowledge domains at a high level of integration. [This is] a critical strategy for solving complex problems and answering complex intellectual questions in emerging disciplines". (National Research Council 2014)

Page (2007) shows that diversity does not always produce better results than ability (meritocracy), but accepts that it does much more often than expected. In this book, we do not seek to support one option over the other, rather both, reminding us that both ability and diversity are important. Knowledge fragmentation is a major challenge for the 4D domain (but not the only one). In disruptive innovation, it seems useful to have people who can successfully and (if possible) skillfully play different functional roles. The success of a team requires trust between partners from different backgrounds and cultures, and also, in interdisciplinary approaches where convergence is called for, different points of view and visions to be shared, in order to build them together. As far as 4D printing is concerned, the approaches to the subject lead us to think that it may be disruptive and change many habits in the future. For all that, the road is neither covered with rose petals nor does it have a very clear objective, but this is what makes it so interesting to the two authors of this work which, hardly finished, will likely need to be deeply reconsidered.

In the field of complexity that can be broadly applied to the 4D theme, Citton (2016) mentions the concept of hyper-objects. He writes:

> Hyper-objects cannot be isolated like classical objects because their existence is inter-objective: their mode of existence is constituted by relations between the things we identify as objects (a glacier, a water level, a crop, a thermometer). It does not matter whether we call them

> "systems", "attractors", "networks", "entanglements", it does not matter: they exist and operate as a web of relations, conditioning the behavior of the objects on which our attention is focused. […] Hyper-objects are viscous: they tend to stick to the beings whose existence is implicated in their own […]

We will have to deal with this, like Captain Haddock not being able to get rid of a sticky substance in Tintin… if the theme keeps its promises. It is a thought of adventure which:

> by the shifting and complex order of knowledge it provides, invites us to face the new and the unexpected. This relationship of transformation, made of tinkering, of adaptation to circumstances and the unforeseen, of forgetting pre-established models and of ingenuity obliges us, in any case, to invent a specific mode of transmission, which is not of the order of the teaching of an abstract knowledge but the learning of an ingenious know-how. (Faucheux and Forest 2011)

This is, far from this present and potential future, the object of this book, which does not only consider examining the links between known modes of actuation and their possible penetration into 4D space, with an essential concept called "printability" (or simple feasibility in additive manufacturing), because we still have to go through the hurdles of additive manufacturing (to varying degrees), as long as we have not mastered self-organizations or organizations activated by thought alone.

"Printability" is defined here as the possibility of making a 4D object using materials that are compatible with each other or whose shapes allow the overall cohesion of the 4D construct to be maintained.

It also means examining other, more autonomous sources of inspiration, and even imagining others. In particular, if we have arrived at digital technology thanks to the mastery of semiconductors and electron transfers, and industrial production at very low unit cost, we must remember that living organisms are assemblies of organic algorithms based on other principles that it might be wise to explore. Finally, beyond the happy idea set to emerge, it is necessary to find ways of integrating knowledge and skills to satisfy societal needs or demands through interdisciplinary convergence actions. Between the dream and the satisfaction of a functional specification, everything has to be done. It is by the yardstick of the mastery of (all) these elements that we will know if this border-object, which we have called a "temporal object", or if this hyper-object, which is the result of 3D printing (which belongs to the old world, whatever we think of it), is going to be introduced into our daily life in punctually or radically, in an ethical, environmental or other framework.

In France, there is expertise in additive manufacturing, because it is an industrial issue. This technology (André 2017a) has been popularized through Fab-Labs. Industrial solutions are yet to be developed, with a large pool of opportunities (hence the continued growth in terms of markets, patents, publications, etc.) such as those of the use of multi-materials and/or the integration of sustainable development imperatives. The existence of open hardware projects such as RepRap, and the possibility of distributed manufacturing, giving individual responsibility back to manufacturers and designers, can be seen as a break from highly centralized production (Ciavaldini et al. 2017). In particular, 3D printers allow for the rapid production of ready-to-use components on an as-needed basis using a single machine, which changes the way companies could design their inventory and supply management. Consequently, by favoring local production, it is possible to limit transport costs and stocks of parts of rare use, but it also allows the production of prohibited parts (weapons, for example) and counterfeits.

For its part, 4D printing is a more complex technology, since it requires the use of active materials in the manufacturing process and processes to stimulate these materials so that the object produced can evolve. Apart from simple demonstrations, it is not possible to imagine the resumption of 4D manufacturing in the FabLab context at the moment. From an industrial point of view, apart from the aspects of "printability" and the definition of stimulation systems to be mastered, the production of real 4D parts in the form of "creative destruction" does not seem much more delicate than certain 3D productions. On the other hand, the potential applications envisaged are much broader than those associated with simple additive manufacturing production, which may raise questions for society or open up new opportunities, in particular for medical applications (drug delivery, active prostheses, organ replacement with bio-printing, etc., all with the supervision due to the use of health products (Raimondeau 2018)) or military. This broad panorama will be presented and discussed later in this book. For many, the search for the "best 4D solutions" desired by the public and/or by society will involve innovations that also aim to reduce the environmental impact of the goods and services produced.

In this qualitative approach, Doucouré (2021) recommends attitudes and behaviors that must maintain the researcher's credibility vis-à-vis the "target" user and industrial audiences (when they know them); he writes: "these include patience, adaptability, empathy and availability". It is through the production of hybrid scientific and socially relevant knowledge that interdisciplinarity or transdisciplinarity contributes to societal transformations (Herrero et al. 2018). Much more than a juxtaposition of knowledge, we speak of the production of novel knowledge enabled by work on 4D printing. "The challenge is not only practical (integration of non-academic knowledge), but also epistemological: in its 'strong' form, transdisciplinarity breaks with classical logic and the assumption of a single reality" (Hermesse and Vankeerberghen 2021). Interactions must lead, thanks to adapted methodologies, to scientific approaches coming from different disciplines and non-scientific actors concerned by the 4D issue, resulting from the dialog of different types of knowledge.

A factor of legitimacy and external recognition, credibility of the strategic approach to a promising subject, new competitive advantages, opening up of new markets, improvement of the image and motivation of the company's partners, etc., together with aspects of social responsibility, or even ethics, make up a story to be constructed. "But innovation is also destructive in that it induces substitution phenomena: it makes existing products disappear, destroys market shares, renders certain skills obsolete, eliminates jobs and dissolves the established order" (Mathieu 2010). But we are not quite there yet… Even if ...

Box P.4. *Toward an ethics of 4D printing?*

This interaction between science and society should serve to develop compromises and place technoscience under an obligation of dependence on society, which finances it! Thus, we must study science and society as both constituents and consequences of action, which will soon lead us to reflect on the social responsibility of the actors and therefore think about the 4D ethical framework respecting society. The question raised thus requires us to think about the "neutrality" of interactions, as some interactors may be "more equal than others"! (Orwell 2020). Indeed, in a world of exponential technological progress, with a society of unstable but growing fears and desires, different subjects are debated and negotiated, and the representatives of different worlds or sub-worlds confront each other and try to manipulate each other. The question of ethics associated with emerging technologies (such as 4D printing) which have a very wide range of applications likely to partially disrupt the social context and induce unacceptable risks would oblige us to respond to Bourdieu (1997), who wrote: "How to create Universes in which people have an interest in virtue"? This is a temporary chance however, we are not there yet!

> But when the know-how is analysable and decomposable to the end, knowledge and doing can become disconnected. Knowledge is then incorporated into a non-human relationship: a book, a treatise, a program, an instruction sheet, a sketch, etc. The great trend in work since the end of the Middle Ages lies in its increasing algorithmization, linked to the impossibility of pushing this algorithmization to its conclusion: new incorporated know-how is born from the very dynamics of algorithmization. (Barel 1977)

Figure P.5 brings together, in a probably very simplistic way, what is happening and what was happening in the "world before", with already enormous changes brought about by the development of digital technology. This has been profoundly reflected in the development of services. Stiegler (2008) reminds us of the possibilities offered by "grammatization", allowing the formalization of links between commands and machines (with considerable simplifications such as the elimination of punched cards, but a writing style that does not easily accept errors and grammatical mistakes). But, to stay on this example, the computer and its software are separable from the 3D printer (and from 4D manufacturing as well). In the projection made for a future that is probably still far off, Human-Systems relations should become more direct and flexible and, as a result, free from the numerous current interfaces and/or filters. This is also an element pointed out earlier in this document: active materials (or their stimulation systems) must be connected.

There seems to be a wide gap between a point-and-tangent approach and a form of divergence from the predictable "technological sublime" (Csicsery-Ronay 2008), "between the marvelous spectacle of science nourished by the scientistic mystique" of Jules Verne and the "sense of wonder" feeding Gernsback's first cultural industry; between the "prototopias" of a Rumpala, inspired by Deleuzian "lines of flight", and "design fiction", inspired by rhetorical models applied to corporate storytelling" (Langlet 2021). In Figure P.5, a number of these concepts, not explored in depth, can be found side by side, without any debate having yet been able to structure or prioritize them. Since we are a little too far removed from the context of this work, we did not want to take the full measure of them, to identify those that could be the subject of consensus, and those that could already be the subject of studies, etc. This work of anticipation and foresight needs to be done and matured.

Foresight is conceived as an articulation, as rigorous as possible, between reason and imagination. Let us recall that for Kant (1994), imagination is a faculty of synthesis ensuring the passage between sensitive intuitions and concepts. When a techno-scientific narrative is constructed and shared by/in an organization, it can form a vision with objectives (breaking with the existing) to be reached, especially if they envisage making sense of unprecedented situations. Not considered, or assumed to be practically impossible, success, which certainly implies great intellectual and organizational agility, self-reflection and confidence, reinforces enthusiasm when the unthinkable happens and the responsible reinforcement of our freedom to undertake (Gauthier 2021).

We can thus imagine that this matter (taken here in a generic vision) or what surrounds it must possess the local and global properties, more or less interdependent, represented on the right-hand side of this figure (for prospective approaches, see National Intelligence Council (2021) and RSIR (2021)). However, as we will show in this book, we are still very far from having a robust way of using any of the properties mentioned on the right-hand side of this figure, which means that we should already be considering new modes of creation and testing them. The idea is indeed to integrate the various scientific and technological convergences within the same cognitive system. But, some difficulties linked to the transition between the over-exponential increase of factual information relative to the integration of knowledge (intensive vs. extensive) and its feeding for conceptual breakthroughs should be overcome (maybe thanks to AI, but not only?). Other "ills" also need to be dealt with, such as resolving the issues associated with contradictory legitimization requirements, the multiplication of possible choices, the proliferation of risks, the uncertainties of the world, societal rejections, innovations with unforeseeable consequences and unanticipated diversions. However, this immense task must be accomplished if we want to create "objects invested with spirit" (Husserl 1996).

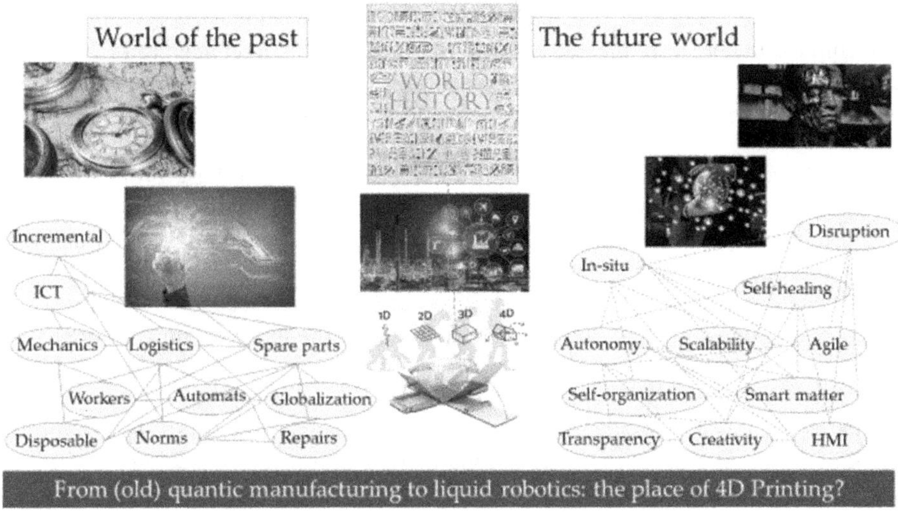

Figure P.5. *Quantum-continuum transition in the development of active material uses and 4D printing. For a color version of this figure, see www.iste.co.uk/demoly/4Dprinting1.zip*

> The future is not an event independent of what we do with it. On the contrary, it depends on our present attitude [...]. Just as the history of past disasters highlights their avoidability, science fiction and its dystopian worlds show that the future is the result of choices made by humans. (Cassely 2013)

P.3. From research to product(s)

A new technique such as 4D printing appears as an opportunistic instrument, with many applications. Let us recall that Breton (1991) wrote that "technology, as a whole, carries within it a conception of the world that its most striking manifestations contribute to shaping and changing. But it has a dynamism of its own, like the sciences that enable its advances. Thus, with new uses, we can expect new social behaviors, with revisited thinking patterns, because technology makes it possible to satisfy new demands".

The invisibility of the knowledge economy could for a long time, because it was "less cluttered", be produced in an apparently mechanical way associated with a main activity devoted to the manufacture of an object or the provision of a service. It was thus produced in the course of action or else as part of apprenticeships, admittedly more elaborate, but still associated with a main activity of production or use (Foray 2018), which gave rise to the economic concepts of learning by doing (Arrow 1962) and learning by use (Rosenberg 1982). As long as everything could be concentrated in one head, the knowledge economy could remain in areas of penumbra. But the inflation of information and the complexity of the systems to be created have changed the situation. Moreover, let us remember what Cowen (2011) wrote:

> We can't figure out why we can't do it anymore. All these problems have a single, but unnoticed, cause: we have lived for at least three hundred years on fruit that was just waiting to be picked [...]. But in the last forty years, these fruits have become scarce, and we have acted as if they were still there. We didn't want to recognize that we had reached a technological plateau and that the tree was much more bare than we wanted to admit.

So, as soon as a guideline seems to bring satisfactory technical progress, there is a risk of continuing the creative action via incremental research, seeking to continue to win without making a clear break. But this quest obviously has its limits.

Our most important collective problems are perhaps not born of a lack of will – of the inability to decide – but of cognitive failures, the ineffective organization of uncertainty, the difficulties that originate from knowledge-deficient organizations, just considered from the angle of their legitimacy. Many managers focus on improvement operations, rather than modifying their competitive positions; this situation can be amplified by the current pandemic (André 2020). Thus, re-engineering, quality management and cost reductions are part of the standard panoply of many companies (Hamel 2000). They have adopted similar strategies, which are a real competitive impasse: "by imitating each other, they focus on incremental improvements in costs and/or quality" (Lehmann-Ortega and Roy 2009). It would then be appropriate to move on to more radical, disruptive, but riskier changes (Kim and Mauborgne, 1999, 2005; Schlegelmilch et al. 2003) leading to profound strategic changes. The quest for quick financial returns for shareholders does not support a radical transformation approach.

Today, many leaders and experts support a return to home ground of material industries. From journalists to experts to political leaders, they all proclaim their fear of the disappearance of local manufacturing (Guillou 2014). Indeed, there are several arguments in favor of this demand: First of all, manufacturing is seen as an area of high employment. "Exports of manufactured goods are one of the keys to escape the external payments imbalance. The high level of public debt and borrowing difficulties, on the one hand, and the upward trend in the price of imported raw materials, on the other hand, underline the need to export" ... But, and this is recent, the crisis linked to the coronavirus pandemic has shown the weakness of developed states which do not sufficiently control certain productions considered recently as strategic (with the associated fragmentations in the value chains).

But, in this restrictive framework where strong initiatives must be supported to imagine a competitive and divergent future, we can recall some very worrying writings on this possibility. In 2004, for example, there were profound renunciations in France of any vision of a conquering future. For example, MEDEF boss Sellière (2004) wrote: "The beginning of this century marks the end of the technological rent for the West. Today, it is on human capital [...] that our response to the new international situation is based. Education, training: the knowledge economy is not an empty slogan" ... In short, let us wait for other countries to do it, get on the bandwagon and agree with Innerarity (2015) when he talks about the disproportion that exists "between our limited knowledge and the problems we are facing – as constituting a society of ignorance"! We may also recall this writing by Nietzsche (1990), which is hardly more reassuring:

> Everywhere there are symptoms of a civilization that is withering away [...], the contemptible thirst for money and pleasure of the learned circles, their lack of love and grandeur. That the learned circles are also caught up in this movement is what I am becoming more and more aware of. They are becoming poorer and poorer in thought and love every day. Everything serves the rising barbarism, art as well as science. Where should we turn our eyes? The great flood of barbarism is at our doors!

In this context of followership where it is, at best, the incremental that will continue to dominate with vigor, we must restrict our ambitions to moderately disruptive (or incremental pushed to its limits)! In a world corseted by finance and its weak tendency to accept risk-taking, the vision of a future close to 4D science fiction can only be achieved if we are already capable of proving the feasibility of much simpler applicative proposals (despite a growth rate of over 40% of annual publications), and this requires not only the emergence of new ideas but also an enlightened coupling between in-depth studies on essential pivotal themes (and the sciences that contribute to their robustness) and realistic applicative actions that are credible. It is based on this understanding of the pooling of interests and skills that an embryonic roadmap will be proposed later in this preamble (which should be the subject of debate) for medium-term actions that could be likely to change the modes of linkage between researchers, the disciplines they carry and society.

> Individuation is the process by which individuals and, with them, the societies they form, are constituted and constantly transformed – and, in this, psychic and collective individuation is the way in which a society forms a body, unites, at the same time as it inherits an experience from the past, what is often called knowledge, and more broadly, wisdom. (Stiegler 2006)

But as long as everything can come out of one head, everything is fine (or almost). When we talk about technology, we call upon a lot of knowledge and it is its good integration that defines the success of an innovation. For the industrial part, it is the risk-taking elements that can limit the adventure because of the large number of criteria to be mastered. On the other hand, the observation that will be analyzed by the two authors shows that a nucleation can take place by continuity (conservatism?), starting from a scientific community bringing enlightenment to a subject (such as that of 4D printing) likely to lead a whole community in the initial direction, supposed to be good: it is then a question of a "true" connection by point and tangent; from there to enabling the escape by inertia of the subject relevant to many titles of its objective of societal usefulness, there is only a step! As a reminder, Figure P.6 (Royal Academy of Engineering 2021) is an essential alarm in the technical processes of transferring know-how to society: if a set of ideas, concepts and skills from research is indeed available, only a small part of the path toward applications useful to society will have been covered. Other equally essential steps must be explored (even if they only very imperfectly fit into the aim of this book).

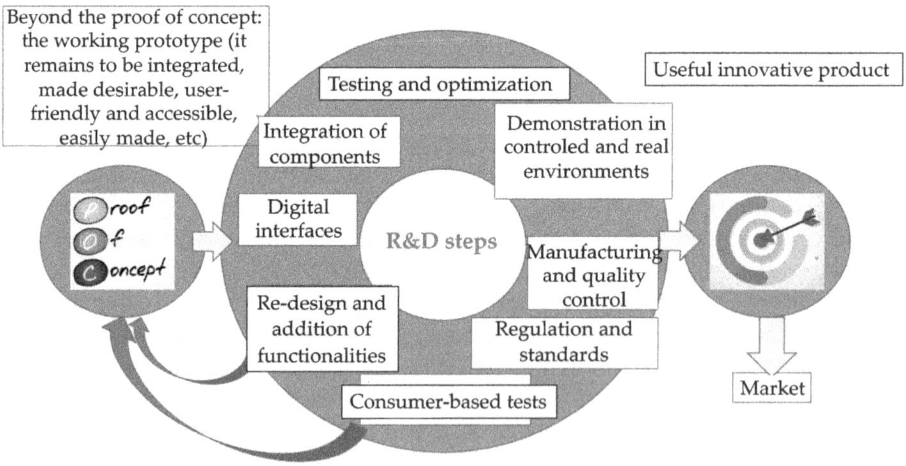

Figure P.6. *From concept to application. For a color version of this figure, see www.iste.co.uk/demoly/4Dprinting1.zip*

Regardless of this inescapable aspect, Lewkowicz (2017) goes further with the domestication, the assimilation of technologies that makes practical, useful, reliable, esthetic and trustworthy artifacts attractive or desirable. What is interesting for him is not only the integration of this model in qualitative approaches (understanding uses, their possible diversion, meanings and values granted), but also the analysis of what the new artifacts and their uses mean for the users (social positioning, usefulness, etc.). This design mode studied by Thouret and Caillette-Beaudoin (2017) in another framework shows that the actors convened in this type of operation must agree on a vision and common values, the basis for mutual trust.

So, to know which directions to go in, we can already try to think of some, but also carry out surveys by questioning more or less enlightened audiences. The idea, as presented in Chapter 2, was to link common sense to imagination, knowing that, in this case, people agree to quickly answer questions they have never or practically never thought of. Stengers (2020) considers that the issuers of this type of survey may have no desire to coerce in order to cynically retrieve the "right" answer, but to bring up, in the face of ignorance, reformulations of the questions asked, stemming from common sense. It has to be said that the feedback is not very useful, as it conforms, without any desire to surprise.

So, to try to get closer to the "real thing", can we find people representing disciplinary "samples" who are likely to accept and propose divergent ideas, each one keeping their own. Also, how do we share a small privilege with those who have thought a bit about it, and who admit that they "know that they don't know"? who having thought a bit about it accept to say that they "know that they don't know". But in a normative academic world where agnotology (the science of not knowing) has no place, will we find experts who are convinced of their initial inability to know what they could say and who would be ready to defer to those who have a head start (but not much more), especially if they confront them with scientific challenges that they may consider as traps. Moreover, each discipline has its specialized jargon and words do not "mean anything" independently of their origin and their use by a community. Would they agree to divert part of their time to an interdisciplinary adventure in which they are not involved, because it is necessary to go beyond simple assemblies? The divergences that could be revealed in open debates on a theme such as 4D printing, on its boundaries, its opportunities, its definition demands, could constitute a collective concern, giving strength to the choices to be made.

> We must then imagine a Socrates who needs the citizens of Athens to ruminate, to accept that there is no need to give authority to the commonplaces that furnish current thought, but without being ready to deny the aspects of existence that their proposals made important.

> On the contrary, their rumination should activate this sense of importance, attach it to an aspect of existence, an aspect that would belong to existence itself, irreducible to what we have become accustomed to referring to as the relativity of the "subjective". (Stengers 2020)

> Ludwig Wittgenstein, who denied the possibility of explaining any fact by reference to another, said that "at the root of the whole modern world view lies the illusion that the so-called laws of nature are the explanations of natural phenomena". Such warnings leave me cold. To say this to a physicist is like telling a tiger in search of prey that all flesh is grass. That we scientists do not know how to explain, with the approval of philosophers, what we do when we seek explanations does not mean that we are engaged in an idle task. Philosophers may help us understand what we do, but with or without them, we will continue to do it. (Weinberg 1997)

Accepting a form of empiricism, of testing in a moving universe, unstable within a certain control of the change induced by the advent of 4D printing as a relay between technologies, certainly refined from the previous technological paradigm, to move to an "intelligent", autonomous, adaptive material, etc., met our wishes. Moreover, everyone knows from Whitehead (1994) that:

> Each profession makes progress, but progresses according to its own routine, tracing its own furrow [...]. The furrow prevents wandering across the landscape and the abstracting of something, to which no attention is paid anymore [...]. The rest of life is treated superficially, with the imperfect categories of a thought that derives from the profession.

So, in the end, based on the bibliography, we thought together (but rather separately). For all that, this outstretched hand translated by this work reminds us that thought needs fixed references to escape certain dogmatisms, forms of confusion or arbitrariness. What a lack of science and technology warfare on such a subject as 4D printing shows is a lack of deliberate will to conquer new scientific spaces in order to integrate them into socially accepted disseminating technologies. So, in the absence of debate, we had to give our opinion in this book, as well-supported as possible to cover, with our means, a small part of an epistemology of 4D printing to be built, or in the course of construction. The acquired knowledge that we present and a little "brainstorming" lead us to propose some bases (but not more) for a roadmap, for the medium term linked to the desired development of 4D printing.

As will be shown throughout this document, and in particular in the analysis of the bibliographic references, the French scientific forces in the 4D field are too modest and above all too strongly dispersed with, despite everything, a strong focus on active polymers (which probably does not help these high-level scientific specialists to broaden the scope of their imagination!). We, therefore, felt that it was not possible to let the research world interested in this field continue to operate solely on the basis of a biased offer, with the number of publications in serious scientific journals as the (sole) objective. It can also, and this is the object of our proposal, function by crossing informed demands, in the short, medium and long term, with generic know-how from 3D (and from all the sciences that can support this development). It is a question of covering the entire value chain, given the very large number of applications (as shown in Figure P.7), and of committing to the development of a coherent and productive future 4D sector today. This is, as we shall show, a risky operation, but surely it is necessary to try? Moreover, Minguet and Osty (2008) have proposed that the organization be "a cooperative structure in which the professions [disciplines] represent the stabilizing anchors and the projects [innovation] the spurs of a profusion of dedicated products, a source of [resourcing and legitimacy]". The medium and long term will make it possible to refine this need and, in particular, to analyze the bibliography, provided that we seek to periodically review and consolidate this approach, summarized in Figure P.6.

> In a civilization saturated by technology, there is no more adventure, only the risk of being late. (Sloterdijk 2006)

> Either the researcher treats the object [4D] in a partial manner, ignoring or concealing all the links that inevitably exist between this partial view and the global biases that he cannot fail to have on the object. Or else, it places its global biases at the heart of scientific questioning: not as interesting elements to look at, but as the motors of a dynamic that then takes the inevitable desire for unity as a stepping stone towards what becomes a requirement for unity. (Alvarez-Pereyre 2003)

This desire to get even closer to real use, with all its difficulties (Nicolescu 1996), is based on cultural background, that of modeling and integration (systemic approach), and that of experimentation (technology centers, for example, to avoid going through a mode of manufacture imposed because it is available), at the same time as an indispensable opening to other scientific and technical fields, by "borrowing" their concepts, methodologies (principles of permanent learning and cross-fertilization) and instruments, by "bringing" subjects of study to them.

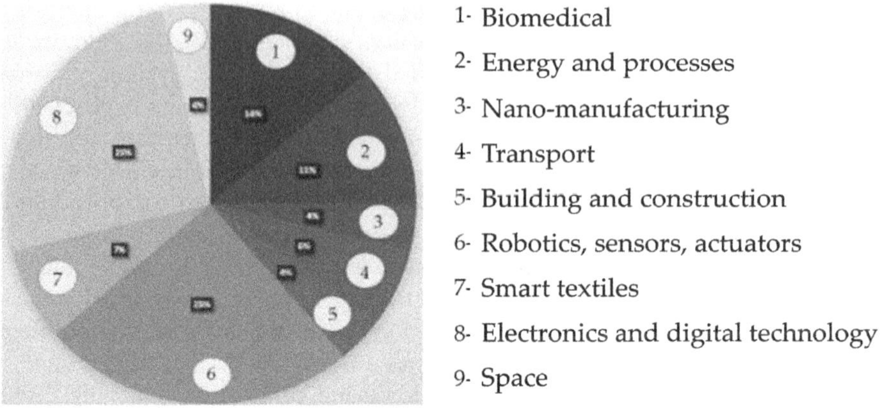

Figure P.7. *Examples of applications of 4D proposed in the literature. For a color version of this figure, see www.iste.co.uk/demoly/4Dprinting1.zip*

The presentation of a vision, associated with lines of action, necessarily plural, is made indispensable in a highly evolving universe involving forms of communication adapted to the promotion of real added values permitted by technoscience in a socio-economic context with systemic causalities (Stengers and Bensaude-Vincent 2003). This situation of reinforcement or rejection of the link between science and society also corresponds to the nature of the social acceptability of new technologies such as 4D printing, which remains rather alien to the scientific approach of a rather monodisciplinary nature. However, reflection on this rapprochement/rejection does not necessarily mean the absence of a boundary that should have been respected in order to separate the external interests of the various stakeholders in scientific and technological research activities. This will indeed lead to a debate on the involvement of scientists with the social body and on the results of research, on a new regime of knowledge production, but little or no debate on the means to achieve them, which is the responsibility of the research structures. Moles (1990) recommends that research activities start with a certain degree of vagueness and "assume it as such", followed by Albaladejo et al. (2009) who write: "The initial specifications are more in the form of a definition of a function or a role that the research group must play than in the form of a product to be delivered [...]. The essential part of the definition of the result to be obtained is as much in the production methods as in the final product, the specifications".

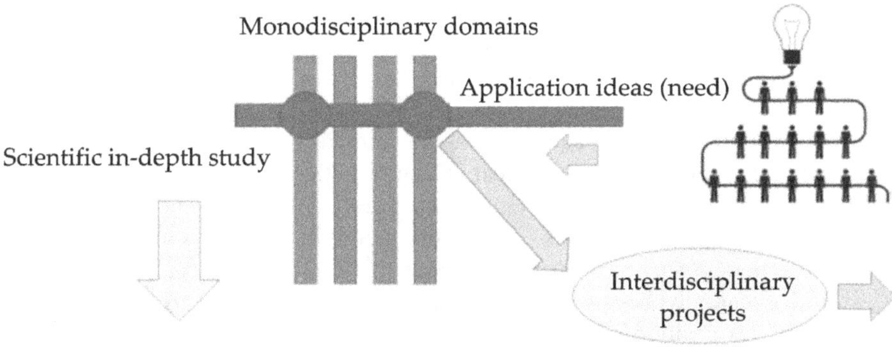

Figure P.8. *Proposed basic action plan for the development of 4D printing: crossing between top-down and bottom-up approaches. For a color version of this figure, see www.iste.co.uk/demoly/4Dprinting1.zip*

However, in the openings that we hope to see, with indispensable deepening to avoid an accounting drought, without any ethical foundation, we thought it useful to recall these sentences by Lemontey (written in 1801 and quoted by Vigoureux in 2020): "Man was no longer considered as a goal, but as a means; and the day he was evaluated in money, all administrative morality was annihilated." Under these conditions, we propose (which will be taken up in our conclusion):

– first, to rely on research units linked to large organizations (ONR) that will be involved in the further development of generic themes associated with 4D printing and bio-printing (more specifically oriented toward "personalized medicine"). They should be concerned with the exploration of generic themes that would serve as pillars for the development of 4D printing;

– to cross the knowledge developed (as well as other complementary knowledge) with application demands from the economic world to manage interdisciplinary projects for industrial and/or medical developments;

– in association, to maintain a scientific and technological watch of the prospective for all the actors, and to consider all the possible openings allowing our country to be revealed in an effective dynamic (such as the one which it should have had if it had not put the first patent on 3D printing in the bin in 1984)!

From then on, dear reader, there will be the possibility of deliberation, even controversy, with its specific heuristic contributions, but in a deferred way. It is a way to enrich a posteriori forms of active participation, "communication that induces reflection on preferences, values, and interests in a non-coercive way" (Dryzek 2000), and is therefore rather oriented toward the collective decision (Mansbridge et al. 2011) that it aims to make mutually acceptable. Will we be able to go beyond the evils that limit or constrain our desire to go further and further, in agreement with Crozier and Tillette (2000): “the statism, the hierarchical elitism and the egalitarianism that it [France] hides under the veil of exceptionality”?

> Disruption cannot be conceived as an internal innovation within the company, even if it is radical. Changing the rules of the game, therefore, implies that the innovation introduces something new for the company itself, for its competitors, but also and above all, for its customers. It is through their contact that the disruptive strategy will eventually be qualified as a success if they adhere to the new value proposition made available to them. (Lehmann-Ortega and Roy 2009)

P.4. References

Albaladejo, C., Hubert, B., Roche, B. (2009). Chercheurs en situation de partenariat : prescrire la subjectivité – réflexion à partir du programme de pérennisation de l’agropastoralisme au Pays Basque. In *Dynamique des savoirs, dynamique des changements*, Béguin, P. and Cerf, M. (eds). Octarès, Toulouse.

Allard, L. (2016). Dans quel monde voulons-nous être connectés ? Trans-humanisme vs companionisme. *Nectart*, 2016/2, 125–132.

Alvarez-Pereyre, F. (2003). *L’exigence interdisciplinaire*. MSH, Paris.

André, J.C. (2006). Elements of consideration on the methodology to be developed in occupational safety and health prospecting and forecasting: Report and final proposals on forecasting on emerging risks related to technical innovations. Report to the EU (Centre thématique “Observatoire des risques”). OSH Bilbao Agency, Bilbao.

André, J.C. (2017a). *From Additive Manufacturing to 3D/4D Printing 1: From Concepts to Achievements*. ISTE Ltd, London, and John Wiley & Sons, New York.

André, J.C. (2017b). *From Additive Manufacturing to 3D/4D Printing 2: Current Techniques, Improvements and their Limitations*. ISTE Ltd, London, and John Wiley & Sons, New York.

André, J.C. (2017c). *From Additive Manufacturing to 3D/4D Printing 3: Breakthrough Innovations: Programmable Material, 4D Printing and Bio-printing*. ISTE Ltd, London, and John Wiley & Sons, New York.

André, J.C. (2020). Engagement solidaire et dé-confinement. *Environnement Risques et Santé*, 19, 238–239 [Online]. Available at: https://www.jle.com/fr/covid19-engagement-solidaire-deconfinement.

Arendt, H. (1968). *Conditions de l'Homme moderne*. Calmann-Lévy, Paris.

Arrow, K.J. (1962). Economic welfare and the allocation of resources for inventions. In *The Rate and Direction of Inventive Activity – Economic and Social Factors*, Nelson, R. (ed.). Princeton University Press, Princeton.

Aubert, N. (2018). Accélération et hyper-connexion à l'ère du capitalisme financier : accomplissement de soi ou dépossession de soi ? In *@ La recherche du temps – Individus hyper-connectés, société accélérée : tensions et transformations*, Aubert, N. (ed.). Erès, Paris.

Bachelard, G. (1966). *Le nouvel esprit scientifique*. PUF, Paris.

Barel, Y. (1977). La ville avant la planification urbaine. In *Prendre la ville : équisse d'une histoire de l'urbanisme d'état*. Anthropos, Paris.

Berthelot, M. (1897). *Science et morale*. Calmann-Lévy, Paris.

Bourdieu, P. (1997). *Les usages sociaux de la science*. INRA, Paris.

BPI France (2021). Rapprocher l'industrie et la tech : Bpifrance lance la plateforme Tech in Fab et annonce avec France Industrie un partenariat pour la Deeptech [Online]. Available at: https://presse.bpifrance.fr/rapprocher-lindustrie-et-la-tech-bpifrance-lance-la-plateforme-tech-in-fab-et-annonce-avec-france-industrie-un-partenariat-pour-la-deeptech/.

Breton, T. (1991). *La dimension invisible. Le défi du temps et de l'information*. Odile Jacob, Paris.

Brown, N. and Sandholm, T. (2018). Superhuman AI for heads-up no-limit poker: Libratus beats top professionals. *Science*, 359, 418–424.

Brunet-Labbez, J. (2013). *Innover avec la génération Y*. Studyrama, Paris.

Cain, C.J. (2017). Go and artificial intelligence: Potential for strategic decision making [Online]. Available at: http://publications.armywarcollege.edu/pubs/3401.pdfb [Accessed 25 March 2022].

Cassely, J.L. (2013). À quoi sert la science-fiction ? A nous expliquer de quoi l'avenir ne sera – peut-être – pas fait [Online]. Available at: http://www.slate.fr/life/80157/science-fiction-recits-apocalyptiques-avenir.

Christensen, C. (2011). *The Innovator's Dilemma: The Revolutionary Book that Will Change the Way You Do Business*. Harper Business, New York.

Ciavaldini, P.A., Roussel, V., Bui, L. (2017). MakerNet : la fabrication distribuée. *Annales des Mines – Réalités industrielles*, 2017/3, 65–69.

Citton, Y. (2016). Accélérer notre attention collective aux hyper-objets. In *Accélération !*, de Sutter, L. (ed.). PUF, Paris.

Clarke, A.C. (ed.) (1962). Hazards of prophecy: The failure of imagination. In *Profile of the Future*. Bantam Books, New York.

Conche, M. (2007). *Montaigne et la philosophie*. PUF, Paris.

Cowen, T. (2011). *The Great Stagnation: How America Ate All the Low-hanging Fruit of Modern History, Got Sick and Will (Eventually) Feel Better*. Dutton, New York.

Cresswell, R. (2010). Techniques et culture : les bases d'un programme de travail. In *Anthologie raisonnée de Techniques & Culture*, Éditions de l'EHESS (ed.). Éditions de la Maison des Sciences de l'Homme, Paris.

Crozier, M. and Tillette, B. (2000). *Quand la France s'ouvrira*. Fayard, Paris.

Csicsery-Ronay Jr., I. (2008). *The Seven Beauties of Science Fiction*. Wesleyan University Press, Middletown.

De Brabandère, L. (2017). *Homo-Informatix*. Le Pommier, Paris.

Doucouré, B. (2021). Crédibilité du chercheur, relation de confiance et éthique en recherche qualitative : l'implexité à la croisée des chemins. *Recherches qualitatives*, 40, 46–60 [Online]. Available at: https://www.erudit.org/fr/revues/rechqual/2021-v40-n1-rechqual05941/1076346ar/.

Dourish, P. (2003). The appropriation of interactive technologies: Some lessons from placeless documents. *Computer Supported Cooperative Work*, 12, 465–490.

Dryzek, J.S. (2000). *Deliberative Democracy and Beyond: Liberals, Critics and Contestations*. Oxford University Press, Oxford.

Du Sautoy, M. (2020). *Le code de la créativité : comment l'IA apprend à écrire, peindre et penser*. Éditions Héloïse d'Ormesson, Paris.

Fabry, P. (2015). L'Ère des robots : transition à hauts risques [Online]. Available at: https://www.contrepoints.org/2015/03/18/201543-lere-des-robots-transition-a-hauts-risques.

Faucheux, M. and Forest, J. (2011). Stimulating a creative rationality to stimulate innovation. *Creativity and Innovation Management*, 20, 207–212.

Feyerabend, P. (1988). *Contre la méthode ; esquisse d'une théorie anarchiste de la connaissance*. Le Seuil, Paris.

Fink, M. (2018). L'innovation : un enjeu culturel. Entretien de Mathias Fink, avec Marie-Françoise Chevallier-Le Guyader, et Paul Maitre. *Raison présente*, 206, 21–29. Union rationaliste, Paris.

Foray, D. (1992). Choix de la technique, rendements croissants et processus historiques. In *La technoscience – les fractures du discours*, Pradès, J. (ed.). L'Harmattan, Paris.

Foray, D. (2018). *L'économie de la connaissance*. La Découverte, Paris.

Foster, C. and Graham, M. (2017). Reconsidering the role of the digital in global production networks. *Global Networks*, 17, 68–88.

France Diplomatie (2019). Annonce de la stratégie intégrée du Japon pour la mise en œuvre de la société 5.0 [Online]. Available at: https://www.diplomatie.gouv.fr/fr/politique-etrangere-de-la-france/diplomatie-scientifique-et-universitaire/veille-scientifique-et-technologique/japon/article/annonce-de-la-strategie-integree-du-japon-pour-la-mise-en-oeuvre-de-la-societe.

Frochot, C., Schaer, E., André, J.C. (2020). Il ne faut pas avoir peur des "nano", c'est juste une nanocrise mal anticipée. *Environnement risques et santé*, 19, 446–466.

Gaudin, T. (1984). Les ingénieurs et l'innovation. *Culture technique*, 12, 133–135.

Gauthier, T. (2021). La prospective, ou l'art de concilier "oui mais…" et "et si…" [Online]. Available at: https://usbeketrica.com/fr/prospective-conciliation-oui-non.

Gendron, C. and Pierssens, M. (2019). L'entreprise vue par la science-fiction : d'aujourd'hui à demain. *Entreprises et histoire*, 96, 8–13.

Géneau, D. and Caulier, S. (2016). Le 3D imprime sa marque sur l'industrie ; construire localement, à moindre coût, avec un service après-vente amélioré [Online]. Available at: http://www.ledevoir.com/economie/actualites-economiques/465475/le-3d-imprime-sa-marque-sur-l-industrie-construire-localement-a-moindre-cout-avec-un-service-apres-vente-ameliore.

Gille, B. (1978). *Histoire des techniques*. Gallimard, Paris.

Guillou, S. (2014). Convergence of EU and US Industrial Policy: "The obsession of competitiveness". *Revue d'économie industrielle*, 145, 181–200.

Gunning, D. (2018). *Artificial Intelligence and Machine Learning to Accelerate Translational Research*. National Academic Press, New York [Online]. Available at: http://nap.edu/25197 [Accessed 25 March 2022].

Halleux, R. (2009). *Le savoir de la main*. Armand Colin, Paris.

Hamel, G. (2000). *Leading the Revolution*. Harvard Business School Press, Boston.

Heidegger, M. (2003). *La question de la technique*. Gallimard, Paris.

Herrero, P., Dedeuwaerdere, T., Osinski, A. (2018). Design features for transformative transdisciplinary research. *Sustainability Science*, 14, 751–769.

Hinks, J. (2015). 5 things you should know about industry 4.0 [Online]. Available at: http://www.techradar.com/news/world-of-tech/future-tech/5-things-you-should-know-about-industry-4-0-1289534 [Accessed 25 March 2022].

Husserl, E. (1996). *Recherches phénoménologiques pour la constitution*. PUF, Paris.

Innerarity, D. (2015). *Démocratie et société de la connaissance*. Presses universitaires de Grenoble, Grenoble.

Kant, E. (1994). Prologue à la quête. In *Sciences et imaginaires*, Maréchal, I.A. (ed.). Albin Michel, Paris.

Kaplan, F. (2012). *La métamorphose des objets*. FYP, Paris.

Kim, C. and Mauborgne, R. (1999). Creating new market space. *Harvard Business Review*, 77 [Online]. Available at: https://hbr.org/1999/01/creating-new-market-space [Accessed 25 March 2022].

Kim, C. and Mauborgne, R. (2005). *Stratégie océan bleu*. Village mondial, Paris.

Knapton, S. (2017). AI implants will allow us to control our homes with our thoughts within 20 years, government report claims [Online]. Available at: http://www.telegraph.co.uk/science/2017/10/15/ai-implants-will-allow-us-control-homes-thoughts-within-20-years/ [Accessed 25 March 2022].

Kuhn, T. (2008). *La structure des révolutions scientifiques*. Champs-Sciences, Paris.

Kurzweil, R. (2011). 2045: The year man becomes immortal [Online]. Available at: http://content.time.com/time/magazine/article/0,9171,2048299,00.html [Accessed 25 March 2022].

Langlet, I. (2021). Théorie(s) de la science-fiction [Online]. Available at: https://www.fabula.org/actualites/resf-20-theories-de-la-science-fiction-cfp_101958.php.

Lazar, P. (2014). Comprendre pour comprendre, comprendre pour agir. In *La science en question(s)*, Wieviorka, M. (ed.). Éditions Sciences Humaines, Auxerre.

Leblanc, D. (2017). *Smart World : comment de simples idées deviennent-elles de grandes innovations ?* Edilivres, Saint-Denis.

Lehmann-Ortega, L. and Roy, P. (2009). Les stratégies de rupture : synthèse et perspectives. *Revue française de gestion*, 197, 113–126.

Lemonnier, P. (2010). L'étude des systèmes techniques, une urgence en technologie culturelle. In *Anthologie raisonnée de Techniques & Culture*, Éditions de l'EHESS (ed.). Éditions de la Maison des Sciences de l'Homme, Paris.

Leonhardt, G. (2018). Technology vs humanity. *FTA2018 Conference – Future in the Making*, 4–5 June, Brussels.

Lewkowicz, M. (2017). De la domestication des technologies. *Annales des Mines – Réalités industrielles*, 2017/2, 33–36.

Manceau, D. and Morand, P. (2014). A few arguments in favor of a holistic approach to innovation in economics and management. *Journal of Innovation Economics & Management*, 15, 101–115.

Mansbridge, J.J., Bohman, J., Chambers, S., Estlund, D., Føllesdal, A., Fung, A., Lafont, C., Manin, B., Luismarti, J. (2011). La place de l'intérêt particulier et le rôle du pouvoir dans la démocratie délibérative. *Raisons politiques*, 42, 47–82.

Mathieu, A. (2010). Développement durable et innovation : dépasser l'antagonisme pour une complémentarité au service de la performance globale. *Stratégies d'entreprises en développement durable*, 159–182.

Mauss, M. (1947). *Manuel d'éthnographie*. Payot, Paris.

Minguet, G. and Osty, F. (2008). *En quête d'innovation : du projet au produit de haute technologie*. Lavoisier, Paris.

Moles, A.A. (1990). *Les sciences de l'imprécis*. Le Seuil, Paris.

Moulonguet, T. (2017). La société 5.0 [Online]. Available at: https://www.revuedesdeuxmondes.fr/societe-5-0/.

Moyer, C. (2016). How Google's AlphaGo Beat a Go World Champion. *The Atlantic*, March 28 [Online]. Available at: http://www.theatlantic.com/technology/archive/2016/03/the-invisible-opponent/475611/ [Accessed 25 March 2022].

National Academy Press (2021). Measuring convergence in science and engineering: Proceedings of a workshop [Online]. Available at: https://www.nap.edu/catalog/26040/measuring-convergence-in-science-and-engineering-proceedings-of-a-workshop [Accessed 25 March 2022].

National Intelligence Council (2021). Global trends 2040 [Online]. Available at: https://iatranshumanisme.com/wp-content/uploads/2021/04/Global-Trends-2040-A-More-Contested-World.pdf [Accessed 25 March 2022].

National Research Council (2014). *Convergence: Facilitating Transdisciplinary Integration of Life Sciences, Physical Sciences, Engineering, and Beyond*. The National Academies Press [Online]. Available at: http://dx.doi.org/10.17226/18722.

Nicolescu, B. (1996). *La transdisciplinarité ; manifeste*. Editions du Rocher, Monaco.

Nietzsche, F. (1990). *Œuvres philosophiques*. Gallimard, Paris.

Nordmann, A. (2004). *Technologies convergentes – façonner l'avenir des sociétés européennes*. Commission Européenne, Luxembourg.

OECD (2021). Collaborative platforms for emerging technology – Creating convergence spaces [Online]. Available at: https://www.oecd-ilibrary.org/docserver/ed1e030d-en.pdf?expires=1619851449&id=id&accname=guest&checksum=3435ED8BC41305D047C2FEBBF8D30F13 [Accessed 25 March 2022].

Paché, G., Saglietto, L., Quillaud, A., Bonet Fernandez, D. (2018). Logistique et technologies disruptives dans les réseaux globalisés de production : le rôle clé des données massives. *Revue d'économie industrielle*, 163, 77–110.

Page, S.E. (2007). *The Difference: How the Power of Diversity Creates Better Groups, Firms, Schools, and Societies*. Princeton University Press, Princeton.

Pau, W. (2014). IoT: 3C's of Internet of Things (aka Welcome to Industry 4.0!) [Online]. Available at: https://www.linkedin.com/pulse/20141104130144-5544582-iot-3c-s-of-Internet-of-things-aka-welcome-to-industry-4-0; http://waynepau.blogspot.ca/2014/11/iot-3cs-of-Internet-of-things-aka.html [Accessed 25 March 2022].

Price, C. (1966). La prospective, ou l'art de concilier "oui mais…" et "et si…" [cited by Gauthier T. (2021)] [Online]. Available at: https://usbeketrica.com/fr/prospective-conciliation-oui-non.

PricewaterhouseCoopers (2016). Industry 4.0; Opportunities and challenges for consumer product and retail companies [Online]. Available at: http://www.strategyand.pwc.com/media/file/Industry-4.0-RC.pdf [Accessed 25 March 2022].

Raimondeau, C. (ed.) (2018). Les produits de santé. In *L'épreuve de santé publique*. Presses de l'EHESP, Rennes.

Roco, M.C., Bainbridge, W.S., Tonn, B., Whitesides, G. (2013). *Convergence of Knowledge, Technology and Society*. Springer International Publishing, Cham.

Rosa, H. (2018). *Remède à l'accélération*. Champs-Essais, Paris.

Rosenberg, N. (1982). *Handbook of the Economics of Innovation*. North-Holland, Amsterdam.

Royal Academy of Engineering (2021). Late-stage R&D: Business perspectives [Online]. Available at: https://www.raeng.org.uk/publications/other/late-stage-r-and-d-business-perspectives [Accessed 25 March 2022].

RSIS (Rajaratnam School of International Studies – NTU Singapore) (2021). Global trends 2040: A hyper-connected future? [Online]. Available at: https://www.rsis.edu.sg/wp-content/uploads/2021/04/CO21061.pdf [Accessed 25 March 2022].

Salgues, B. (2018). *Society 5.0*. ISTE Ltd, London, and John Wiley & Sons, New York.

Sarazin, B. (2019). Mener à bien une innovation de rupture. In *Innovation de rupture – Guide pour disrupter votre marché*, Sarazin, B. (ed.). EMS Éditions, Paris.

Schaer, E. and André, J.C. (2020a). *Process Engineering Renewal 1: Background and Training*. ISTE Ltd, London, and John Wiley & Sons, New York.

Schaer, E. and André, J.C. (2020b). *Process Engineering Renewal 2: Research*. ISTE Ltd, London, and John Wiley & Sons, New York.

Schaer, E. and André, J.C. (2020c). *Process Engineering Renewal 3: Prospects*. ISTE Ltd, London, and John Wiley & Sons, New York.

Schlegelmilch, B., Diamantopoulos, A., Kreuz, P. (2003). Strategic innovation: The construct, its drivers and its strategic outcomes. *Journal of Strategic Marketing*, 11, 117–133.

Schwab, K. (2015). What is the fourth industrial revolution? [Online]. Available at: https://agenda.weforum.org/2015/12/the-fourth-industrial-revolution-what-it-means-and-how-to-respond/ [Accessed 25 March 2022].

Seillière, E.A. (2004). *Les dirigeants face au changement*. Éditions du huitième jour, Paris.

Serres, M. (1991). *Le Tiers-Instruit*. François Bourin, Paris.

Simondon, G. (2012). *Du mode d'existence des objets techniques*. Aubier, Paris.

Sloterdijk, P. (2006). *Le palais de cristal : à l'intérieur du capitalisme planétaire*. Moren Sell, Paris.

Stengers, I. (2020). *Réactiver le sens commun*. Les Empêcheurs de penser en rond, Paris.

Stengers, I. and Bensaude-Vincent, B. (2003). *100 mots pour commencer à penser les Sciences*. Les Empêcheurs de penser en rond, Paris.

Stiegler, B. (2004). *De la misère symbolique*. Champs/Essais, Paris.

Stiegler, B. (2006). *Ré-enchanter le monde : la valeur esprit contre le populisme industriel*. Champ Essais, Paris.

Thouret, A. and Caillette-Beaudoin, A. (2017). Le co-design en continu : un facteur clé de la réussite de la prise en charge des patients dans leur parcours de soins. *Annales des Mines – Réalités industrielles*, 2017/2, 10–15.

Twain, M. (2018). *Une vie sans fin*. Grasset, Paris.

Valero, J. (2016). Industry 4.0: The future is here [Online]. Available at: http://www.euractiv.com/section/innovation-industry/linksdossier/industry-4-0-the-future-of-manufacturing-is-here/#ea-accordion-further-reading [Accessed 25 March 2022].

Van Andel, P. and Bourcier, D. (2009). *De la sérendipité dans la science, la technique, l'art et le droit ; leçons de l'inattendu*. L'ACTMEM "Libres Sciences", Paris.

Vigoureux, J.M. (2020). *Détournement de sciences*. Ecosociété, Quebec.

Virilio, P. (1996). *Un paysage d'événements*. Galilée, Paris.

Weinberg, S. (1997). *Le rêve d'une théorie ultime*. Odile Jacob, Paris.

Whitehead, A.N. (1994). *La science et le monde moderne*. Éditions du Rocher, Paris.

Yeung, H. (2018). The logic of production networks. In *The New Oxford Handbook of Economic Geography*, Clark, G., Feldman, M., Gertler, M., Wójcik, D. (eds). Oxford University Press, Oxford.

Zimmermann, J. (2021). Y-a-t-il des gens créatifs et d'autres pas ? L'auteur de La baignoire d'Archimède répond et c'est surprenant [Online]. Available at: https://www.ladepeche.fr/2021/04/20/y-a-t-il-des-gens-creatifs-et-dautres-pas-lauteur-de-la-baignoire-darchimede-repond-et-cest-surprenant-9500960.php.

Introduction

> History teaches us that the "pioneers" always opened for others paths that they themselves did not take. (Löwith 1969)

> At the heart of this standardization process, the training of young researchers, which is crucial to creative science, has been particularly targeted. Doctoral students and thesis directors are now supervised like children by doctoral schools corseted by numerous decrees and monitored by "monitoring committees" summoned to ensure the orthodoxy of their course. (Plagnol 2021)

> Do you want on the greatest and most important of all things, I mean freedom, to see caprice and liberty reign? No my friend: freedom consists in wanting things to happen not as you please, but as they happen. (Epictetus 2015)

> [In France], we live and think in the element of this moderately optimistic presenteeism, colored by worries, without strong beliefs, with a limited, impoverished imagination. The most derisory claims have replaced projects, social jealousy and resentment have driven out enthusiasm. (Taguieff 2021)

In 1871, Auguste Blanqui, a French theorist of revolutionary socialism, was imprisoned and wrote "*Eternity By the Stars*" in prison. For him, if the universe is sufficiently vast, there is a chance that any object will be reproduced somewhere identically. All we have to do is search the universe for the desired object! But what do we do if time is short? Will we be able to find the object? Undoubtedly, Blanqui was not intimately familiar with Maxwell and his demons? At the same time, with the development of automata and the exploitation of new forms of energy, there arose the fantasy of an ideal mutation, where humans would have, outside of themselves, another almost-human, artificial clone, which would act with a speed, a precision, a regulation and a limitless force to produce the material goods they think they need. Indeed, the world of technology has undergone a major transformation in recent decades. Elements that previously existed in a purely mechanical or electrical (i.e. physical) form, and, in particular, those elements describing logic, control and decision-making, are increasingly taking the form of embedded systems and software (i.e. cyber-elements), albeit linked to materials and using energy. However, it is obvious that there is hardly any manufacturing that does not involve digital technology. "The image of a moment when everything would be transferred, transposed, gesture, force, thought, into a perfect double" (Guièze 1983). The engineer or designer, in charge of the efficient, economical and robust production of objects, including those produced via digital technology or for dedicated computing, is the intermediary for achieving this objective.

Industrial revolutions have, since antiquity and despite wars, made it possible for the citizens of the planet to lead a better practical life, and in developed countries, science and technology have made it possible to weave a particularly large number of digital links between humans and machines throughout our planet. From an industrial point of view, a new revolution has recently emerged, that of Industry 4.0, characterized by a whole set of new technologies that merge different worlds: physical, chemical, biological and digital. This new development makes room for different promises such as ubiquity, mobile supercomputers, artificial intelligence embedded in robots, self-driving cars, augmented brains, genetic modification and the death of death (Schwab 2016; André 2019). Within this emerging new world lies, much more calmly, without too much expressed fantasy, additive manufacturing (see Figure I.1), but at its heart, what is it? On this reflexive basis, where does one of its children, 4D printing, fit in?

The adoption of "digital" manufacturing technologies is expected to make actors more efficient, speeding up the time-to-market for innovative new products, helping companies optimize assets and processes, reducing waste and rework, and better matching supply and demand, which should lead to significant productivity impacts for companies, those that have successfully integrated digital technologies into their operations. This is the credo… Regardless of these technological aspects, for Philbeck (2017), we should instead think of production systems, not just technologies. For this author, individual technologies are interesting, but it is their systemic impact that counts. Emerging technologies challenge our societal values and norms, sometimes for the "good", but sometimes also in a negative way. But with a global market of a few tens of billions of €/year, additive manufacturing is still a very small economic market and this consideration, which can be applied to it, will have little effect on society.

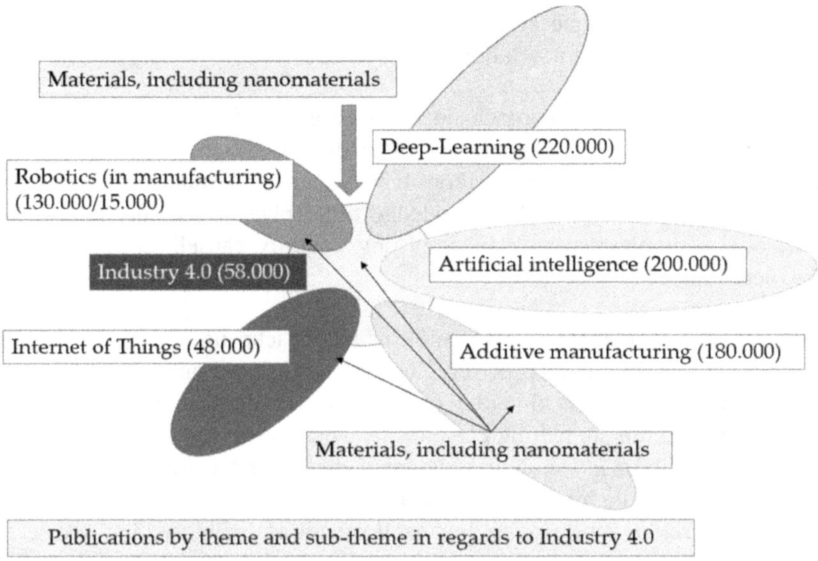

Figure I.1. *Characteristic elements of Industry 4.0 (with the number of scientific publications by elementary "bricks", according to the library of the University of Lorraine). Note that the matter and materials component, which is very "horizontal", is not presented as such (including the nanomaterials)*

An intentional artifact (related to the work of the engineer and/or designer) can be seen as a means of linking an "internal" environment, the substance, functioning and organization of the object itself, and an external environment, the surroundings in which it is implemented. If the two environments are compatible, the artifact meets the specifications. As H.A. Simon (2004) pointed out in another context, knowledge of an artifact such as a machine "has an advantage over knowledge of nature, because it is based on valid prior foundations, the purposes of which will be deviated from with a certain amount of new intentionality to give the projects intelligibility and openings to society". Without us realizing it, the development of digital technology is made up of alternations between centralized infrastructures at the beginning, then distributed with the advent of nomadic computing, then mixed today with increasingly externalized processing and storage services. Therefore the 3D machine is (and this is one of its strengths) a useful peripheral in this field.

> The engineer's job is to mobilize the technical knowledge at his disposal to imagine innovations that represent a compromise acceptable to all stakeholders. Technical perfection is the poison of the economy! One must be prepared to tinker to satisfy the greatest number of people. The resources of technology are such that we cannot believe an engineer who assures us without batting an eyelid

> that there is only one efficient and profitable technical solution and that we must choose it. (Callon and Lacoste 2011)

Thus, in the cultural approach in engineering sciences, the search for an understanding of reality relies largely on "strategies of detour, in other words to the construction of models or the production of virtual facts" (Granger 1995). However, these possible detours exploit the knowledge of a robust possibility and, as far as possible, of the simplest possible by imposing majority causalities and the greatest possible determinism.

It was on this basis that, in 1984, additive manufacturing emerged (André et al. 1984). The idea was to place precisely "at the right place" and "at the right time" material (and/or energy) so that the organization could lead by addition to the desired object (and not something close to it). To achieve this deterministic goal, it was necessary to look for situations where it was possible to take control of the complexity of the whole by choosing specific processes: polymerization induced by light or by electromagnetic waves, localized fusion of fusible wires, powders, etc. (see André (2017a)). The notion of voxel (elementary volume transformed both numerically and physically) intervenes in this process: an object is thus defined by the addition/aggregation of *n* cohesive voxels.

According to Tibbits[1] (2013):

> An unprecedented revolution is underway. It consists of the ability to program biological and physical materials to change shape, change properties [...]. The idea behind 4D printing is to do 3D printing with multiple materials, so you can put multiple materials in, and add a new capability: transformation. As soon as they come out, the parts can transform from one shape to another, directly, by themselves. It is like robotics without wires or motors. You print this part, and it can be transformed into something else.

Who has not dreamed of this after seeing the movie "Terminator 2"? Tibbits (2016) thus defines a new objective to 3D printing with, in anticipation, its inclusion in the context of Industry 4.0 (Jardim-Goncalves et al. 2017; André 2019), with the expression of a new need, because one degree of freedom is added to 3D manufacturing. In Tibbits' highly visual demonstration, "scientific construction is a matter of a logical process from facts, induction, but there is also the work done by the unexpected, which will suddenly allow a thread to be pulled as if a string were

1 Not all early references to the origin of 4D printing come from Tibbits. Elizabeth Yinling Tsai of MIT might have been the first with her 2013 Master of Science report (Tsai 2013) or, somewhat earlier, Bowyer in 2012 at the University of Bath (Jones et al. 2012).

found in a tangled skein" (Vincent 1994). It is with clear messages that carry the imagination that we arrive at a radically new vision with 4D printing.

However, in spite of Tibbits, can we forget these premonitory sentences of Dagognet (1989) who wrote:

> Matter, although it spreads out before us, immediately presents us with a challenge: is it possible to leave the prison of our consciousness in order to explore it, invade it, and then master it? Can we export ourselves? Can we really hope for this fabulous journey and on which vessel? [...] But, because matter can be defined as solidified energy, it is not surprising that an energetic treatment transforms its order or manages to insert a little disorder in its order...

By reducing the cost of the complexity carried by the digital component of a product or a machine with a standardized base, 3D additive manufacturing is already part of this logic. However, with a complimentary opportunity, which is to have original possibilities linked to the resolute evolution in time and space of the form and functionality of the objects created, 4D printing can have specific assets to better cooperate with the other elementary bricks of Industry 4.0. As a reminder, these building blocks are, apart from artificial intelligence, (flexible) robotics and the Internet of Things (IoT). The development of the specific axis constituted by 4D printing can therefore be an important strategic element of integration in the technical progress associated with the Industry 4.0 concept. By design, 4D manufacturing generates a technological inseparability when it comes to producing a 3D part in the printer, made active by energy stimuli, rather than by activities that are organized in different successive steps. This is therefore a major disruptive integration that should revolutionize the 4.0 domain if it can meet the required industrial specifications. But, as Dufourcq (2021), the CEO of the *Banque Publique d'Investissement*, reminds us, "While this [Industry 4.0] transformation may appear complex for industrialists, many start-ups, meanwhile, do not have the keys to penetrate the industry biotope. To succeed in this challenge, we need to create transversality." The debate on the integration of knowledge that is too separate is therefore not over!

However, before we talk about 4D printing, do we know what we are talking about when we talk about 3D printing or additive manufacturing? It seems clear to everyone, but if you dig a little deeper, there are no clear boundaries between the so-called additive processes and many other assembly processes. Indeed, many researchers forget that manufacturing technologies are only a means to an end and not an end in themselves. It is on this initial aspect of digital manufacturing that we wanted to come back before trying to identify what should represent this legitimate, still fragile, child of 3D printing, the 4D. In fact, the questions asked serve to present singular elements that will be dealt with in the various chapters of this book, illustrating, in particular, the locks and the search for a possible "user manual" that is attractive to the user.

> The difference between the spirit of geometry and the spirit of finesse: In the former the principles are palpable, but far from common usage, so that one can hardly turn one's head that way, for lack of habit. But, as soon as we turn our head, we see the principles in full, and we would have to be quite wrong-minded to reason badly about principles so large that it is almost impossible for them to escape. But, in the spirit of finesse, the principles are in common use and before everyone's eyes. All one has to do is turn one's head or do violence to oneself; it is only a matter of having good eyesight. But it is necessary to have good eyesight, because the principles are so loose and in such great number that it is almost impossible for them to escape. (Pascal 2004)

I.1. Attempt to define 3D printing

> The authenticity of a thing integrates all that it contains that is transmissible by virtue of its origin, its material duration, and its historical testimony. This testimony, based on materiality, is called into question by reproduction, from which all materiality is removed. No doubt only this testimony is affected, but in it the authority of the thing and its traditional weight. (Benjamin 2000)

If A.C. Clarke is possibly at the origin of the idea of the invention, in his numerous forecasts, of a machine to replicate any object, in 1964 (https://www.openculture.com/2014/06/in-1964-arthur-c-clarke-predicts-the-internet-3d-printers-and-trained-monkey-servants.html), other authors thought of this same type of operation, like Hergé (in 1972, according to http://tpebac-impression3d.e-monsite.com/pages/l-imprimante-3d/histoire-de-l-imprimante-3d. In the cartoon *Tintin and the Lake of Sharks*, Professor Calculus invents a three-dimensional photocopier immediately coveted by Rastapopoulos in order to make fakes by duplicating works of art stolen from major museums) and Franquin in France (published in 1960, according to https://fr.wikipedia.org/wiki/Les_Petits_Formats, in the daily newspaper "*Le Parisien Libéré*" and then in the form of a comic book in Belgium. However, it was not until 1984 that the first patents on this increasingly topical subject were issued (17% annual increase) (André et al. 1984; Hull 1984)).

For "l'internaute" (https://www.linternaute.fr/dictionnaire/fr/definition/imprimante-3d/), the proposed definition is:

> Printer with the capacity to produce three-dimensional objects from a model previously defined in a computer document.

Futura Sciences (https://www.futura-sciences.com/tech/definitions/imprimante-3d-impression-3d-15137/) give the definition:

> 3D printing is a so-called additive manufacturing technique that proceeds by adding material, unlike techniques proceeding by removing material such as machining. 3D printing can be used to produce everyday objects, spare parts or prototypes for testing. The starting point is a computer file representing the object in three dimensions, broken down into slices. This information is sent to a 3D printer which will manufacture the object by adding successive layers.

According to Wikipedia (https://en.wikipedia.org/wiki/3D_printing),

> 3D printing, or additive manufacturing, is the construction of a three-dimensional object from a computer-aided design (CAD) model or a digital 3D model. The term "3D printing" can refer to a variety of processes in which material is deposited, joined, or solidified under computer control to create a three-dimensional object, with material being added together (such as plastics, liquids, or powder grains being fused together), typically layer by layer.

The 3D file of the part is processed by a specific process planning software that organizes the slicing of the part in terms of layers. The layers and related printing parameters are sent to the 3D printer, which deposits or solidifies the material layer by layer until the final part is obtained. The principle remains close to that of a classic 2D printer with this major difference: it is the stacking of layers that creates the volume. It is defined by the AFNOR (2011) as "all the processes allowing to manufacture, layer by layer, by adding material, a physical object from a digital object". Finally, for Cult 3D (https://cults3d.com/fr/blog/articles/definition-impression-3d):

> Every 3D printer builds parts according to the same principle: a digital model (called CAD file or 3D model) is transformed into a three-dimensional physical object by adding one layer of material at a time. This is where the alternative term additive manufacturing comes from. 3D printing is a fundamentally different way of producing parts than traditional subtractive (CNC machining) or formative (injection molding) manufacturing technologies. In 3D printing, no special tools are needed (for example, a cutting tool with a certain geometry or a mold). Instead, the part is manufactured directly on the platform in a layer by layer manner, which leads to a unique set of advantages and limitations.

We could have continued this small exercise, but we would have quickly noticed a certain continuity in the remarks: a computer, a program of realization of an object, a printing by layer (e.g.) using a transformable matter thanks to or because of an energy contribution. With mono-materials, the manufacturing elements (the voxels – elementary volumes) are assembled in the continuity of the manufacturing process (case of stereolithography). The generic designation of "additive manufacturing" allows us to differentiate this type of process from classical manufacturing processes, which work by elimination of material (subtractive manufacturing), but not really from the manufacturing of molds. This ambiguity on the definition of additivity is amplified by the existence of recent 3D manufacturing processes without going through the layers step (see de Beer et al. (2019), Kelly et al. (2019), van der Laan et al. (2019), Mayer et al. (2020), Loterie et al. (2020), Regehly et al. (2020) and Doualle et al. (2021)). It continues the continuous aspect of manufacturing when several materials are involved. With processes using, for example, dynamic masks to polymerize an entire surface, the voxel size is not unique! Moreover, the rivets of the Eiffel Tower are now replaced by welds without loss of material, welds managed by computer in the automotive industry… Moreover, the BeAM process (https://www.beam-machines.fr/) is a 3D process based on this principle.

So what is a reasonable definition of 3D printing processes? Basically, what is clear is that we have a digital process, a transformation of a material in a machine to produce an object normally in a single operation. This can eliminate assembly processes, multi-material production methods, post-processing operations (annealing, apodization, etc.), etc. Table I.1 provides a comparison between different manufacturing approaches, with some reminders of their advantages and disadvantages.

Process	Digital	MT/ Machine	Item in 1 step	Advantages	Disadvantages
Additive manufacturing or 3D printing	YES	YES	YES	Re-design, customization, material savings, flexibility	Voxel size and resolution, not so green, specialized method; availability of materials
Mass construction	YES	YES	YES	Robotization, welding assembly	Fixed structure
Industrial molds	YES	YES	YES	Material gain	Fixed structure; specific materials
Building construction (see Mayer (2018))	YES	YES	YES	Either cemented bricks or liquid cement deposit	Question of design to realize the supporting structures; design
Machining	YES	NO	YES	Proven and well-controlled processes	Material loss (subtractive manufacturing)

Table I.1. *Some elements of comparison between different computerized assembly processes (MT: Molding Technique)*

For their part, Alghamdi et al. (2021) consider 4D printing as a technology that produces shapes whose properties evolve on a time basis. They write,

> 4D printing has demonstrated its potential to fabricate self-repairing, self-assembling and self-adapting intelligent components [...]. 4D printing faces many obstacles that it must overcome to reach its full potential in manufacturing technology. Some of the main obstacles include reduced mechanical strength of materials, longer response time to stimuli resulting in slow shape changes.

These authors propose four main approaches to 4D printing: self-assembly of elements, bi-stability, strain shifting and shape memory. This provides a foundation that will be expanded upon in this book.

However, despite this proposal, it is still very difficult today to propose a clear definition of what 3D printing represents, even if everyone has integrated in their imagination the low-cost molten wire 3D printer (which is a material extrusion process) placed on their desktop or in their garage. However, as André (2017a) reminds us, 3D technologies are numerous and meet the criteria presented in Table I.1. For all that, we can distinguish the production of parts in mono-materials, which correspond to the main applications (homogeneous 4D printing), then hybrid fabrications, using a single machine, with various materials (hybrid 4D printing) and finally heterogeneous processes, which use 4D manufacturing principles in part (heterogeneous 4D printing). Figure I.2 shows these three definitions.

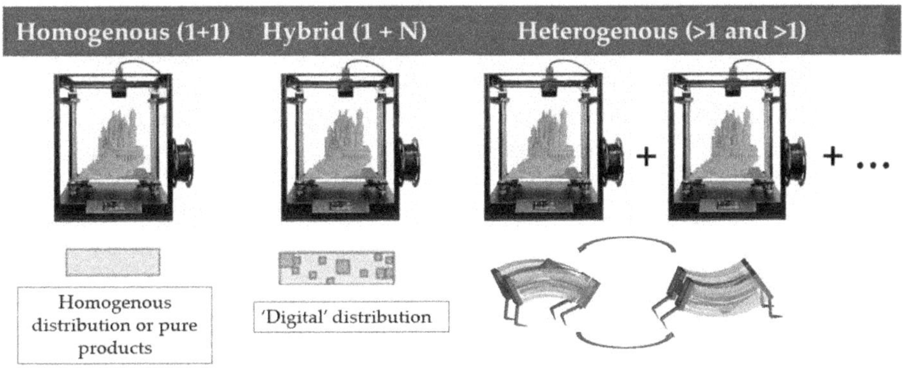

Figure I.2. *The different forms of 4D printing (e.g. it is possible to do hybrid 3D printing with a multi-material machine and heterogeneous 3D printing when we couple different processes). For a color version of this figure, see www.iste.co.uk/demoly/4Dprinting1.zip*

For Magistrelli, heterogeneous manufacturing processes combine additive and subtractive manufacturing processes.

> These are implemented sequentially or incorporated, with proper control of component attachment and orientation. Such a process is used both to improve dimensional accuracy and to speed up the overall manufacturing process. [Heterogeneous] processes can be used to solve complex production problems where a single manufacturing method (subtractive or additive) is not sufficient. (Magistrelli 2021)

Robots can be used for this combination to produce complex parts.

André Mailfert, a very talented cabinet-maker (and a bit of a forger), wrote a famous book in 1968 (for those who love antiques) in which he wrote that a piece of furniture remains authentic as long as it contains more than 50% of the materials present at the time of its manufacture! It seems difficult to define such a limit here, because it would be necessary to know what these 50% represent. The vagueness will therefore be maintained.

Karry Stevenson (2021) citing a survey reminds us that respondents consider 3D printing, in its "fuzzy" definition, to be of interest for the following reasons:

– high performance of 3D parts (46%);

– reduction in design and production costs (46%);

– speed of execution (45%).

This is a change related to the non-reproduction of parts made from other proven machining technologies, which illustrates more redesign. With a radically new design, it is possible to make a part that is as strong, but less heavy, leading to an improvement in the performance of the mechanical elements that can largely compensate for the possible extra costs of 3D printing. Moreover, as already reported in André (2017a), there is the possibility of reducing the number of steps between design and production, which can lead to considerable time savings.

The Essentium (2020) survey, cited by Stevenson, also shows other insights:

– 98% of managers in the manufacturing industry consider an improvement in the quality of the materials used in their products.

– But 37% are held back by the high cost of materials used in 3D printing.

– 24% of respondents said that 3D printing materials are unreliable (see also André (2020)).

This finding illustrates that the future of 3D technologies lies in materials, if only because of the modest, even limited supply of materials available for producing 3D objects, although efforts by materials producers (with partnerships with machine manufacturers) are moving in the right direction. However, for Stevenson (2021), "it is a tedious process, as certification of materials requires a lot of testing, especially since there are so many possible 3D printers and processes. The good news is that this work is well underway, and will eventually remove the material barrier."

Furthermore, while there is a wide range of rules for taking advantage of conventional manufacturing processes, according to Cohen (2021), 3D printing breaks many of these rules by introducing production capabilities that are not compatible with conventional manufacturing. For 3D printing, conventional manufacturing parameters should be available for comparison with those of additive manufacturing. According to this author, design engineers make an initial selection of criteria by induction. A database of components and assemblies should then be available to evaluate the potential of 3D printing in order to select the best design and production methods to lead to robust decision support. Figure I.3 represents a theoretical framework for defining the suitability of 3D as a production method (Cohen 2021).

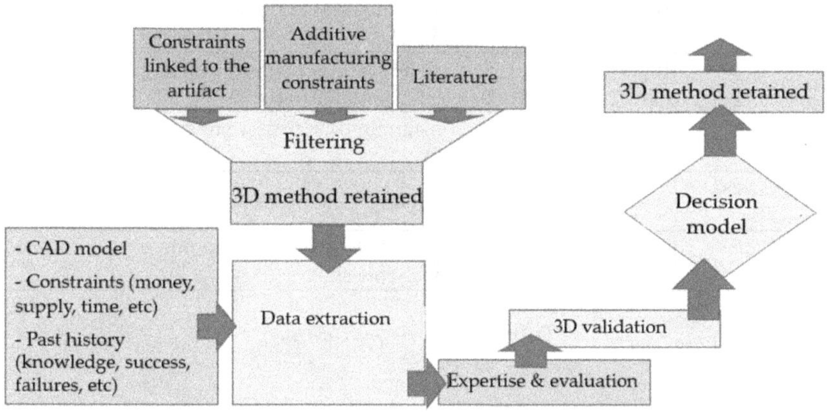

Figure I.3. *Theoretical framework for the suitability of additive manufacturing in industrial production*

Other methodologies are also proposed, such as design analysis, which consists of identifying the parts where 3D printing can be used effectively, based on the knowledge of the following criteria: function integration, customization, lightweight design and operational efficiency. Cohen (2021) considers this procedure to be limited because it does not take into account all the operational factors of the production process.

Finally, he introduces Senvol, a company that provides data in order to implement 3D technology; it defines seven supply chain scenarios where 3D can be profitable. Cohen writes:

> Verifying the [3D] candidacy through this approach is fairly straightforward: if a component meets the requirements of one or more scenarios, then additive manufacturing should be explored further. However, if none of these scenarios are satisfied, then [3D] is probably not the ideal technology process to exploit. (see Table I.2)

Box I.1. *Is this lack of definition detrimental to development?*

Scenario	Description/Comments
Manufacturing cost	The cost of the parts because they are complex, have high fixed costs (e.g. tooling) or are produced in low volumes. In this case, 3D can be more profitable.
Turnaround times too long	Is the production time too long? Are the capital costs too high? Should the time to market be accelerated? 3D can help get parts to market faster.
High inventory costs	If the question of storage is an issue, 3D can allow on-demand production that reduces stocks.
Exclusive	If parts are proprietary, there may be a supply risk. By qualifying a 3D part, it is possible to avoid sourcing from a single supplier.
Remote locations	If it is difficult, time-consuming or expensive to ship parts, 3D can enable on-site production.
High import/export costs	If import/export costs are high, local production can limit these costs.
Performance improvement	3D printing allows an object to be redesigned to improve performance beyond what was previously possible.

Table I.2. *Senvol program*

This approach is easy to use; it is not very objective and it does not differentiate between the different additive manufacturing techniques, allowing for informed choices. In this situation, it is therefore difficult to drive a fabrication laboratory with robustness toward one of the additive manufacturing methodologies, with the right material.

These economic, process and performance situations do not involve the question of definition, but they do highlight important difficulties in finding the "right" material or the "right" process for a given specific application (when there is already a market of several tens of billions of euros/year) and in knowing what the ideal path is for a given production. With a market that is about a hundred times smaller and a need for new "material" performance, new manufacturing processes, various ways to change the form or the functionality of a 4D object, these difficulties are likely to be deeply amplified.

> Good old chance; occasionally chance guides our hand, and the wisest providence could not invent music more beautiful than what our foolish hand then produces. (Nietzsche 2007)

I.2. What about 4D printing?

> If Men could regulate all their affairs according to a settled opinion, or if fortune were always favorable to them, they would never be in the grip of any superstition; but they are often reduced to such an extremity that they cannot stop at any opinion, and most of the time, owing to the uncertain goods of fortune, which they desire without measure, they float miserably between hope and fear; that is why they have souls so prone to believe anything. (Spinoza 1997)

Similar to 3D printing, 4D envisages displacements or functionalities induced by stimulations by considering the parameter of time, in order to continuously transform an object made by 3D printing, subjected to the will of its creator by the control of the material and its stimulation. This is all well and good, but is there not a risk of falling into the trap of a teleological illusion in which we decide to designate the ultimate goal of the entire evolutionary process in advance, while masking its more or less radical unpredictability?

Published in 2020, the CNRS's objective and performance contract confirms the strategy chosen for this work:

> Faced with a growing demand for multifunctional materials with optimized, combined, or even extraordinary properties (metamaterials), the classic "materials by design" approach consisting of playing only on the chemistry and nano-structural refinement of materials or on their macroscopic form may reach its limits. In order to overcome these limits, it is necessary to exploit an intermediate scale, the "meso-structure", in which the architecture of materials is controlled by geometric optimization (topology and morphology) of heterogeneities and/or hybridization: cellular, composite or fibrous materials, sandwich structures and structured surfaces, entangled materials, materials with active meso-structure, programmable materials, etc. The use of concepts from database mining and artificial intelligence coupled with robotic or automated systems is an avenue that deserves to be explored. This new "materials by design" approach must be closely linked to the optimization of the processes for the elaboration and shaping of these materials: shaping of composite materials, 3D and 4D printing, etc. It gives the mechanic specializing in materials and structures a central role.

The five-year objective is to design new (meta)-architectured materials that break with existing solutions, making substantial progress on:

– optimization algorithms for meso-structures to better understand and integrate their instabilities and strong material nonlinearities;

– processes for the preparation of these new materials.

Box I.2. *The CNRS vision (2020)*

Between materials, metamaterials, their shaping and design, even if the expression "4D printing" is not mentioned, it is clear that this field is fully in line with the strategy of the CNRS in its priority orientations – engineering themes. Would we have succeeded in convincing this research organization to which we belong or are we attached of the merits of an appropriation of the 4D theme, or even more? We believe that this work will be able to support this national strategy.

Moreover, in another framework, Stiegler (2019) reminds us that, for Malthus, all problems are rooted not only in change, but also in differences in rhythms that she calls "heterochrony". This concept is interesting in that it highlights the fact that an active matter, subjected to a stimulus, can react with temporal gradients of behavior. Furthermore, it is possible to activate only one part of the object or to have a multi-material system (e.g. one active, the other passive) combined in a particular way. The active and activated component acts while the non-active or non-activated one is affected, which, in principle, leads to a great fertility of possibilities. However, in an object where all the elements are linked together, the interactions between the elements mobilized in the transformation affect the others (although today we have no confirmed and robust control over the modeling of transformations). Is there not a risk of being condemned to dealing with simplistic systems or, rather like Blanqui, or, in another framework, Darwin, to trial and error in examining the results? The question of having a liberated intelligence to guide the optimal design of a "missioned" object has been and will continue to be raised.

Lippmann (1937) wrote, on this type of subject: "Intelligence must cut out abstract, isolated and artificially simplified patterns from the infinite complexity of the real world". However, are we sure to find cuttings that will allow us to manage, for example, movements or changes in functionality (chemical reactivity, color, etc.)? Figure I.4 shows the general principle associated with the transition from a traditional additive manufacturing process to 4D printing: first of all, it is necessary to be able to build, at least in part, an object by one of the known 3D processes with adaptive or rendered adaptive materials, and secondly to be able to make its shape evolve from one or more specific stimuli. There is thus a complex coupling between materials and processes in this operation. This principle naturally leads to the realization of relatively flexible (or locally flexible) parts that must be able to deform or change their functionality in the presence of a stimulus, on the one hand,

and homogeneous or heterogeneous parts that are manufactured, if possible, using "classic" 3D printers, on the other.

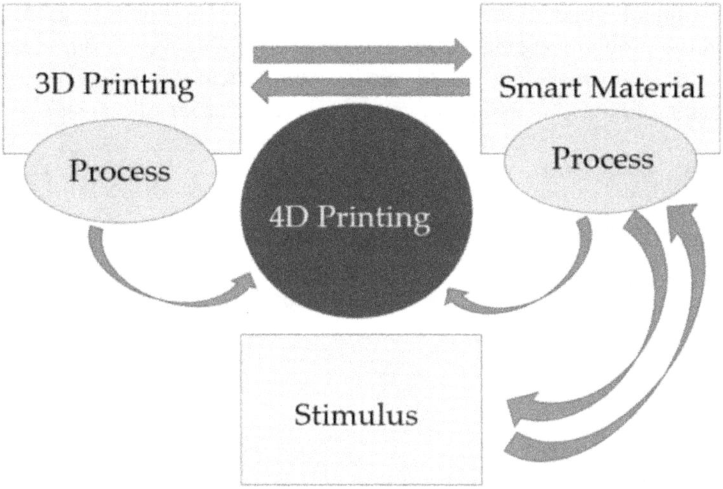

Figure I.4. *Process-material coupling in 4D printing. For a color version of this figure, see www.iste.co.uk/demoly/4Dprinting1.zip*

Indeed, most of the 3D processes, commercial ones in particular, use, as shown in Figure I.2, only one type of matter or only one material (or even a few, with the technique of deposition of molten wires) with a single digital 3D manufacturing machine, which has led to 4D processes respecting the same rules of use or possibilities as classical 3D processes. Figure I.5 shows different ways to "play" with the material by judiciously positioning materials with different properties so that the object created can evolve over time, or by creating metamaterial structures with specific deformation properties under specific stimuli.

DEFINITION.– *Metamaterials*: this term denotes artificial media, generally structured in a periodic (or even aperiodic) manner, with specific mechanical properties.

In 4D printing, it is essentially a question of working with materials that are already in use and whose properties are known (for reasons of ease of construction and use of commercial machines), distributed in an optimized manner in either their pure form or not in the 3D process. However, we will find the same difficulties as for 3D printing processes. Indeed, in the publications dealing with new materials which could be successfully applied to new scientific directions, the authors of these scientific presentations have, in general, manufactured materials for 4D printing, in sufficient quantities to carry out their proofs-of-concept and characterize their original materials. It is therefore difficult to obtain these materials, which imposes radical choices that are likely to be detrimental to the development of the theme. It is not a question of good or bad will, but of technical impossibility; designers of original active products are not in the business of producing custom materials, and there are no companies that could manufacture them (at reasonable prices) in quantities suitable for tests that go beyond the level of proof-of-concept.

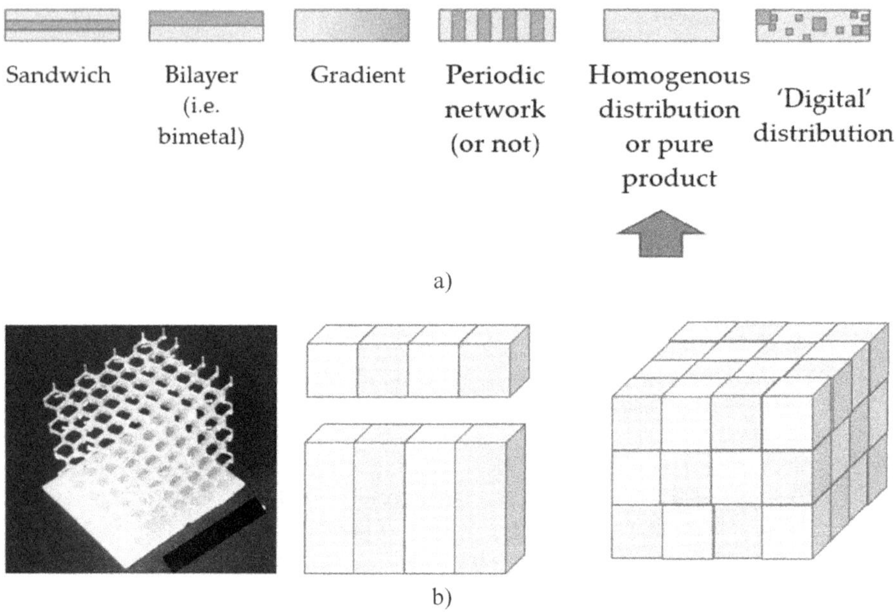

Figure I.5. *How to change the spatial properties of a structure over time: a) 2D system; b) 3D system according to Castles et al. (2016). For a color version of this figure, see www.iste.co.uk/demoly/4Dprinting1.zip*

> Evolution is not a very imaginative mechanism capable of inventing what it needs, but it is extremely efficient at cobbling together by recycling what it has. (Durand 2018)

With the same or different materials, having variable responses in the presence of stimuli of various origins, the shapes of the structures can evolve spatially and over time. Figure I.6 collects some typical effects sought for in 4D printing (André 2017c; Yao et al. 2018).

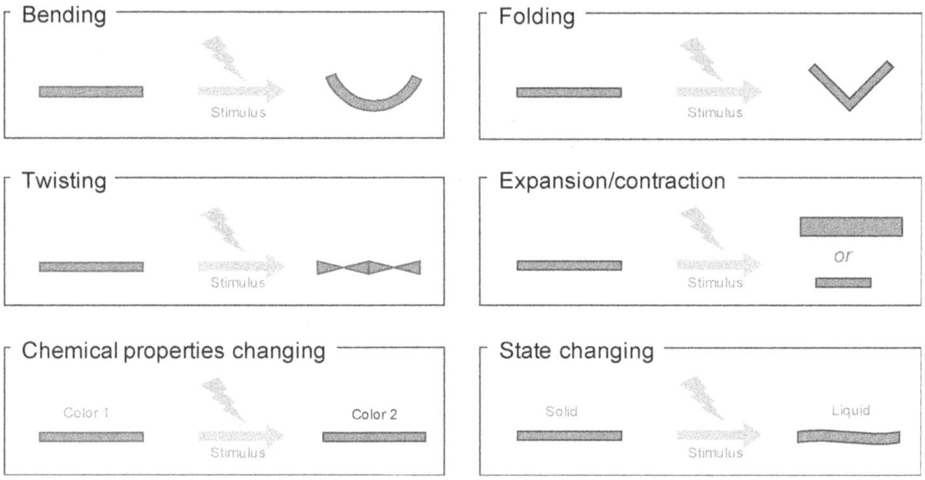

Figure I.6. *Change opportunities of pure and/or associated materials that can be exploited in 4D printing (from Dimassi et al. (2021)). For a color version of this figure, see www.iste.co.uk/demoly/4Dprinting1.zip*

If, for example, we look at an electric motor, it is made up of an assembly of elements of various shapes, materials and origins; it is then powered by electrical energy to produce the desired rotation. If we manufacture the different parts that make up this motor using 3D printers, can we talk about 4D printing after assembly, since we will be respecting the criteria of additive manufacturing, active materials and stimulation? The same will be true for many actuators? Then how can we define a set of formal or even informal (but accepted) rules and standards, whose aim is to ensure the production and rapid dissemination of reliable fundamental knowledge on a subject that we have managed to define? In what way does the organization of science around clear norms, operationalized by concrete principles of realization, offer a coherent answer to this problem of definition? To return to the spirit of the definition of 4D printing, we find the same difficulties in framing the subject as for additive manufacturing. In this book, when we talk about homogeneous 4D printing, it means that we have made an object with a single material leading to a single active material. If several materials are involved in a 4D printing process using a single machine, it will be called hybrid 4D printing; finally, if several manufacturing processes are involved, including, necessarily, a 3D printing process, it will be called heterogeneous manufacturing.

NOTE I.1.– With FFF (fused filament fabrication) or SLA (stereolithography) type processes, the material used to make the object can be pure or filled. For us, it will be a question of bringing the manufacturing into the hybrid processes, even if the distribution of the fillers is homogeneous in the polymeric matrix (FFF) or monomeric (SLA).

NOTE I.2.– If other methods of manufacturing contribute to the realization of the 4D object, for example, subtractive manufacturing machine tools and digital connections of these machine tools for different purposes, the authors will not present them in this book. They may be mentioned, but representing domains considered as disjoint, their contribution to the 4D domain is considered as subsidiary.

In this concern of a certain coherence, it seemed necessary to us to have a fundamental knowledge base or a corpus of refocused generic knowledge, allowing a global comprehension of the phenomena involved in 4D printing.

I.3. An "explosion" of complexities in 4D printing

> To search is to invent the world, to set new rules for an ephemeral world. Not like tyrants who also invent another world, but impose it on others. The researcher does not remake the world, he undoes it to make it. He imagines one, then confronts it with the real world to enlighten it, not to exhaust it. Research is a never-ending quest. (Rose 2001)

We have already mentioned the difficulty of finding the right materials for 3D printing. However, we have gone from a certain self-sufficiency (even if progress is required) to deep and complex interdependencies. Indeed, to this first problem, which will be dealt with in the chapters of this book, we can add other critical elements, that of performance, that of stimulation and its implementation (with the associated heterochronies), the design and usage software, etc. In short, with all these obstacles the die is far from being cast! In this open setting, the fragmentation of disciplines undoubtedly allows for participation in scientific competition, but even if we think that we have a possible reserve of creative and adaptive potential, this dynamic must be integrated into a responsible interdisciplinary convergence approach in order to move the 4D subject forward toward social utility.

I.3.1. *Stimulation process*

For the moment, we have not expressed ourselves on the stimulation processes; these can be global or localized, each with its own laws of interaction with the active materials inducing variable temporalities (heterochronies). These are typically effects related to heat, humidity, chemistry, light, mechanical effects, electric fields and magnetic fields. By playing between local and global, there are already 14 opportunities to "play" with.

I.3.2. *Materials*

> The ornament is no longer really separated from its support: the one and the other are produced together, by the same procedure. It is impossible to discern two layers, one simply supporting the other. (Dagognet 1989)

In homogeneous 3D printing, it is possible to produce objects in polymers and metal (see, for example, Pinho et al. (2020)). If we move to hybrid, the number of candidates is practically infinite as long as we know how to bond the materials together (by different cohesive techniques). The number of candidates is even greater if, as illustrated in Figure I.4, objects of any shape (but precisely defined by the designer) can be made with different materials, active or not, distributed over the surface or in the volume of the object. This radical increase can be further amplified if one seeks to realize 4D objects from a heterogeneous process where one can integrate parts of the object made outside 3D printers. This can involve materials with a concentration gradient (already possible in hybrid 4D printing), and also those with surface treatments. Moreover, as for the electric motor used previously as an example, the choice of passive or active materials, their spatial distribution, is difficult. Difficult, in fact, because there are no databases or manufacturing habits on a subject as recent as 4D printing to guide the designer to make the "right" choice.

To go a little further in the enumeration of the questions associated with the materials domain, we have to consider the existence of metamaterials, which hold their specific properties from their form and their structure.

I.3.3. *Controlling deformations*

Apart from simple structures such as origami (which produces spectacular results), as soon as we create a 4D printed object that is a little more complex than these basic elements, interdependencies will appear between inert matter and active matter, between the aspects of heterochrony already expressed and the avoidance of collisions between mobile elements (e.g. in flexible robotics), taking into account criteria of mechanical performance and temporal responsiveness, etc.

In short, trying to discuss 4D printing is a very risky operation because in this complexity the links between all the influencing factors, the selection of materials, stimuli, their form and their temporalities of action constitute a game whose rules are not known (which induces risks, but a very open field of action). In this spirit of trying to find causalities, Momeni et al. (2017) propose possible applications of 4D printing in Figure I.7.

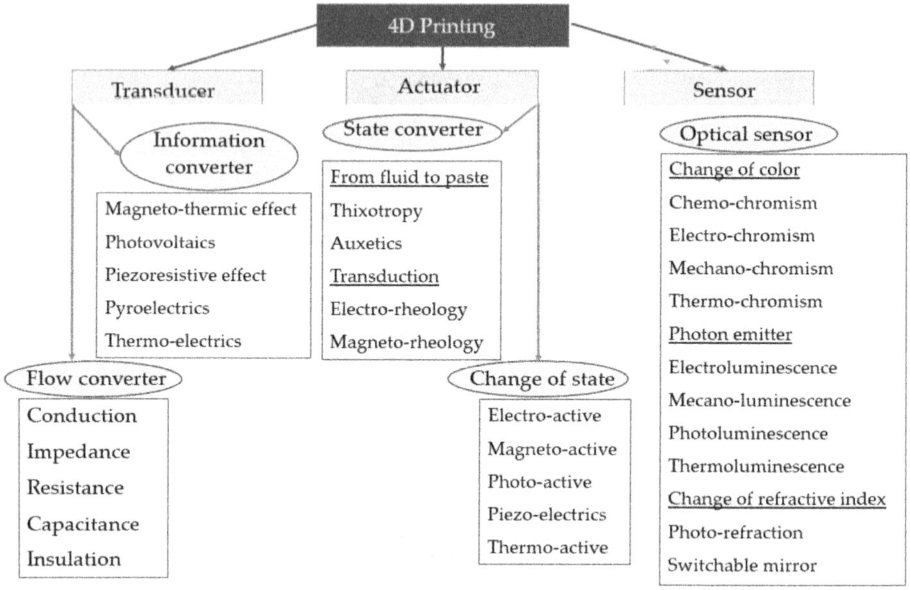

Figure I.7. *Potential application of 4D printing in terms of functionality (according to Momeni et al. 2017)*

The potential for application, even if it is based on "inverse problem" type processing principles that are generally difficult to implement, is important, because it makes it possible to deal with relatively simple problems in certain niches, in order to make progress on a very open topic.

> Impulses are pivots around which the reorganization of activities revolves, they are the agents of deviation, giving new directions to old habits. (Dewey 2008)

I.4. Conclusion

> In cultural terms, no enterprise is built on dreams alone, and none is built apart from them. Action, successful action, is of necessity the result of practical considerations. But the goal of all action is explicitly defined by the deep nature of the human being, his dreams,

> his vision of life, his culture. The dynamics of life, the challenge of risk and uncertainty require a new effort of creativity that will lead us to the reconstruction of the notion of progress, the one that philosophers and ideologists of certainty have undermined and almost destroyed. (Giarini and Stahel 1990)

Are the principles of causality that link the different phenomena quickly mentioned in this introduction (on which science is normally based) usable here because they are mastered, or are they the result of an inductive approach (one that is still modest because of the youth of 4D technology), or are they simply fiction? Perhaps we should wait for a fortunate but uncertain chance in its occurrence to find original and useful solutions. "Theory is when you know everything and nothing works; practice is when everything works and nobody knows why. Here, we combine theory with practice: nothing works and nobody knows why!" (Einstein 2007). But this situation of expected opportunity would certainly not have satisfied an empiricist philosopher like David Hume; so one of the aims of this book is to try to find regular and robust sequences for moving forward and innovating on this topic. "We too often forget that specialists are produced from amateurs, as the military is produced from civilians" (Latour 2007). However, in the teleological approach that must be taken for the exploitation of a concept with a view to its societal usefulness, will these rational chains suffice to find the right path(s) in the logic of an "inverse problem"?

Unlike research that is aimed at incremental innovations, as is increasingly practiced in 3D printing, there are forms of empiricism, learned tinkering, heuristics and management of uncertainty in the approach associated with 4D printing, which are intended to lead the reader toward a permanent, ethical questioning of all progress, of all radical evolution. So, for tomorrow, it will undoubtedly be possible to envisage new ways of creating 3D/4D objects. Today, we have a fairly good command of synthesis, of the way in which the object is constructed. However, we could also ask ourselves if it would not be possible to develop systems in which the object would be allowed to have an intentionality, and thus to entrust it with the choice of looking for what it needs to make itself, and thus to move on to self-organization with the selection of the necessary elements that it would extract from a "bank" for the construction of the final object.

> This would go beyond 4D printing, which tackles the assembly and installation of products that should be able to assemble themselves and their installation in extreme environments, all of which could be facilitated by "Programmable Matter". Programmable materials and objects that build themselves would thus make assembly plants and cumbersome installation procedures superfluous [...] Robotization, the heart of the 20th century's productivity gains, would thus be

> integrated into the products themselves, with, one can imagine, a few ethical problems to consider. (France Stratégie 2017)

Basically, for one of the authors (the older one), the technology does not discover anything new and is, perhaps, not so admirable as that, with aspects of “angling” (even if it is exciting and one can be tempted to introduce a little rationality in this field); what was essential, and this remains true today, was to conquer the material with intensifications on extensions of new ideas and tensions in the competition, whereas currently 3D technologies are accepted elements of economic development with fewer and fewer breakthroughs and more and more incremental innovations. The adventure is summed up, resolved, and it is not finished in a singular sequence of hermeneutic movements, carried out as a journey inside its laboratory(ies).

There is indeed much room for creativity and trial and error (so long as it does not cost too much). As will be shown, changes in thinking and action will be required. They correspond to the conceptual shift from the economy enabled by emerging and desirable technologies to a new economy of creativity which, in a second phase, will have to take better account of emerging social, economic, organizational, geopolitical and even environmental constraints. It is a form of “design thinking” that should be taken into consideration. With 4D printing and its promises, what would be desirable would be to show, through changes negotiated with the stakeholders concerned (some of whom are mute), the path to a better exploration of complexity, that we can do well, if not better, perhaps with less equipment, but differently. Thus, this fertile situation does not result in a praise of interiority, quite the contrary. This book, because it is part of the exercise, could constitute a turning back on oneself. We owe it to the reader to avoid this pitfall!

Let us dream about this future to be built together. Paradigmatic changes will certainly have to take place in the near future. Scientific initiatives that are marginal today and remain aporetic in the current world of research, which is plunged into uncertainty, if not in crisis, could turn into a new scientific era, which is less restricted, that could be applied to 4D printing. The approach attempted in this book is therefore to try to question a present (whose likelihood of achieving all its objectives is impossible to know) and to determine the conceptual elements that could lead to an original future, based on new ways of learning about and accessing knowledge, with the proactive discovery of new application niches, all by exploiting revisited paradigms and departures from work habits as far as possible. As Barbara Stiegler (2019) reminds us, in a quote applicable to 4D printing:

> In the crisis of drift created by the crisis of absolutes and traditions, but also by epistemological fallibilism for which no science can ever conquer definitive certainty, the only guide that can restore some

> control over events is precisely science, but understood in a new way, as a collective work of cooperation and pragmatic testing of experimentation.

It is followed by this comment by Lippmann (2015): "Mastery [of the question], whether you like it or not, is an immense collaboration"! Cheers!

> Our mind, which seeks solid points of support, has as its principal function, in the ordinary course of life, to represent states and things [...]. It substitutes the discontinuous for the continuous, mobility for stability, and fixed points for the changing tendency. (Bergson 2009)

> Imagination is awesome in that it produces images that magnify reality and properly invent it. (Guérin 2015)

> What is called union in a body [...] is a very equivocal thing: the true one is a union of harmony, which makes all the parts, however opposed they may seem to us, contribute to the general good of society, as dissonances in music contribute to the total agreement. (Montesquieu 1968)

I.5. References

AFNOR (2011). Fabrication additive – Vocabulaire. NF E67-001, October 2011, Paris.

Alghamdi, S.S., John, S., Choudhury, N.R., Dutta, N.K. (2021). Additive manufacturing of polymer materials: Progress, promise and challenges. *Polymers*, 13, 753.

André, J.C. (2017a). *From Additive Manufacturing to 3D/4D Printing 1: From Concepts to Achievements*. ISTE Ltd, London, and John Wiley & Sons, New York.

André, J.C. (2017b). *From Additive Manufacturing to 3D/4D Printing 2: Current Techniques, Improvements and their Limitations*. ISTE Ltd, London, and John Wiley & Sons, New York.

André, J.C. (2017c). *From Additive Manufacturing to 3D/4D Printing 3: Breakthrough Innovations: Programmable Material, 4D Printing and Bio-printing*. ISTE Ltd, London, and John Wiley & Sons, New York.

André, J.C. (2019). *Industry 4.0 – Paradoxes and Conflicts*. ISTE Ltd, London, and John Wiley & Sons, New York.

André, J.C. (2020). Recyclage des déchets d'impression 3D. *Techniques de l'Ingénieur*, BM7975.

André, J.C., Le Méhauté, A., De Witte, O. (1984). Dispositif pour réaliser un modèle de pièce industrielle. French Patent no. 84 11 241, 16 July 1984.

de Beer, M.P., van der Laan, H.L., Cole, M.A., Whelan, R.J., Burns, M.A., Scott, T.F. (2019). Rapid, continuous additive manufacturing by volumetric polymerization inhibition patterning. *Sciences Advances*, 5, eaau8723.

Benjamin, W. (2000). L'œuvre d'art à l'époque de sa reproductibilité technique. In *Walter Benjamin Œuvres 3*, Benjamin, W. (ed.). Gallimard, Paris.

Bergson, H. (2009). *La pensée et le mouvant*. PUF, Paris.

Blanqui, A. (2012). *L'éternité par les astres*. Impressions nouvelles, Paris.

Callon, M. and Lacoste, A. (2011). Défendre l'innovation responsible. *Debating Innovation*, 1, 5–18.

Castles, F., Isakov, D., Lui, A., Lei, Q., Dancer, C.E.J., Wang, Y., Janurudin, J.M., Speller, S.C., Grovenor, C.R.M., Grant, P.S. (2016). Microwave dielectric characterization of 3D-printed BaTiO3/ABS polymer composites. *Scientific Reports*, 6, 22714.

CNRS (2020). *Contrat d'objectif et de performance 2019–2023 entre l'Etat et le Centre National de la Recherche Scientifique*. CNRS, Paris.

Cohen, J. (2021). Comment savoir que la fabrication additive idéale est la technologie idéale pour une production ? [Online]. Available at: https://3dadept.com/comment-savoir-que-la-fabrication-additive-ideale-est-la-technologie-ideale-pour-une-production/.

Dagognet, F. (1989). *Rematérialiser*. Vrin, Paris.

Dewey, J. (2008). Human nature and conduct: An introduction to social psychology. In *The Middle Works of John Dewey*, Boydston, J.A. (ed.). Southern Illinois University Press, Carbondale.

Dimassi, S., Demoly, F., Cruz, C., Qi, H.J., Kim, K.Y., André, J.C., Gomes, S. (2021). An ontology-based framework to formalize and represent 4D printing knowledge in design. *Computers in Industry*, 126, 103374.

Doualle, T., André, J.C., Gallais, L. (2021). 3D-printing of silica glass through multiphoton polymerization process. *Optical Letters*, 46, 364–367.

Dufourcq, N. (2021). Industries : la French Fab en marche vers le 4.0 [Online]. Available at: https://www.entreprendre.fr/industries-la-french-fab-en-marche-vers-le-4-0/.

Durand, T.C. (2018). *L'ironie de l'évolution*. Le Seuil, Paris.

Einstein, A. (2007). *L'espoir économique : vers la révolution du quaternaire*. Bourin, Paris.

Epictetus (2015). *Le manuel*. Garnier-Flammarion, Paris.

Essentium (2020). Trends in 3D printing at scale [Online]. Available at: https://f.hubspotusercontent20.net/hubfs/5442420/Trends%20in%203D%20Printing%20at%20Scale%20Essentium%202020.pdf.

France Stratégie (2017). 2017/2027 – Répondre à l'innovation disruptive – Actions critiques [Online]. Available at: http://www.strategie.gouv.fr/publications/20172027-repondre-linnovation-disruptive-actions-critiques [Accessed 18 March 2022].

Giarini, O. and Stahel, W.R. (1990). *Les limites du certain : affronter les risques dans une nouvelle économie de service*. Presses Polytechniques et Universitaires Romandes, Lausanne.

Granger, G.G. (1995). *Le probable, le possible et le virtuel*. Odile Jacob, Paris.

Guérin, M. (2015). *La croyance de A à Z ; un des plus grands mystères de la philosophie*. Encre marine, Paris.

Guièze, G. (1983). Les énoncés de Mind and machines comme image de l'Objet technique. In *Pensée et machine*, Anderson, A.R. (ed.). Champ Vallon, Paris.

Hull, C. (1984). Method for production of three-dimensional objects by stereo-lithography. US Patent 5,762,856, 1 August 1984.

Jardim-Goncalves, R., Romero, D., Grilo, A. (2017). Factories of the future: Challenges and leading innovations in intelligent manufacturing. *International Journal of Computer Integrated Manufacturing*, 30, 4–14.

Jones, R.O., Iravani, P., Bowyer, A. (2012). Rapid manufacturing of functional engineering components [Online]. Available at: https://purehost.bath.ac.uk/ws/portalfiles/portal/216470/RapidManufacturingFunctionalComponents.pdf [Accessed 18 March 2022].

Kelly, B.E., Bhattacharya, I., Heidari, H., Shusteff, M., Spadaccini, C.M., Taylor, H.K. (2019). Volumetric additive manufacturing via tomographic reconstruction. *Science*, 363, 1075–1079.

van der Laan, H.L., Burns, M.A., Scott, T.F. (2019). Volumetric photo-polymerization confinement through dual-wavelength photo-initiation and photo-inhibition. *ACS Macro Letters*, 8, 899–904.

Latour, B. (2007). *L'espoir de Pandore ; pour une version réaliste de l'activité scientifique*. La Découverte, Paris.

Lippmann, W. (1937). *The Good Society*. Little, Brown and Company, Boston.

Lippmann, W. (2015). *Drift and Mastery – An Attempt to Diagnose the Current Unrest*. The University of Wisconsin, Madison.

Loterie, D., Delrot, P., Moser, C. (2020). High-resolution tomographic volumetric additive manufacturing. *Nature Communications*, 11, 852.

Löwith, K. (1969). *De Hegel à Nietzsche*. Gallimard, Paris.

Magistrelli, G. (2021). Comment la robotique aide-t-elle à construire une usine de fabrication additive smart et efficace ? [Online]. Available at: https://www.a3dm-magazine.fr/magazine/toutes-industries/robotique-fabrication-additive-usine-futur.

Mailfert, A. (1968). *Au pays des antiquaires – confidences d'un maquilleur professionnel*. Flammarion, Paris.

Mayer, N. (2018). Ce robot peut construire votre maison en 3 jours ! [Online]. Available at: https://www.futura-sciences.com/maison/actualites/ce-robot-peut-construire-votre-maison-3-jours-73917/.

Mayer, F., Ryklin, D., Wacker, I., Curticean, R., Čalkovský, M., Niemeyer, A., Dong, Z., Levkin, P.A., Gerthsen, D., Schröder, R.R., Wegener, M. (2020). 3D two-photon microprinting of nano-porous architectures. *Advanced Materials*, 32, 2002044.

Momeni, F., Mehdi Hassani, S.M., Liu, X., Ni, J. (2017). A review of 4D printing. *Materials & Design*, 122, 42–79.

Montesquieu, de Secondat C.L. (1968). *Considérations sur les causes de la grandeur des romains et de leur décadence*. Garnier-Flammarion, Paris.

Nietzsche, F. (2007). *Le gai savoir*. Garnier-Flammarion, Paris.

Pascal, B. (2004). *Pensées*. Le livre de poche, Paris.

Philbeck, P. (2017). *Member of the "National Academies" Committee (2017) "The Fourth Industrial Revolution"*. National Academies Press, New York [Online]. Available at: http://nap.edu/24699 [Accessed 18 March 2022].

Pinho, A.C., Buga, C.S., Piedade, A.P. (2020). The chemistry behind 4D printing. *Applied Materials Today*, 19, 100611.

Plagnol, A. (2021). Covid-19 et science officielle : un désastre evidence-based [Online]. Available at: https://up-magazine.info/decryptages/analyses/80499-covid-19-et-science-officielle-un-desastre-evidence-based/?utm_source=ActiveCampaign&utm_medium=email&utm_content=RSS%3AITEM%3ATITLE&utm_campaign=ACTION%21+Quotidienne+lu-01%2F02.

Regehly, M., Garmshausen, Y., Reuter, M., König, N.F., Israel, E., Kelly, D.P., Chou, C.Y., Koch, K., Asfari, B., Hecht, S. (2020). Xolography for linear volumetric 3D printing. *Nature*, 588, 620–624.

Rose, J. (2001). *Profession quasi-chercheur*. L'Harmattan, Paris.

Schwab, K. (2016). *The Fourth Industrial Revolution*. World Economic Forum, New York.

Simon, H.A. (2004). *Les sciences de l'artificiel*. Folio-Essais, Paris.

Spinoza, B. (1997). *Traité théologico-politique*. Garnier-Flammarion, Paris.

Stevenson, K. (2021). Barriers and benefits with additive manufacturing [Online]. Available at: https://www.fabbaloo.com/news/barriers-and-benefits-with-additive-manufacturing?utm_source=newsletter&utm_medium=email&utm_campaign=fabbaloo_daily_3d_printing_news&utm_term=2021-02-04 [Accessed 18 March 2022].

Stiegler, B. (2019). *Il faut s'adapter – sur un nouvel impératif politique*. NRF Essais, Paris.

Taguieff, P.A. (2021). *Les Nietzschéens et leurs ennemis*. Editions du Cerf, Paris.

Tibbits, S. (2013). The emergence of 4D printing [Online]. Available at: https://www.ted.com/talks/skylar_tibbits_the_emergence_of_4d_printing.

Tibbits, S. (2016). *Self-Assembly Lab: Experiments in Programming Matter*. Routledge, New York.

Tsai, E.Y. (2013). 4D printing: Towards biomimetic additive manufacturing [Online]. Available at: https://dspace.mit.edu/handle/1721.1/91821 [Accessed 18 March 2022].

Vincent, J.D. (1994). Le cerveau est une métaphore. In *Sciences et imaginaires*, Maréchal, I.A. (ed.). Albin Michel, Paris.

Yao, Y., Waters, J.T., Shneidman, A.V., Cui, J., Wang, X., Mandsberg, N.K., Li, S., Balazs, A.C., Aizenberg, J. (2018). Multi-responsive polymeric microstructures with encoded predetermined and self-regulated deformability. *Proceedings of the National Academy of Science*, 115, 12950–12955.

1

Is 4D Printing Disruptive or Incremental, or a Bit of Both?

> The fall of France is not a new theme, but it is a reality on many levels, and the most diverse explanations abound. (Silverzahn 2021)

> Richelieu [...] said he recognized from experience that, all other things being equal, rebels are always half as strong as the defenders of official power. (Weil 2021)

1.1. Introduction

A certain malaise is developing, despite the prevailing discourse on the decline of France in terms of technological development and scientific relevance. Because of a lack of will, a lack of vision of the future and therefore a lack of risk-taking, 3D printing has not been French. Moreover, as Philippe Silverzahn (2021) points out, "our country and its industry are absent in almost any of the major current technological races, whether in Bio-Tech, artificial intelligence, robotics, digital, or energy". For all that, the government is not stingy when it comes to launching proposals, rather than catching up, within well-funded projects such as the "*Programmes d'investissements d'avenir*" (see, e.g. Alcimed (2021)). As a reminder, they were set up by the State to finance innovative and promising investments on the national territory, in order to allow France to increase its growth and employment potential. Thus, for innovation, for this bibliographic source, the national criteria for innovation are as follows:

"Let's make no mistake about it, even if some projects will be financed at fairly low TRLs, the priority is on applications with high market potential and leverage effect to boost French competitiveness", with the following criteria:

– Can the project differentiate itself from ongoing translational research projects in the field worldwide? How does it add value?

– Is the preliminary data on which the project is based sufficiently robust?

– Can they justify an ambitious translational research project?

> When the State, in recent years, became aware of the delay accumulated by our country, it could only rely on the administrative system as it is, i.e., a system that continues to give priority to the control of the form rather than to the relevance of the action. (Crozier and Tillette 2000)

DEFINITION.– *Translational research*: initially, this transfer-oriented research was situated between basic, in-depth research in biology (e.g. understanding the mechanisms behind the development of a cancer) and application-oriented research aimed at evaluating the effectiveness and tolerance of new treatments in humans. By extension, it is a transfer science exploring TRLs 3–4.

DEFINITION.– *TRL "Technology Readiness Level"*: this English-language concept is translated by a scale called TRL, which evaluates the level of maturity of an activity aimed at technological applications. It starts at 1 for basic research and reaches 9 for applications.

In the undoubtedly naive minds of many researchers (including our own), entrepreneurship and innovation contribute significantly to the success of firms and the economic growth of nations (Schumpeter 1942; Drucker 2007). Companies create value, directly and indirectly, for themselves and their customers. Entrepreneurship and innovation are often linked directly or indirectly to the notions of creativity and design. Starting from the needs expressed or the offers resulting from research, the design processes define, step by step, the artifact which must eventually meet the objectives retained (needs and constraints) by successive decisions with increasingly detailed approaches (and therefore increasingly irreversible in the limited duration of the action). It may therefore be useful to take stock of what has already been achieved and to examine future courses of action.

In the underlying race of the current major calls for tenders, in an attempt to catch up with the leading pack (which invests much more in research than our country in many scientific fields), we are exploiting the paths defined elsewhere and by others, hence this blind following. In 1984, for 3D printing (André et al. 1984), the financing supporting an economic development could have been modest, because it was merely a slightly sophisticated form of tinkering; the emergence of a child of the additive manufacturing technologies. However, 4D printing concerns a much higher field of complexity and risk-taking. This is one of the major issues of the messages we wish to convey in this book.

So, a nagging question preoccupies the authors: Is 4D manufacturing translational research (therefore incremental) or, on the contrary, despite its filiation, a breakthrough technology? In the first case, this would mean that the paradigms and methods would be stabilized and that we would participate after other foreign research groups in this follow-up of good scientific quality, because we would have already paved the low TRLs. Then, for example, work in 4D printing could be included in calls for tenders for research support accepted by the system. On the other hand, if lakes of uncertainty exist with methodologies yet to be invented, even if we have a certain vision of the future and certain means of achieving it, it is this vision of incompleteness that risks being favored by a country that exploits too much a bias of preference not to do, because it avoids the responsibility of political or non-political contributors.

On the other hand, in the context of university education, and especially in the context of the Grandes Ecoles d'Ingénieurs, mastery of the future, apart from the basic theoretical teachings, is important for two reasons. First, for future tasks over decades, useful skills for the future must be planned and taught. Second, teacher-researchers need to select the most promising pedagogies, tools and technologies to apply in their teaching work. The problem is that there is no robust method for assessing the importance of alternatives – how important will a technology like 4D printing really be in the near and especially long-term future? A useful question to answer would be: "What should every teacher-researcher know about the coming technological changes and how do we prepare for them? In these changes, where would 4D printing fit in?" (Jaakkola et al. 2017).

Moreover, according to Bessen et al. (2020), the decline in disruptive innovations since 2000 is linked to a factor related to the ability of already well-equipped companies to invest massively in intangible assets. They invest in their proprietary software, mega-data and information technology (IT) capabilities, which in the short term can enable competitive advantages, at relatively low cost (marginal cost). Again, the risk is minimized. In both extreme situations, it seems difficult to bring about innovations or breakthrough technologies. However, it is commonly known that information at all levels is expressed on the accessibility of innovations that will radically change the way consumers, industry or companies operate. This means that only startups and SMEs should be in charge of this uncertain future if they hope one day to compete with large companies (like GAFAM).

On the basis of the comments presented in the introduction to this chapter, we propose to take a look at this position, which will be further developed in the other chapters of the book. This reflection makes it possible to appreciate the position of

4D printing in this setting where, for the truly new, risk-taking is left to fragile structures and practically personal initiatives. So, in these conditions, the place of science and technoscience can be strongly questioned: How then to show that it can play a proactive role by developing new processes, illustrating the possibility of a convergence between the expression of an applicative need (more or less formulated) and scientific production. This phase of emergence of new processes, as long as it is understood by the hierarchy, and supported as such, is a possibility to envisage applications that were inconceivable less than 10 years ago, while knowing that these trends can still evolve strongly. This is what is at stake in this chapter, which voluntarily separates intelligent and/or stimulable materials from the 4D printing that uses them.

1.2. Prospective approach

> Learning to admit and exploit the seemingly incongruous means practicing new ways of perceiving and ordering experience that will allow one to maintain an inclusive peripheral vision through an essentially plastic attitude of mind. (Gordon 1965)

It is possible, before embarking on new research, not only to make a bibliography on a subject; especially since, when it is emerging, the number of publications to be read can be relatively modest (and also, it avoids wasting time). Science fiction can be another way; finally, a third way is a foresight, which is not always adapted to the sudden emergence of events or innovations. And yet…

> As anticipation adapted to an unpredictable future, studies on the future remain essential to the organization of the future passage to action: without anticipation, no objective; without objective, no coordination of actions; without coordination of actions today, no human, material, legal means [...] to be mobilized tomorrow to implement a strategy, be it military or corporate. (Roubelat 2017)

But, if this situation is well conceived by the company that needs a certain continuity to function, even if adaptability, flexibility and an ever-increasing capacity to adapt are demanded, what happens when a disruptive innovation that is not foreseen, not easily assimilated, not perceived as relevant by decision-makers (see Kodak syndrome) occurs? This question highlights a real difficulty in a world constrained by major trends and subject to unbridled competition (McKay and McKiernan 2010).

In a prospective approach where it is the uncertain that is explored, generally over longer periods than those associated with European or national plans, there is the possibility of being a little original, without any major constraint apart from motivation and conviction, which makes it possible, faced with multiple and undetermined futures, to question the comfort of the mind, to try to awaken consciences that have fallen asleep on false or old certainties. It is then a question of conceiving possible desired futures and supporting the real means to try to achieve them (Ackoff 1973); it is indeed an approach to the future that is not based solely on the extrapolation of past trends, leading to the highlighting of certain breaks to be made.

Absolute discontinuities occur without having been preceded by warning signs, and this is the whole debate on weak signals, which bring together creativity focused on an object (but in all its richness of possibilities) and foresight (insofar as they have been anticipated! The transistor and its digital sequels are good examples of non-anticipation). But as Fernand Braudel (1993) pointed out: "any effort against the current is doomed in advance"... Some prospective analyses come from strong independent personalities, such as D. Bell (1999), A. Toffler (1991) and P.F. Drucker (2007), who can, thanks to their discriminating power, issue divergent and original opinions, as is the practice in creativity (no need to seek a consensus). The interest of these personalities is that they force us to think about the future differently: Prigogine and Stengers (1979) introduced the notion of order by fluctuation, showing that weak fluctuations can orient systems by bifurcation toward one branch of evolution or another. This highlighting of weak signals can then force original conclusive propositions through in-depth investigation of the issues. Obviously, in this world where economic competition defines the (current) destiny of a consumer society, it is the technological destiny that is the object of attention of many futurists. All highly developed countries agree on the decisive importance of knowledge and industrial innovation as factors of independence and economic competitiveness. But do they not take insufficient account of technological breakthroughs because they are averaged in groups of processes or because of a lack of discrimination (see section 1.1)?

> Creativity is the impulse that invites thinking something new, something amazing, something of value. (Du Sautoy 2020)
>
> Every inventor, every scientist, has been so because they have known and been able to open themselves up to a field far from their own, from which they have drawn inspiration. They have erased the unfortunate divisions. (Dagognet 1989)

But, if it is not necessary to go too far into this subject, it may be useful to remember the history of 3D printing. Before 1984, there were scientific attempts reported in André (2017a), but there was no foresight about the emergence of this technology (apart from some comics from André Franquin in the "small formats" (Franquin (1973) or d'Hergé ...). For 4D printing, Sculpteo (2020) considers that this technology could be used, giving a few examples, ranging from self-assembly of elements to self-repair, then to uses in extreme conditions, finally in medicine and fashion. For his part, Moigne (2020) talks about applications in soft robotics and actuation, citing Barnes et al. (2020). The manufacture of flexible robots that would swim like jellyfish or the transport and delivery of drugs are thus envisaged. This position is supported by Staedter (2019) and Future Bridge (2020). Finally, Tibbits' (2014) prediction of combining 3D printing and programmable matter is a breakthrough that has not really been anticipated outside of science fiction movies like Terminator 2.

Vision, discovery, learning and advancement are among the fundamental missions of the academic body involved in research; they are mutually reinforcing. The transfer of knowledge to those in society who can use it for the general good contributes to each of these missions. These transfers of knowledge and learning take place through scientific publications, student training, conferences, cooperative ventures, etc. Not everything is accessible, and when it is, it may be with delay, as in the case of innovations protected by patent (intellectual property or IP) and by the granting of licenses to private companies. These means of knowledge sharing that have contributed to the history of science–society relations can be exploited to access information on the 4D subject. This is how an analysis of the bibliography was carried out and collected in Table 1.1, which shows the elements expressed for the near future (indeed, it was not possible to have long-term visions in the documents examined). In these future visions, Figure 1.1 from Ergene and Yalçın (2019) gives indications of the cause (stimulus)–consequence relationships.

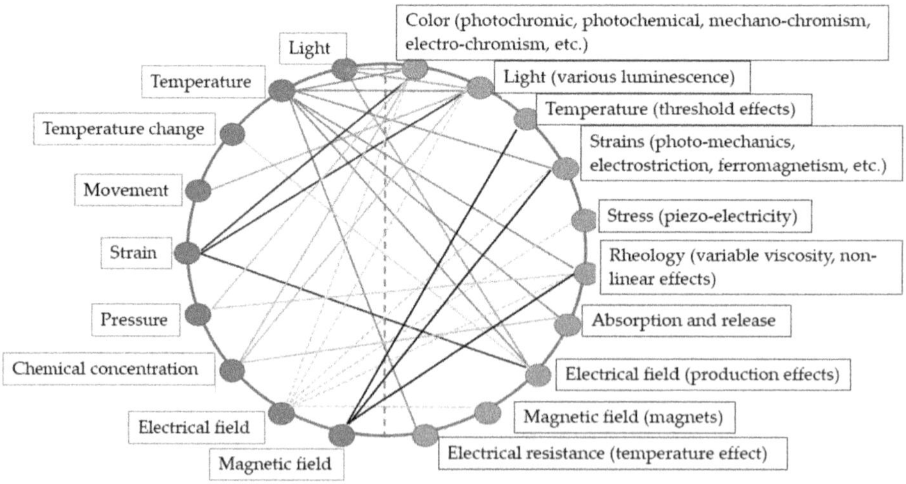

Figure 1.1. *Examples of cause-effect relationships in 4D printing. For a color version of this figure, see www.iste.co.uk/demoly/4Dprinting1.zip*

> Every profession makes progress, but progresses according to its own routine, tracing its own furrow [...]. The furrow prevents wandering through the landscape and abstracting something, to which no attention is paid anymore [...]. The rest of life is treated superficially, with the imperfect categories of a thought that derives from the profession. (Whitehead 1994)

Application	Comments	References
Biomedical		
4D bio-printing	Different biological tissues and organs	Wan et al. 2020; Lui et al. 2019
4D stents	4D structures of adaptive materials introduced into blood vessels (biocompatibility)	Duerig et al. 1999; Wei et al. 2017; Xue et al. 2019; Cohn et al., 2020; Siddharth et al. 2020; Yang et al. 2020
Implants	Shape memory alloys	Elahinia et al. 2012
Personal protections	3D elastomers	Kumar and Pumera 2021
Bio-printing	Adaptive scaffolds (biocompatibility); roadmap	Miao et al. 2016; André 2017c; Sun et al. 2020
Vascular repair	Use of adaptive materials to assist in the repair of blood vessels (biocompatibility)	Kuang et al. 2018
Intelligent sutures	Imitation of tendon mechanics	Ma et al. 2021
Dispensing of drugs	Timely delivery of orally administered drugs to the right place (biocompatibility)	Novakovic et al. 2018; Lamichhane et al. 2019; Lui et al. 2019; Melocchi et al. 2019; Shie et al. 2019; Municoy et al. 2020; Li 2016; Inphonlek et al. 2020; Jia et al. 2021
Medical treatment	Swimming robot	Wang et al. 2020b
Drugs	Nanoscopic materials sensitive to an electromagnetic potential (e.g. local hyperthermia)	Li and Su 2018; Kirillova and Ionov 2019; Lui et al. 2019; Bajpai et al. 2020
Odontology	Consideration of tissue displacement in dental repair	Hamza 2018
Orthopedics	Organic stimulable materials	Semba et al. 2020
Energy and processes		
Storage	Loading by thermomechanical means	Roudbarian et al. 2019
Production	Piezoelectricity; thermoelectricity	Cholleti 2018; Chen et al. 2020; Rösch et al. 2021

Solar	Solar concentrators	Svetozarevic et al. 2016; Momeni and Ni 2018
Micro-fluidics	Adaptive systems	Robertson et al. 2016; Cesewski et al. 2018; Apsite et al. 2020
Microparticles	Deformation induced by nanoparticles	Sun et al. 2012
Metamaterials	Design	Fraternali et al. 2019
Adaptive optics	Integrated and self-adapting optics	Jeong et al. 2020; Apsite et al. 2020
Electrochemical reactor	Several materials including metals	Arenas et al. 2017
Catalysis	Metamaterials and composites	Ludwig et al. 2018; Papetti et al. 2018
Atomic scale		
Manufacturing	Displacement of atoms	IBM 2012
Nano-machines		
Nano-machines	Molecules transformable by the influence of stimulation; catenanes	Sauvage 2003; Lemouchi 2010; Medina-Sanchez et al. 2018; Bandari et al. 2020; Novotný et al. 2020
Nano and micromotors	Idem	Ning et al. 2018; Wang and Pumera 2020
Transportation		
Adaptive structures	Spoilers in adaptive aerodynamics; speed-dependent deformations (but according to Audi's Wiese et al. (2020), this demand does not appear)	Sellitto and Riccio 2019; Das and Dujana 2020; Siddharth et al. 2020
Morphing	Adaptive systems	Daynes and Weaver 2013; Papoutsis-Kiachagias et al. 2019; Chillara and Tapino 2020; Le Good et al. 2021
Tyres	Wheel-tread torque	Michelin 2016; Monino 2017
Avionics	Active elements	Daynes and Weaver 2013; Su et al. 2018b
Mechanical insulation	Auxiliary/meta-material structures Damping/meta-materials	Liu et al. 2020b; Xue et al. 2020 ; Clough et al. 2019

Architecture		
Design and construction	Adaptive structures	Thomas et al. 2014
Structures	Adaptive structures	Beites 2013; Grassi et al. 2020
Esthetic facades	Reconfigurable shapes adapted to heat and/or light	Abdel-Rahman and Tafrihi 2018; Morteza Hosseini et al. 2019
Robotics, sensors and actuators		
Automation	Association of active elements	Shen et al. 2019; Decroly et al. 2020
Self-organization	Self-organization of robots	Ren et al. 2020
Origami and mobile kirigami	Development of active 3D structures from a flat base on which active materials are printed	Ge et al. 2014; Modes and Warner 2016; Liu et al. 2018b; Apsite et al. 2020; Liu et al. 2020a; Shen et al. 2020a; Alderete et al. 2021; Lin et al. 2021
Flexible robotics	Replacement of structures and joints by deformable elements (actuators)	Gul et al. 2018; Schaffner et al. 2018; Zolfagharian et al. 2019, 2020; Duriez 2020; Scalet 2020; Shen et al. 2020b; Zhu et al. 2020; Schara et al. 2021; Tyagi et al. 2021
Micro-robots	Co-bots, bio-bots; self-organization	Blackiston et al. 2021; André 2017c
Swimming robots	Submillimeter scale displacements	Bumjin 2018; Ceylan et al. 2019; Wu et al. 2021
Mechanical closures	Realization of a 4D object under mechanical tension	Su et al. 2018a
Deformable structures	Temperature and humidity deformable shapes; shape memory systems	André 2017c; Mouzakis 2018; Lopez-Valdeolivas et al. 2018; Wang et al. 2018c; Zhang and Xiao 2018
Sensors	Strain-dependent strength; mechanical effects	Surjadi et al. 2019; Garces and Ayranci 2020; Verpaalen et al. 2020
Engines	Activated deformable shapes	Peng et al. 2020; Rubio et al. 2021
Clamps and actuators	Deformation of a 4D object serving as an actuator thanks to an external stimulation	Miriyev et al. 2017; Chen and Shea 2018; Liu et al. 2018a; Yuan et al. 2018; Akbari et al. 2019; Lui et al. 2019; Schreiber et al. 2019; Liu et al. 2020b; Milana et al. 2020; Shao et al. 2020; Ze et al. 2020

Bilames	Stimulation-induced changes	Hagaman et al. 2018
Bistables	Shape memory alloys	Rostaing 2004; Jeong et al. 2019; Jiang et al. 2019; Boni and Royer-Carfagni 2021
Shape memories	Alloys and polymers	Nicolas et al. 2000; Wagg et al. 2007; Ould Moussa 2011; Simoneau 2013; Yao et al. 2021
Optics	Various elements	Willis et al. 2012
Magnetism	Insulating coupling, magnetic materials	Huber et al. 2016
Bio-mimicry	Active auxetic structure	Lee et al. 2020; Wang et al. 2020c
Color change	Transformation induced by an electromagnetic field or light or temperature	Rai et al. 2017; Salmon et al. 2019
"Programmed" self-organization	Robotic voxel assemblies; catomes	Moskvitch 2015; Rus and Tolley 2015; Poty et al. 2017; Su et al. 2020
Fashion and clothing		
Smart clothing and textiles	Adaptive clothing (changing colors, permeability, size, etc.)	Chan Vili 2007; Hu and Lu 2014; Rayate and Jaemi 2018; Zhang et al. 2018b; Ren et al. 2019; Yu et al. 2020; Biswas et al. 2021; Koch et al. 2021; Miao et al. 2021
Military applications	Camouflage (color change)	Puren 2013; Wang et al. 2019a
Color change	Color changes with deformation	Vatankhah-Varnosfaderani et al. 2018
Shoes	Adaptive materials	Konakovic et al. 2016; Rastogi and Kandasubramanian 2019
Jewelery	Mobile jewelry	Zarek et al. 2016b
Electronics		
4D and structural electronics	Integration of electronics into objects (4D) including actuators; flexible printed electronics	Espalin et al. 2014; McDonald et al. 2014; Lee et al. 2017b; Oh et al. 2018; Lu et al. 2018; Liten-CEA 2019; Wang et al. 2019b
Printed electronics	Printing with conductive inks (including those containing nanoparticles)	Zarek et al. 2016a; Kwon et al. 2020; Deng et al. 2017; Lou et al. 2017; Jasiuk et al. 2018
Switch	Photo-activation	Gu et al. 2018
Circuitry	Photo-deposition	Tabridzi et al. 2017

Transducer	Several successive operations	Cesewski et al. 2018
Biochemical analysis	Miniaturization, integration, automation and parallelization of biochemical processes	Zhang et al. 2018a
Wave absorption	4D acoustic wave absorption structure	Le et al. 2020
Digital		
Synapses	Adaptive materials	Park et al. 2020
Memories	Idem	Zhang et al. 2019
NAND gates	Idem	Zhang et al. 2019
Information storage	Three-dimensional microscopic indentation layers	De Angelis 2019
Space		
Antennas, solar collectors, etc.	Spatial structures that can be deployed	Mitchell et al. 2018; Ntouanoglou et al. 2018; Tofail et al. 2018; Melli et al. 2020
Various		
Food	Pasta	Tao et al. 2021
Self-repair	Poly-methyl siloxane elastomers	Idumah et al. 2020; Liu et al. 2020c
3D animation	3D TV; 3/4D cinema	Patel et al. 2017
Waterproofing	Functionalization of textiles	Puren 2013; Yu et al. 2018
Packaging	Smart packaging	Kuang et al. 2019
Alphabet	Letters of the alphabet made in 4D printing	Hoa and Rosca 2020
6D printing	The consumer will use his sixth sense; the data will be transferred to the 6D printing interface Other view	Moonmoon 2019; Simeral et al. 2021 Georgantzinos et al. 2021

Table 1.1. *Examples of application targets*

The partitioning in this table is relatively artificial; thus, in Figure 1.2, we have represented, for a garment, a concept of energy production and storage in association with other sought-after qualities which will promote the attractiveness of a device. There can be several 4D operations in the same object.

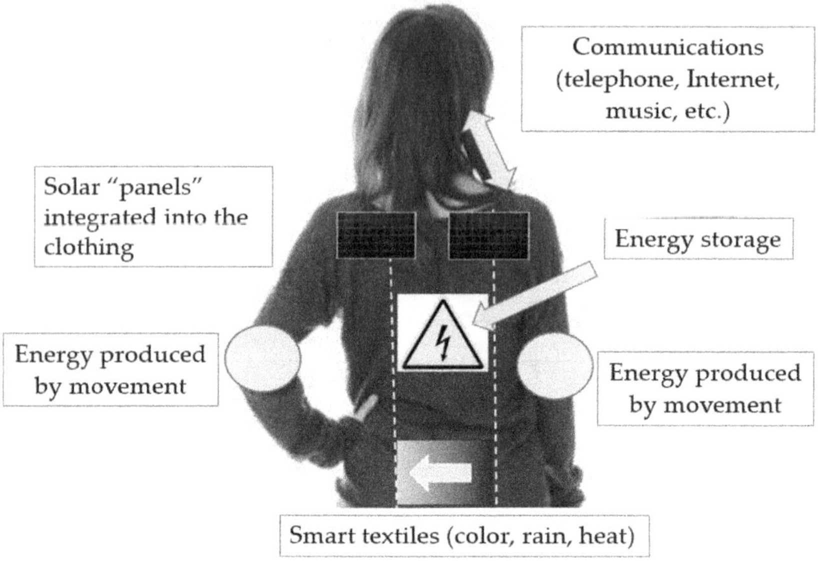

Figure 1.2. *Integrations enabled by 4D printing; the example of smart clothing. For a color version of this figure, see www.iste.co.uk/demoly/4Dprinting1.zip*

DEFINITION.– *Structural electronics*: generic technologies for integrating electronics into the heart of objects, that is by combining electronic functions with the structure of the parts themselves.

For Melli et al. (2020), 4D printing is a revolutionary invention; it is expected to have a significant impact, provided the challenges it faces are overcome, on the fabrication of dynamic structures for various applications. This assumes, at a minimum, that the varieties of tunable, 4D-printable smart materials increase in number and performance. Thus, 4D printing remains for the moment at a proof-of-concept stage that has yet to show its capacity to pre-program the material, which must react appropriately to various stimuli localized in time and space. An axis emerges in 4D actions, that of space, because it is quickly conceivable to realize simple adaptive structures allowing weight savings (as a reminder, each kg launched from the Earth costs, according to Mitchell et al. (2018), about 10,000 €). For these same authors, a step should be taken when it will be possible to have large printers combining additive manufacturing and robotics, with multi-material printing. However, in this picture, these ideas and concepts remain in line with what can be envisaged; there are few "revolutionary" contributions (apart from the idea of 6D printing). Has science reached the peak of wisdom or does it no longer know how to question itself?

This situation is recent, and one of the origins can be defined, in agreement with Gordon (2016), by these sentences from Cowen (2011):

> We can't figure out why we can't do it anymore. All these problems have a single, but unnoticed, cause: we have lived for at least three hundred years on fruit that was just waiting to be picked [...]. But in the last forty years, these fruits have become scarce, and we have acted as if they were still there. We didn't want to recognize that we had reached a technological plateau and that the tree was much barer than we wanted to admit.

So, as soon as a guideline seems to bring scientific and/or technical progress, we pursue the action via incremental research (as is the case today with 3D printing) by seeking to win without making a clear break. But this quest obviously has its limits.

For Vinsel (2020), to complement Cowen's analysis, interest in technological innovation (and its benefits) is overestimated (see also Vinsel and Russel (2020)). He concludes that "this situation invites us to work on technological situations that are not real and are not even realistic projections of where science and technology are going" (Guillaud 2021). This is somewhat what was observed in the case of nanotechnologies, where considerable promises were made more than 20 years ago, with a rather mediocre application result (Frochot et al. 2020). Will the same be true for 4D printing?

1.3. A tectonics of paradigms

Figure 1.3, which is a bit provocative, reminds us of the old Christopher Columbus syndrome: when he left, he did not know where he was going; when he arrived, he did not know where he was and all the while, he did not do it with his own money!

> But another world exists. Today, at the micro and nano scales, an unprecedented revolution is underway. It is the ability to program biological and physical materials to change shape, change properties, and even perform computation in a non-silicon based material. (Tibbits 2013)

The *technology-push* model has long been dominant and is still applied (Badillo 2013). It has inspired the development of technologies such as 3D printing. It is a linear model with an innovation that is the result of science and technology. This "*top-down*" innovation respects the "E-C-R" (emission, communication, reception) communication model inspired by Shannon's work and is situated in a

Schumpeterian logic (Schumpeter 1942, 1952; Von Hippel 2013), that of the innovating entrepreneur (the emitter). In this simple situation, users are considered, to a first approximation, as passive receivers. Schmookler (1966) introduced the concept of "*demand-pull*" or "*market-pull*" involving the user. Rothwell and Zegveld (1982) then combined these two models with feedback effects. "The innovation process can be thought of as a set of communication paths through which knowledge is transferred" (Trott 2008). More recent models, such as network models or open models of innovation, are based on a systems approach to innovation (Chesbrough 2003; Chesbrough et al. 2006). While the situation is relatively clear for 3D printing, we discuss below the situation for 4D printing and its possible applications.

Figure 1.3. *Peter Breughel's The Blind Leading the Blind. For a color version of this figure, see www.iste.co.uk/demoly/4Dprinting1.zip*

1.3.1. *3D printing*

> Collective intelligence is an open approach based on the freedom to belong to a process. Because they find benefit in collaborating, individuals voluntarily and autonomously engage in a community bound in time and space. The structure thus developed is perfectly horizontal, without any hierarchy. (Novel and Riot 2013)

The concept of 3D printing was demonstrated and operationalized in 1984 (André et al. 1984) by a patent concerning stereolithography consisting of the spatially resolved polymerization of a light-sensitive resin. The coordinates of the object to be produced were stored in a computer which controlled galvanometric mirrors to transform a liquid into a solid by polymerization of a fluid layer, voxel by voxel, hence the concept of additive manufacturing. Adding a second layer, then a third, and so on, would in principle create the desired prototype part, which finally had to be extracted (and washed) from the non-photo-transformed fluid (André 2020). The other manufacturing principles are based on the same foundation of additivity, but as shown in Figure 1.4 and Table 1.2 the processes differ essentially in the nature and modes of adhesion of the material to the object under construction (André 2017a; Kalyan et al. 2021). The rule of the game is indeed to have either localized energy or a localized positioning of material or both.

When Dagognet (1989) wrote: "Didn't the clothiers catch a glimpse of the future digitalized image, since they worked on the one hand in a binary mode (a presence or absence) and with the help of a few strands, dyed in different ways?", did he not introduce this mechanization as an eminent ancestor of additive manufacturing? Digital additive manufacturing techniques that exploit elementary volumes called voxels (and not strands) are now well known to have many advantages (and naturally limitations expressed in André (2017a)). Recently, a new technology called volumetric 3D printing has been developed (Amra et al. 2017; The Chinese University of Hong Kong 2019; De Beer et al. 2019; Loterie et al. 2020; Mayer et al. 2020; Regehly et al. 2020; Doualle et al. 2021).

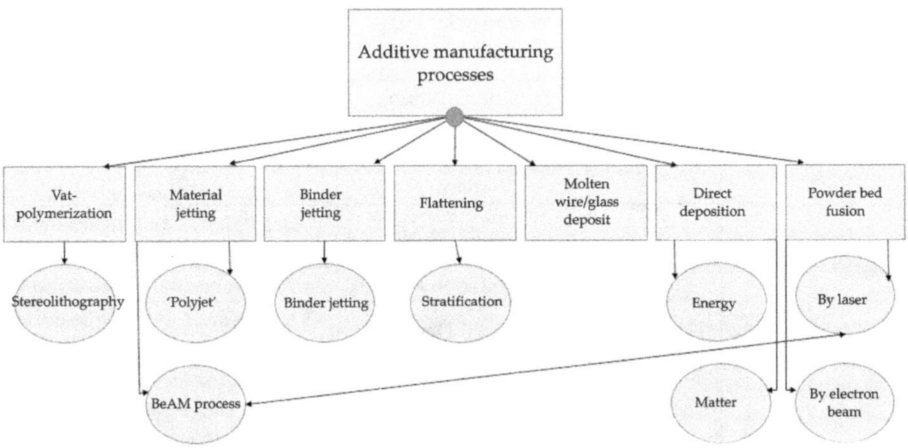

Figure 1.4. *The different additive manufacturing processes*

<table>
<tr><th>Theme of the chapter</th><th>Process name</th><th>Official name</th><th>Comments</th></tr>
<tr><td>Stereolithography</td><td>SLA</td><td>Stereolithography</td><td>Light-induced polymerization of charged and unfilled resin
Different irradiation systems</td></tr>
<tr><td>Strand fusing</td><td>FDM/FFF</td><td>Fused Deposition Modeling/Fused Filament Fabrication</td><td>Wire fusing: layer by layer deposition of a thin layer of molten wire. The material cools in contact with the air and solidifies
From molten polymer to metals and glass</td></tr>
<tr><td rowspan="5">Bonding of sheets or powders</td><td>SDL</td><td>Selective Deposition Lamination</td><td>Sheets with activatable adhesive; uses inkjet system (photosensitive or UV-reactive material)</td></tr>
<tr><td>SC</td><td>Strato-Conception</td><td>Sheet cutting – “mechanical” assembly; at each layer, a laser or a cutting tool removes the contours of the desired object</td></tr>
<tr><td>3DP</td><td>Three-Dimensional Printing</td><td>Powders combined with an adhesive; uses inkjet system (photosensitive or UV-reactive material) or thermal bonding</td></tr>
<tr><td>BJ</td><td>Binder Jetting</td><td>Ditto; particles are deposited on the printing plate and bound by the addition of an aggregating liquid, possibly colored</td></tr>
<tr><td>SUR</td><td>Soluble Unsoluble Reaction</td><td>Change in solubility of an organic material</td></tr>
<tr><td rowspan="6">Melting/Sintering of powders</td><td>SLS</td><td>Selective Laser Sintering</td><td>Laser-induced sintering of flat surfaces</td></tr>
<tr><td>SLM</td><td>Selective Laser Melting</td><td>Laser-induced melting on a flat surface</td></tr>
<tr><td>EBM</td><td>Electron Beam Melting</td><td>Electron beam induced fusion</td></tr>
<tr><td>LMD</td><td>Laser Metal Deposition</td><td>Simultaneous metal deposition and melting</td></tr>
<tr><td>CLAD</td><td>Direct Additive Laser Construction</td><td>Simultaneous supply of material and light energy: material melted during its deposition by a highly directed thermal source</td></tr>
<tr><td>MPA</td><td>Metal Powder Application</td><td>Ditto, particle kinetic energy</td></tr>
</table>

Table 1.2. *The different additive manufacturing processes with some elements of differentiation*

With a number of manufacturing techniques limited to less than 10, the existence of specialized machines and knowledge on the qualities of 3D parts that can be produced are known. If there was a breakthrough in 1984, in 2021, it was rather in the logic of supply. It is only when new 3D machines covering new energy-material interactions are put on the market that users will make their choice. It is basically a machine like any other to be chosen from a commercial set, with its defects and its qualities. This is in fact what is shown in Figure 1.5, from Gao et al. (2015), concerning 3D methods and associated offers.

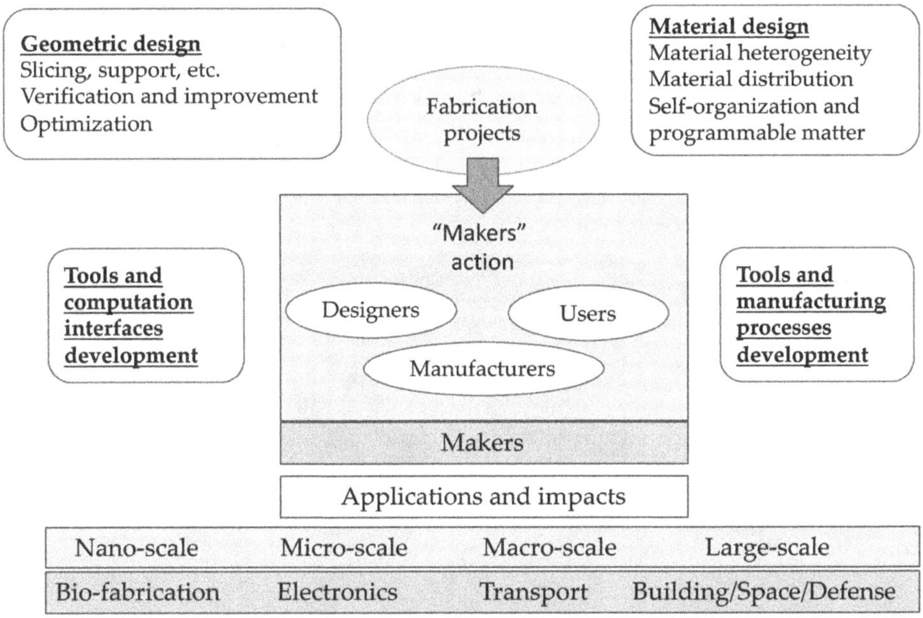

Figure 1.5. *Offerings resulting from the mastery of 3D technologies. For a color version of this figure, see www.iste.co.uk/demoly/4Dprinting1.zip*

1.3.2. *4D printing*

No matter how we look at the material world, there are many visible or hidden materials that frame our everyday lives. These can be inert objects as well as living or formerly living objects. We have shown that scientists are developing tools and ideas, some of which are inspired by nature (bio-mimicry), to create materials that have life-like properties; in other words, materials that can feel, move, change shape or function and adapt to their environment to fulfill specific purposes. These are the foundations on which 4D printing is being developed.

1.3.2.1. *Double disruption*

> Technology can be defined "as the set of procedures and means that a society uses to understand the nature and scope of scientific change, technological development and capabilities, and to assess economic utility and feasibility, social value and relevance". (Petrella 1992)

However, as André (2017a) reminds us, 3D technologies are numerous (see section 1.3.1) and meet precise and now well-known criteria. For all that, we can distinguish the fabrication of 4D parts with single materials which correspond to the main applications targeted by scientific publications (homogeneous 4D printing), then hybrid fabrications, using a single machine, with various materials (hybrid 4D printing) and finally heterogeneous processes which use in part 3D manufacturing principles (heterogeneous 4D printing). Figure 1.6 recalls, with different modes of distribution of passive and active materials (Rafiee et al. 2020), these three definitions.

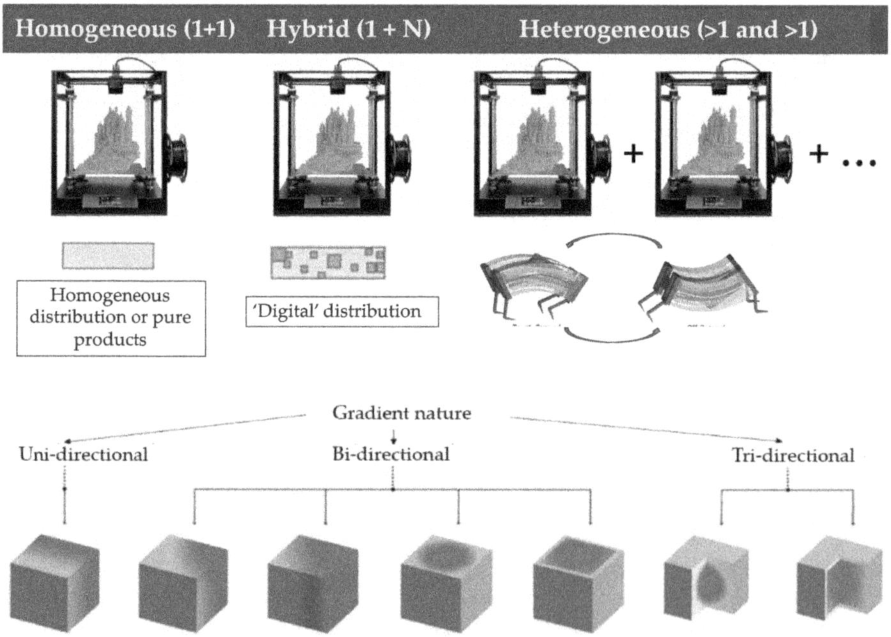

Figure 1.6. *The different forms of 4D printing. For a color version of this figure, see www.iste.co.uk/demoly/4Dprinting1.zip*

COMMENT ON FIGURE 1.6.– *The upper part of this figure has already been presented in a previous chapter; this situation repeats itself for a small number of other figures. They have been kept in the chapters where they belong to avoid the reader having to look for the pages where they were introduced for the first time.*

For Magistrelli (2021), heterogeneous manufacturing processes combine additive and subtractive manufacturing processes.

> These are implemented sequentially or incorporated, with proper control of component attachment and orientation. Such a process is used both to improve dimensional accuracy and to speed up the overall manufacturing process. [Heterogeneous] processes can be used to solve complex production problems where a single manufacturing method (subtractive or additive) is not sufficient.

Robots can be used for this combination to produce complex parts. A first combinatorial analysis would show a great diversity of choices: variable voxel sizes, modes of action and stimulation. However, the excess of possibilities risks, by accumulation of possibilities, not to serve the morphic creation and its evolution. A principle of parsimony should probably be used to optimize the realizations of the 4D devices.

Furthermore, 4D printing requires an association with an energy stimulation mode: light, pH, chemistry, humidity, heating, mechanical, pulsed or continuous electric and magnetic fields, global or localized, single or coupled with a time resolution, etc. Thus, with respect to the 10 families of 3D printing technologies, it is possible to consider 10^3 (homogeneous, hybrid and heterogeneous technologies) x 8–10 (stimulation modes) x 2 (continuous or pulsed) x 2 (couplings), that is approximately 10,000 possibilities, which prohibit any operation based on supply logic (unless one restricts oneself to a small niche). 4D printing is already a breakthrough by associating 3D printing with active material, which means that we must not operate on the basis of supply, but on the basis of demand, from needs expressed by possible users, which is a radical change. This context of double scientific and organizational breakthrough is not yet perceptible, but it will have to be taken into consideration if 4D printing is to develop.

1.3.2.2. *Demand-driven management*

> Why is it beautiful? [...] Because it is a constrained movement, because the profound movement of dance lies precisely in absolute and ecstatic obedience, in the ideal lack of freedom (Zamiatine 2020).

The public is beginning to be familiar with the development of self-healing materials targeting many applications: asphalt, concrete and self-healing fiber-reinforced polymers. But beyond this sensible notion of rejecting the vision of a throwaway society, there are many spaces that can be covered, from the nanometer to the decameter. Clearly, as the understanding of materials can be animated with the development of shape and stimulus technologies advances, large numbers of applications can emerge in all industrial sectors for the production of consumer goods, as shown in Table 1.1.

Even if it is only to develop activatable materials from an idea, knowledge from different disciplines must be drawn upon. While these fabrication, analysis and testing techniques are based on the already broad fields of materials science and engineering, shaping and stimulating them to become active require input from other scientific fields ranging from synthetic chemistry to physics to engineering and digital sciences. For example, in its report, the Royal Society (2021) imagines building walls that could absorb ambient carbon dioxide and act "as bioreactors, using inputs such as light, water, heat, algae, bacteria, nutrients and gases to generate a range of products such as purified water, energy, oxygen, recoverable biomass and heat". In their report, the authors imagine clothing or bandages that react, for example, to changes in a person's body temperature to detect possible illnesses (Wang et al. 2020a). Robotic-type devices could be developed with elements or organs that could incorporate some degree of adaptation or even swarm intelligence, all using ambient energy (heat, light, mechanical, etc.).

4D structures have the potential to fundamentally change the way material goods are designed. These possibilities in the medium term may envisage the exploitation of molecular nano-machines, for example in medicine, "living robots" that would perform high-risk tasks, from storing and processing data in DNA to creating battery-less devices powered by electromagnetic waves (Royal Society 2021). Eventually, replicas of living cells could combine to form a new evolutionary tree with life forms different from those that have evolved on the Earth so far.

1.3.2.3. *Paradigm shift in design*

> Our time is the time of the dispersion of specialized knowledge and the incoherence of values; at least that is how our time becomes aware of itself. Our culture is that of the malaise of culture. From then on, the task of the philosopher seems clear: to propose a conception of the world in which each discipline finds its place; and to give back to our culture cohesion and confidence in itself. (Lefranc 2016)

The construction of artifacts involves the use of different types of materials. So, the design has a fixed goal of providing a robust result for a device that may be static (a table, a bridge or a house) because it is supposed to keep the same shape during its life. With 4D printing, it is possible to incorporate change into design methods, which will force method designers to propose new ways of making and using. There are probably considerable differences between the realization of an adaptive garment and structural electronics, taken as examples. So, these new approaches to be created should be mainly positive, offering applicators a multitude of new possibilities dedicated to each envisaged use.

In any case, the development of 4D printing requires cooperation between researchers in the fields of process, digital, materials science and engineering, mechanics, etc., combined with those who have ideas. They must work together to exchange data and share, invent approaches and methods that meet a set of specifications. The problems highlighted in these interdisciplinary activities should be taken out of the field of project management for further scientific study and then returned to it. This work, cycling between scientific convergence and advancement at different scales, requires associations between engineering sciences and advanced sciences such as physics, a generally difficult marriage.

1.3.3. *The potential development of 4D innovations*

> The dreamer always has a cloud to transform. The cloud helps us to dream the transformation. (Bachelard 1943)

In fact, current industrial developments consist of producing elements in additive manufacturing and gradually introducing them into the usual manufacturing/production processes. 4D printing is still timid on the industrial level because of its low maturity and the time needed for interdisciplinary research. What is shown in this book, moreover, is the very great complexity of the relationships between a wide variety of scientific and technical actors in the development, integration and proposed use of the artifacts created. As discussed elsewhere (André 2017c), best practices are described by organizational and practical solutions associated with the lifecycle stages of 4D systems:

– development;

– integration and organizational anchoring;

– validation of the use of 4D systems in an operational environment.

However, the specific nature of 4D active objects can give rise to interdisciplinary development and concept testing processes (Proofs-of-Concept or POCs), in which the successive phases of convergence (project management, epistemological approach, development, acquisition, initial knowledge and use) form a spiral development process, as schematically represented in Figure 1.7. During these phases, the development of the concept and the testing of successive versions of the instrumental models serve to improve the 4D system. Similarly, during the use phases, lessons learned may influence the functionality and design of the system through feedback, thus necessitating possible radical or modest modifications. This is an iterative process with risk-taking at every stage that blurs the boundaries between the developer engaged in interdisciplinary activity, the producer of the 4D device and the end-user. The notion of the "client-supplier" relationship can no longer be easily satisfied in a deterministic and causal activity, because all the different stakeholders are involved and interconnected in all the stages of the lifecycle of the realization of a 4D object (apart from simple school cases).

> It should be noted that interdisciplinarity certainly calls for the overcoming of disciplinary boundaries, but that the disciplines still remain at the root of the process that activates its decompartmentalization: interdisciplinarity is cyclically linked to disciplinarity and vice versa, insofar as any interdisciplinary practice is likely to make the disciplines involved evolve, or even transform them from within. (Darbellay 2020)

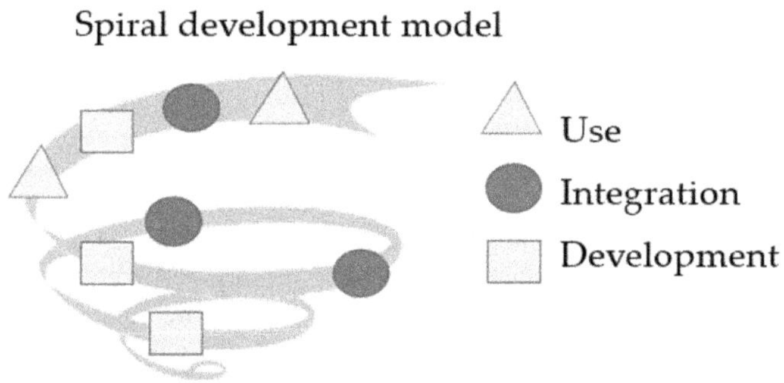

Figure 1.7. *Spiral development process and stages involved in the system design lifecycle. For a color version of this figure, see www.iste.co.uk/demoly/4Dprinting1.zip*

Thus, it is necessary to establish adequate cooperation between the relevant stakeholders already in the initial phase of developing an application using 4D principles. This design research requires an interdisciplinary approach with, as far as possible, good cooperation between all stakeholders. It is on this basis of necessary trust that interdisciplinary actors can meet the demand for convergence through the integration of specialized scientific and technological knowledge.

Independently of these central aspects aiming at the realization of POCs, or even more functional mock-ups, a certain number of questions must be asked by the users (who have worked since the beginning in connection with the designers) before arriving at an operational device (The Hague Centre for Strategic Studies 2021):

– Reaction speed: is it compatible with the intended application?

– Mechanical resistance: idem.

– Reliability: is the result in line with the degree of confidence envisaged by the applicant? Are there any fatigue phenomena making the device less accurate or reliable (Mirzaali et al. 2019)?

– Adaptability: are 4D systems adaptive and reconfigurable (scaling, extending, upgrading, etc.) to meet new requirements?

– Flexibility of the device: ease of modification of the 4D system according to the evolution of the R&D project?

– Volume and mass of the stimulation device: is it compatible with the envisaged application (e.g. in a mobile application)?

– Robustness and sensitivity to real-life failures: can the 4D device operate in degraded mode, and with what effects on the application?

– Safety and hygiene risks for humans? If so, is encapsulation possible? Effect of failures?

– Maintenance: is it necessary? At what cost (training of staff, materials, difficulties in extreme environment activities, etc.).

– Price relative to a positive reception by the market.

– Ethical compliance, etc.

This reflection translates for the authors into the following top-down generic propositions:

1) *4D manufacturing in a short-term approach* (<5 years before producing application paths)

It is necessary to return to what is achieved in mechanics by the association of elements from different "classic" additive manufacturing or robotic methods (for mechanical assemblies that can be disassembled, welding, gluing, etc.) with the specific component of the programming of the manufacturing process, which must take into account the materials, their shapes, their stowage and the effects of the stimuli on the desired deformation. It can be envisaged that the assembly and disassembly of the elements are possible to authorize the repair, even the reuse of the elements. Another advantage, cultural this time, must be noted in this preliminary approach, that of a manufacturing mode largely included in the current manufacturing processes, real factor of integration of innovations and novelties.

2) *Incremental approaches*

> Bio-mimicry: "An approach to innovation that involves the transfer and adaptation of principles and strategies developed by living organisms and ecosystems to produce goods and services in a sustainable manner and make human societies compatible with the biosphere." (Benyus 2017)

In the exploratory approach by scientific advancement (>= 5 years before producing application paths), the potential of "bottom-up" manufacturing remains an element of improvement of certain technologies, or even a way to promote new ones, even if "top-down" manufacturing processes should not be put aside in the production of high-tech devices. Cross-fertilization will already be considered. But, this currently rather virgin field of study (which could exploit ideas from bio-mimicry (Mathijsen 2020; Poppinga et al. 2021; Tahouni et al. 2021)) in the field of manufacturing may have great potential for spontaneous (or stimulated) self-assemblies to broaden the accessibility of some technologies and capabilities to new markets (Islam et al. 2021). In particular, work associated with changes in scale, nonlinear dynamical systems, or the study of behaviors can be enriched by cooperation in the following directions:

– Modeling the behavior of active materials under stimulation.

– Moving from nano-/micro- to centimeter-scale studies (with fatigue studies related to self-organizations under stress).

– Applications in centimetric 4D printing (modeling of the desired deformations and programming of the manufacturing of the constituent elements and the active object): actuators, flexible robotics, etc.

– Use of the knowledge acquired for adaptive optics involving active materials.

– Realization of active microsystems of the optode type, the stimulation, advantageously optical, allows modifications of forms (release of chemical or biochemical substances) or of functionalities (lab-on-fiber concept). In addition, the fiber can act as a stimulating energy carrier and as a feedback system. The fabrication of active deposits and their stimulation is a current topic; the stimulation can advantageously exploit two-photon photochemistry technologies.

– If chemists can contribute their knowledge on the materials selection (Pinho et al. 2020), it is essential in the process, whatever the scale, to characterize the materials and their structures (e.g. metamaterials). Thus, the thermoelastic, acoustic and optical properties of heterogeneous materials (metals, liquid, phase change, piezoelectric, gain, self-organized, etc.) should be studied.

– Finally, innovative adaptive materials can find applications in energy (electricity storage, fuel cells in particular), etc.

In an upward approach, this work (not exclusive) should lead to original scientific research, to a better mastery of research on new frontier objects via interdisciplinarity, likely to lead to industrial protection. These options imply creativity and require the management of projects that bring together a number of scientific disciplines and professional skills in an interdisciplinary approach. It allows the creation of skills and the introduction of working methods that will be useful in upstream-downstream activities.

1.3.4. *Note: example of 4D printing in structural electronics (SE)*

> In the field of participatory research, we do not know a priori the state of knowledge at the end of the process, because this will not result exclusively from the explanatory capacity of the research experiments, but also from the interactions between the knowledge of the researchers and the "lay" knowledge of the other actors around a problem with a view to solving it. (Hubert 2005)

Von Neuman was one of the pioneers of computer design based on the presence of several elements: a memory (programs or algorithms and digital data), the calculation unit for processing, the control unit (computer operation) associated with inputs and outputs. Even today, it is thanks to silicon with its extraordinary performance that the digital field is developing. However, already in 2013, Puren wrote that Microsoft had filed a patent that aimed to develop 3D electronic objects by providing "consumables" consisting of electronic components (chips, LEDs, processors, etc.). This promise, which however did not exclude this metalloid, has, to the authors' knowledge, remained just a promise. The aim here is to examine how 4D printing sees its contribution more widely than through the use of classic (and let us say particularly efficient) semiconductors in the general field of electronics. The aim is to develop new active or intelligent media that will be able to receive and process information in a minimum amount of time, leading to a variety of uses that meet human needs.

In Table 1.1, we have cited this theme among others. To try to go further, we thought it would be useful to consider a more in-depth study of this field to show, with a generic example, how it can enrich the innovations resulting from 4D printing (and vice versa). Figure 1.8 shows the performances sought in structural electronics (Saleh 2020). Even if we are still a long way from the collective manufacturing that is used in traditional electronics, this field of possibilities allows great freedom in its principle and, potentially, very numerous applications. It is in continuity with the Internet of Things and robotics in the development of the Industry 4.0 concept.

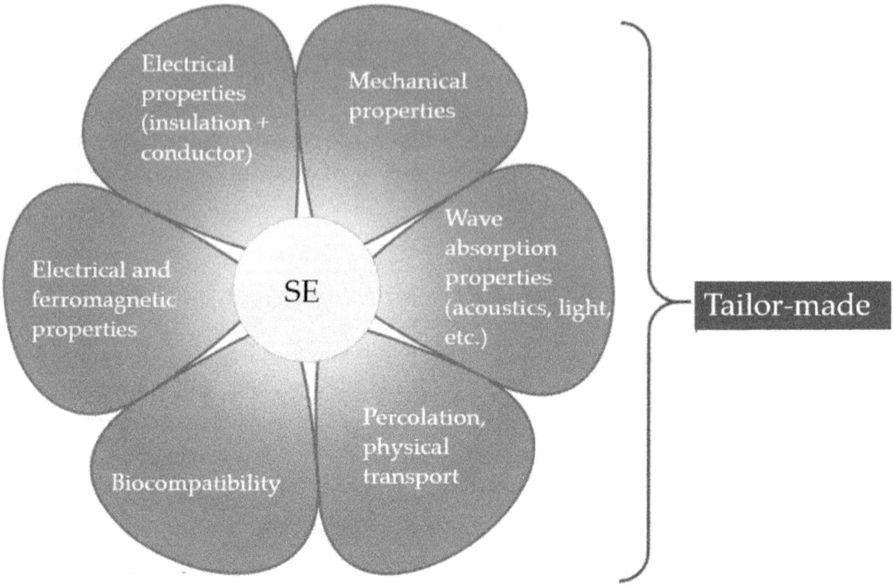

Figure 1.8. *Properties sought in structural electronics (SE). For a color version of this figure, see www.iste.co.uk/demoly/4Dprinting1.zip*

1.3.4.1. *Structural electronics*

> When I opened a Physical Review Letters in 1960, I found a revolutionary idea every time. These days, I come up with two or three a year, in a journal that has become five times as thick. (De Gennes 2020)

As recalled in Table 1.1, structural electronics aims at the integration of electronics in the heart of (4D) objects including actuators; for a significant part it concerns soft, flexible or portable printed electronics, with the objective of an easy link with connected objects (Fresneau 2020). More precisely, structural electronics aims at placing sensors and their amplification circuits as close as possible to objects, whether in industrial equipment, everyday objects or on/within the human body. To further explore future needs in structural electronics, it is desirable to conduct research on the possibility of printing organic transistors on complex substrates to interact with sensor arrays printed on/in the same substrate. In parallel with this energy/information orientation, the integration of devices requires the formulation of materials for the matrices containing the components, the development of printing, shaping and assembly processes using flexible or stretchable substrates of a polymer or textile nature (Liten-CEA 2020). This CEA center (in French: *Commissariat à l'Energie Atomique et aux Energies Alternatives*) has already developed "sensors for physical parameters such as pressure, temperature, elongation and actuators based on organic materials for applications in the automotive, health, intelligent packaging, and building sectors".

In fact, all electronics are concerned with the following angles: energy supply, information intake, information transfer, actuation and possibly information storage. With 4D printing, it could be envisaged to "subvert" the current industrial models by giving a prominent place to imagination, to freedom, if only through the existence of new systems to study, of new forms of integration, and also by allowing a return from the digital to the continuous. On the University of Lorraine website, the theme "*structural electronics*" corresponds to 1,135,000 publications (and to 15,154,000 for the keyword "electronics" alone). Table 1.3 constitutes a magnifying glass effect concerning the themes involved in this recently emerged field (compared to what is presented in Table 1.1).

Structural electronics	**Integration of electronics in the heart of (4D) objects including actuators**	**References**
4D and flexible and structural electronics	Integration; flexible printed electronics	McDonald et al. 2014; Lee et al. 2017b; Lu et al. 2018; Oh et al. 2018; Fernandes et al. 2019; Liten-CEA 2019; Wang et al. 2019b; Agarwala et al. 2020; Kwon et al. 2020; Rodriguez et al. 2021; Wang et al. 2020a
Bioelectronics	Flexible; hybrid biocompatible materials	Herbert et al. 2018; Jung et al. 2019
Electronic skin	Sensor; self-healing; environmental analysis	You et al. 2018; Gu et al. 2019a; Yang et al. 2019
Drivers	Nanoparticles; metallic deposits	Kamyshny and Magdassi 2019; Joshi et al. 2020; Stano et al. 2020; Zhu 2020
Electrodes	Porous polymers/metal nanoparticles	Wang et al. 2019b
Transistors	Material transfer in inorganic materials; organic transistors	Huang et al. 2019; Shi et al. 2020
Switch	Photo-activation	Gu et al. 2018
Electromagnetic shielding	Coupling of carbon nanoparticles and lamellar carbon	Liu et al. 2021
Sensors	Hydrogel base; silicone elastomer and magnetic powder; 3D printing	Ni et al. 2017; Kotikian et al. 2018; Marasso et al. 2018; Kamat et al. 2019; Mirzanejad and Agheli 2019; Zhang et al. 2020

Circuitry	Photo-deposition; electro-deposition	Tabridzi et al. 2017; Chen et al. 2019b
Transducer	Several successive operations	Cesewski et al. 2018
Communication	Chemical route	Chen et al. 2018
Liquid crystal	Liquid crystal elastomers/CNT nanoparticles	Donovan et al. 2019; Kim et al. 2019
Biochemical analysis	Miniaturization; integration; automation and parallel structures of biochemical processes	Zhang et al. 2018a; Muzzafor et al. 2020
Absorber	4D acoustic wave absorption structure	Le et al. 2020
Synapses	Adaptive materials; electrochemistry	Fuller et al. 2017; Sharbati et al. 2018; Park et al. 2020
Diode	μ-fluidic pathway	Lee et al. 2021
Memristor	Adjustable synaptic plasticity; neural networks	Yang et al. 2016
Memories	Adaptive materials	Yu 2018; Zhang et al. 2019
Liquid neurons	Adaptive neural networks	Hasani et al. 2020; Przyczyna et al. 2020
Logic doors	Meso-porous silica; organogels	Bradberry et al. 2015; Zhang et al. 2019
Power generation	Triboelectricity	Di Giacomo et al. 2016; Chen et al. 2019a; Shi and Lee 2019
Energy intake	Use of existing electromagnetic waves	Eid et al. 2021
Energy storage	Super-capacity (nanoparticles); triboelectricity	Maurel 2018; Reyes et al. 2018; Song et al. 2018; Xiong et al. 2018; Chen et al. 2019a; Gu et al. 2019b; Wang et al. 2020b; Wang and Xu 2020
Solar cells	Composites	Burela et al. 2020
Atomic storage	Information about a holmium atom	IBM 2017
Nano motors	Biochemical potential	Abdelmohsen et al. 2016
Micromotors	Various forms of actuation	Medina-Sanchez et al. 2016; Ning et al. 2018; Ren et al. 2020

Table 1.3. *Some themes corresponding to structural electronics*

However, as shown in Figure 1.9, taken from Joshi et al. (2020), the use of charged polymers does not lead to the realization of conductive materials of the same quality as the metals conventionally used in electronics. Progress on this topic is therefore expected before competing with current electronic devices.

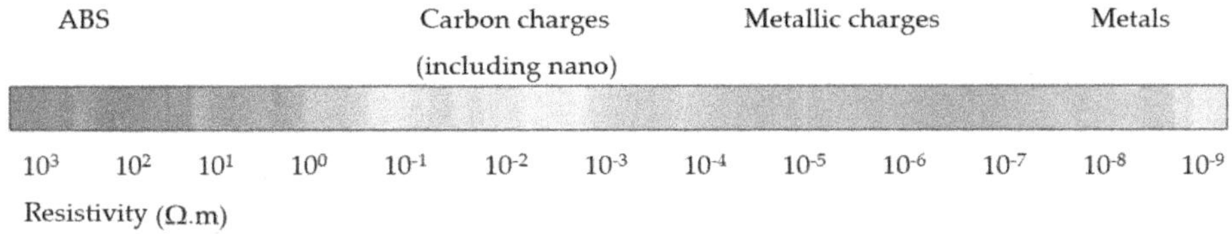

Figure 1.9. *Resistivity of conductive/insulating materials used in 4D printing (illustrating the interest of hybrid manufacturing). For a color version of this figure, see www.iste.co.uk/demoly/4Dprinting1.zip*

As can be seen from Table 1.3, there are a large number of topics that need to be explored to cover the field. This observation highlights the need to work at several scales, to use different types of materials: organic, mineral and semiconductor. Thus, by choosing an element from Table 1.1, we are confronted with the same situation: the existence of a very large number of topics to be dealt with through concepts of physics, chemistry, biology, cleanliness, scale, integration, etc., which are very different from one another. This makes it, a priori, difficult to realistically engage in top-down approaches (e.g. Lee et al. 2017a; Espera et al. 2018; Lu et al. 2018; Fernandes et al. 2019).

According to Wang et al. (2020b, 2020d), flexible electronics is being developed with functional materials, flexible energy devices and versatile sensors, which allow for functionalities such as self-healing, stretching or compression, self-powering, environmental sensing, etc. Some critical issues remain in the development of this particular field of electronics associated with aspects of the integration of flexible electronic systems, power supply, etc. Flexible integrated circuits based on thin film transistors (TFTs) are the focus of wearable electronics for their ability to be fabricated on a variety of flexible substrates, thus creating ultra-thin electronics, with applications in the Internet of Things being targeted. Other activities are in their infancy, such as the exploration of organic field effect transistors or nature-inspired semiconductors (Irimia-Vladu et al. 2010). Multifunctional flexible electronic components such as energy storage devices, sensors and electrical circuits with improved performances continue to be developed, up to the "electronic skin" (e-skin), which should allow applications in artificial intelligence, health monitoring and, more generally, on human-system interactions. However, progress remains to be made in integration, multi-scale approaches, resolution, sensitivity, energy self-powering and more (Chen et al. 2019a; Gu et al. 2019a).

1.3.4.2. *Structural electronics and 3D printing*

> Scaling up is an event regularly encountered by successful startups and it is common that it be also the cause of their failures. (Le Blanc 2017)

Recent developments around new classes of printable materials, such as conductive or semiconducting polymers or those containing metallic nanoparticles, have enabled the emergence of printed electronics (Ganet 2018). Taking advantage of the maturity of conventional printing processes, such as screen printing or inkjet, the realization of new portable and flexible devices has thus been made possible, at least in the form of a POC. Unlike traditional subtractive technologies in microelectronics, 3D and 4D technologies offer new opportunities by enabling the additive production of a structured pattern, reducing the number of manufacturing steps and the amount of materials consumed. Moreover, printing is applicable to flexible or non-planar substrates for the realization of flexible electronic devices or 3D parts.

Following what was presented above, the number of published articles corresponding to the combination of the themes "structural electronics" and "additive manufacturing" is, for the same database, 14,491; it decreases to 2,327 when dealing with *4D printing*. For example, Kuang et al. (2019), in a general article (rather 4D than electronics), report the need to print several types of materials (metals, insulators, composites, inorganic or organic semiconductors) and different resolutions for integrated devices (from μm to mm).

For these authors, high-resolution, high-speed, multi-material 3D printing technologies (yet to be invented) will be needed for the rapid production of materials with complex geometry and devices at multiple scales. Second, new inks (also to be invented) should enable the production of multifunctional materials. They write: "Stimuli-responsive material properties have a profound influence on the method of actuation, robustness of a part, speed of shape change, etc., of 4D printed parts". This is already an issue of control and improvement of 4D processes in general. But for all that, 4D processes, resulting from 3D, do not have the extraordinary properties of materials used in microelectronics, which are generally very pure and/or with perfectly controlled concentrations of impurities, with precise crystallographic conditions, etc. Major efforts will undoubtedly be required if we want to bring the two fields closer together, since they are currently quite separate, with one culture stemming from learned tinkering and the other from a solid rigor with precise quality approaches (but with very different machine costs).

This "material and process" aspect must be combined with theoretical work and design methodologies to accurately predict and optimize the effects on the 4D printing objective. The modeling of geometries, the determination of interactions at different scales for the states of change of shape and the calculation of energy (from heat, mechanical stress, ultra-sound, light, pneumatics, gravity, electromagnetism, etc.) must be undertaken. For the (very simplified) electronics component, Figure 1.10 shows the question posed by scale changes (even if, in electronic and 4D coupling, the interaction and assembly modes have no reason to respect current manufacturing models).

Figure 1.10. *Scaling in electronics. For a color version of this figure, see www.iste.co.uk/demoly/4Dprinting1.zip*

However, today with 3D/4D printing, it is not yet possible to play on very important scale changes (from the atom to the printed circuit (or what will take its place)). From this scale, the many obstacles to be overcome really exist but, obviously, depend on the desired objective to reach. As a reminder, it is the size of the voxel that determines the precision. If we want to get closer to the field of microelectronics, we either accept prohibitive manufacturing times or we develop "multi-voxel" manufacturing processes with variable resolutions, or...

> How is it that philosophy had to wait several centuries to ask that we wait a little while for sugar to melt in a glass containing water? How is it that, when such evidence was presented, we did not immediately associate the mixing and fusion of one body in another with time itself? Two flows, however, poured their component together. (Serres 1985)

Thus, by taking an example, barely touched upon for the moment (about 10^{-4} of the scientific contribution to electronics), many specialized issues are expected; however, this approach largely confirms the point made in the previous paragraph: application development is demand-driven. However, a certain number of generic issues concerning 4D printing could advantageously be taken into consideration, such as the multi-material approach, changes in scale, stimulation modes, variable resolutions and modeling (enough to cover several researchers' lives!). However, the promises are numerous: flexible robotics, actuators, medical applications, home automation, transportation, etc. But, to this day, other pending questions limit this intrusion of an original manufacturing mode into the real world:

– choice of processes and associated creativity;

– design;

– optimization of structures;

– materials with important transformations under stimulation;

– stimulation systems (optical, thermal, mechanical, chemical, ultrasonic, electromagnetic, etc.);

– materials with acceptable mechanical qualities (up to 100 MPa);

– shaping of materials: desired anisotropies; local self-organizations;

– metamaterials and Voronoi structures;

– stimulation modes; software, artificial intelligence;

– deformation modeling;

– treatments of the inverse problem;

– robust applications in soft robotics, actuators, prostheses, "smart" drug delivery, transportation, home automation, etc.

The French fragmentation is similar to that observed on a global scale. Although our country lost the battle for leadership in additive manufacturing in 1986, the CNRS (and INRIA) has disciplinary capacities which, when pooled, could help create a strong dynamic in this field. But, when reading the references in Tables 1.1 and 1.3, France does not seem to shine too brightly in those cited!

1.3.5. *Partial conclusion*

> An individual's compliance with a constraint can also be analyzed as the result of a desire to make their behavior intelligible in the eyes of the group to which they belong, their interlocutors or, more generally, the public in relation to which they act. By conforming to expectations and prescriptions, they make their behavior anticipatable, thus facilitating interaction with other individuals and integration into the group. (Boudon 2003)

What we have shown is a beautiful vision, translated in terms of "reasonably creative" but possible applications. However, the bibliographic approach reveals a very broad spectrum of action with a small number of generic themes to be explored, and this with cultures of sophisticated "bricolage" associated with the origins of 3D printing, the use of materials without major purification, etc. Not only do tectonics of scientific and technological paradigms emerge, but also of cultures, and this with an impressive list of hurdles to be jumped over. Figure 1.11, inspired by Daynes and Weaver (2013), illustrates the difficulties (or hurdles) to be considered in the case of a 4D object composed of passive structures and active structures (without taking into account the stimulation). And these are not the only ones!

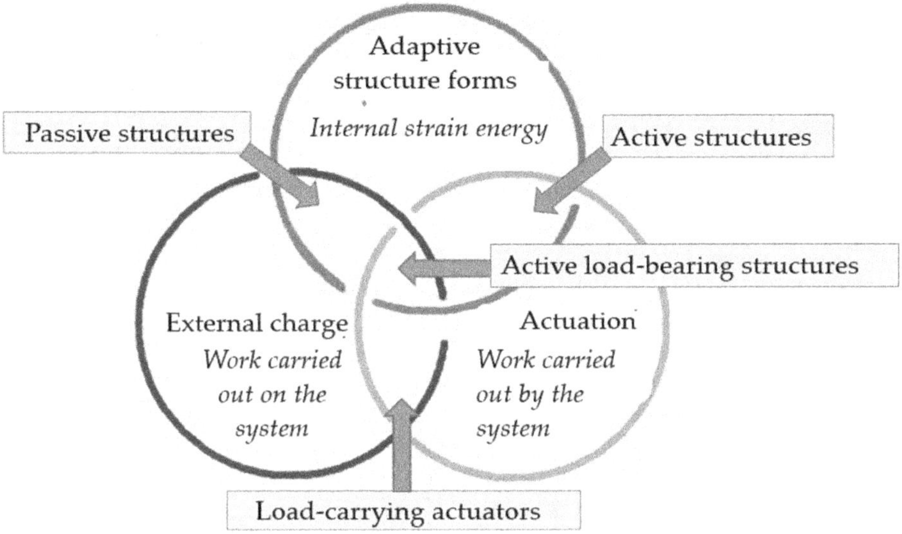

Figure 1.11. *Interdependencies between passive structures and active structures. For a color version of this figure, see www.iste.co.uk/demoly/4Dprinting1.zip*

1.3.5.1. *Upstream-downstream co-piloting*

> What is difficult to accept is that no idea is good at first. Our natural tendency is to imagine and judge at the same time when we think, yet we must learn to separate these two times of creative thinking. (De Brabandère 2019)

According to Bouët and Duréault-Thoméré (2011), the increasing complexity of the questions we have to solve between 3D and 4D printing and an application axis, such as structural electronics, imposes the idea that we cannot find the answers and solutions alone. Thanks to our own creativity, each of us contributes of course to the enrichment of this research, in particular on well-identified axes. But, far from being the addition of each of these intelligences, upstream-downstream collective intelligence methods can stimulate the production of ideas, which should then be enriched exponentially, at least initially. They are the foundation of what are called "positive sum economies".

The proposal to link the concepts of creativity and collective intelligence poses the hypothesis of a possible link between causes and effects. The promoters of collective intelligence and its methods postulate that they are factors of creativity and innovation. Giving birth to a new form of intelligence, which goes beyond individuals and their egos, collective intelligence methods should encourage the letting go of the defense or authorship of ideas. They would offer the possibility to go where no one thought possible by significantly improving the processes of intellectual collaboration. These upstream-downstream co-production approaches would allow for an appropriation of the subject matter and a recognition of the people who facilitate their involvement.

> It is easy to understand that training a human being to compete by making them understand that life is a struggle where only the strongest survive and educating them to cooperate by making them discover that they can be both fully singular and fully connected are two very different options. (Viveret 2012)

They are, however, sensitive to the personalities involved in the process (batch), to a form of learning about friendship and exchange and to their common history and dynamics, and echo a concrete and living reality. But at the same time, subliminal forms of soft coercion (the mission) and awareness (e.g. a facilitator who, in order to move the operation forward, might be tempted to make their contribution a priority) are intertwined. Apart from the work on oneself that could have been carried out by each member of the upstream-downstream group to appropriate the proposals, to make them their own, there is probably no way to go too far in the time allotted (Grojean 2015).

On the other hand, in groups exploring specific topics by associating a theme associated with an applicant and experts from different scientific backgrounds, the operation is generally of the "project" type with interdisciplinary approaches. On the subject of interdisciplinarity, Mahy (2014) writes:

> It is above all a question of sharing a desire, an intention, of wanting to create a solidarity that invites one to adopt a generous and trusting posture and to return to inhabit it when it is momentarily lost from sight, for a moment replaced by the conquering warrior. One can imagine the dynamics of such a group, the maturity it must demonstrate, its relationship to knowledge, to its self-image, and the empathy required to establish and preserve a very special quality of trust…

In short, the die has not yet been cast!

In any case, communities that share the same disciplinary passion are innovative, but tend to close in on themselves. From a scientific point of view, the opportunity of the Convergence Sciences is to develop an interdisciplinary space that is open to society, allowing for a more collective operation of the creation and exploitation of new concepts. In this way, by associating with traditional scientific disciplines on truly open subjects that concern the technological future of the city, discontinuities will emerge that lead to reconfigurations and adaptations. It is these components that must constitute the specificity of interdisciplinarity, aiming at new approaches, other links between researchers and artifacts, between users and production systems. This is why it is necessary to account for the way in which the objects studied by scientists and technicians return to the social world. Thus, techno-scientific disciplines need to be analyzed partly away from their possible relation to scientific "truth" alone, and should be more involved in approaching the horizon of social practices, artifacts and the power relations with which they are associated. This is a set of central questions that strongly links research to the economic, social and political, and puts techno-scientific research in direct contact with society (but this concern is clearly outside the scope of monitoring, foresight and even strategy projects).

> We observe a strong hierarchical prescription of the company, a vertical coordination of its organization. This verticality is characterized by rigid, formalized and hierarchical processes. In contrast, web communities claim a freer, less ordered organization, without formal hierarchy; a more horizontal organization leaving the activity of Internet users free. (Cardon 2005)

1.3.5.2. *Interdisciplinarity*

> Interdisciplinarity leads to marginalization. (Algoud 2014)

Camacho-Hübner (2007) provides an element of definition of interdisciplinary research, he writes:

> Fields which, through mutual contamination, find themselves on an unexplored or even new track, could be considered as belonging to this interdisciplinarity. Nevertheless, it is important to know that the mere bringing together of two or more fields of investigation does not confer on the research in question a true interdisciplinary value. In the same way, any change of direction within the same discipline is not necessarily identified with an interdisciplinary opening, even if this change has been induced directly or indirectly by the appropriation of a research object traditionally belonging to another discipline. (See also Henry (1998), Vinck (2000) and Alvarez-Pereyre (2003))

There is therefore a need to engage in the search for "bifurcations", for synergies between disciplinary fields (bissociations for A. Koestler), a search that conditions the object of the action and its relations between Science, Technology and Society (Ollivier 2002), "not totally predictable by the knowledge of the prior art", as is claimed in the patentability of innovative work. Moreover, for Gramaccia (2008), it would aim for "the social integration of knowledge for more social and cultural mediation". However, if we have to select the few interdisciplinary actors, if we do not try to "recognize" our own deviant "children", it will never be possible to define the concept of interdisciplinarity more precisely.

Risk-taking, which must be supported by the hierarchy (with, let us remember, "candidates" willing to participate in the experimentation), and which is implied by the fact of not anticipating the common project, should be one of the essential driving forces of collective creativity. New ideas, based on partial knowledge of what is real and what is possible, should be born and take shape in the course of a process that has yet to be built. Going through phases of disorder, accepting detours, not knowing, not judging, not excluding, leaving room for chance and freeing one's imagination, it is in these conditions, sometimes playful, sometimes insecure, that divergent thinking and creativity should express. Thanks to very diversified methodologies, made up of work processes and situations, collective intelligence should allow for real interdisciplinary innovation.

Researchers are free to choose their subject, but it is necessary to know how to approach and analyze it, and in these conditions of a project coupling upstream and downstream, it is not possible to do just anything because one's own activity has an effect on that of the other partners. Each member of a group works according to the theoretical presuppositions and methodology offered by each discipline (paradigms), which does not prevent them from questioning and working with other disciplines in a spirit of interdisciplinarity. The work of the group is therefore not the place for the expression of personal opinions.

> It is necessary to begin by ploughing a field, searching the archives, constructing an object of study, structuring a corpus, carrying out an initial classification of observables, then proceeding to a detailed description of what has been collected in the field or constituted into a corpus, or even carrying out experiments, then establishing new classifications according to discriminating criteria (what is similar, what is different), and finally engaging in interpretations. Interpretation is one of the most delicate and distressing moments for a researcher. In a process of constant back and forth between inductive and deductive reasoning, the researcher must engage in internal interpretations, those made within the theoretical framework of the discipline (which can lead to a questioning of the latter), and external

> interpretations in terms of what other disciplines can say on the same subject. (Charaudeau 2021)

> We take the opinions and knowledge of others into account, and then that's it. We must combine them with our own. We are like those who, in need of fire, go to their neighbor's house to get some, and upon finding a large and beautiful one there, stop to warm themselves, without remembering to bring any home. (Montaigne 2020)

Interdisciplinary research, transversal thinking, collective intelligence... these are current trends that challenge traditional conceptual tools, which are robust and validated, but rather conservative and incremental, while drawing attention to the limiting power of methodology. The academic discourse, which aims to broaden the boundaries of intellectual reflection (and this is part of its mission), seems at the same time to set its limits, a process in which a non-stabilized (unnstable?) methodology of discovery inevitably participates. In the age of interdisciplinarity (at least in its political discourse), theory and methodology are facing new challenges for which creativity normally represents the keystone of a collective drive. It is difficult to reason about few cases, but it seems, after an earlier experiment with CNRS-INSIS researchers in 2015, during more than a year of operation, that it is useful to explore (to find answers to) the following general questions: To what extent and under what conditions is methodological innovation in collective intelligence possible in academia today? Can it be envisaged for 4D printing? Today, it is a question of going far beyond the timid roadmap of Utela et al. (2008) by introducing trial and error methods precisely because the logic of supply is no longer credible. Will we be able to change the minds of researchers and especially decision-makers in time for the 4D field to take its rightful place?

> It is necessary to pay attention to what emerges, to what is emerging, but there is no reason to forget resilience [...], the weight of rigidities, of friction. It is important to keep the processes of hybridization, crossbreeding and irreduction in mind, but also to think about the perpetuated vulnerabilities and the impossibilities of doing, to analyze for themselves the dominations, their implementations, and their experiences. (Pestre 2014)

1.4. 4D printing: breakthrough or increment?

As has already been pointed out, entrepreneurship and innovation contribute significantly to the success of firms and the economic growth of nations (Schumpeter 1942, 1952; Drucker 2006). Firms have a mission to create value, directly and indirectly, for themselves and their customers. Entrepreneurship and innovation are often linked directly or indirectly to the notions of imagination, creativity and industrial design. Starting from the needs expressed or the offers resulting from research, the design processes define, step by step, the artifact that must eventually meet the objectives selected (needs and usage requirements) by successive choices with increasingly detailed approaches (and therefore increasingly irreversible in the limited duration of the action). It may therefore be useful to take stock of what has already been achieved and to examine future courses of action, as has already been undertaken.

For professionals in the field, what has been envisioned is that 4D printing can fundamentally change the design and manufacture of objects through combination/association with other digital capabilities (Internet of Things, artificial intelligence and robotics). By making shapes evolve through various stimuli, many applications are targeted. For example, Airbus would develop programmable carbon fibers as fuel savers and ambient air purifiers (Rander 2020). For Garcia (2020), robotics is today limited by the "hard" materials used; "soft" robotics should solve this problem by using some stimulable soft materials. Similarly, in the textile industry, 4D printing would make it possible to manufacture clothing that adapts to the shape and movement of the body. For example, cited by Garcia (2020), the U.S. military is reportedly testing clothing that can change its color depending on the environment or regulate people's perspiration based on the outside temperature. According to Market Analysis Report (2017), in China, the main efforts are in military, space, aerospace, automotive, "smart" textiles and health applications. According to Market Intellica (2019), four approximate quarters outline the 4D landscape: space, aerospace, automotive and miscellaneous. The growing need for soft or flexible objects in various applications, such as folding packaging, adaptive footwear, controlled origami-robots, adaptable wind turbines, etc., has supported the emergence of 4D printing through the use of different materials that form material anisotropy and allow the object to modify the 3D structure by bending, elongating, twisting and corrugating along its axes (Future Bridge 2020). More generally, functional developments are expected from 4D printing. In fact, apart from simple examples, it is more about promises than concrete, demonstrable achievements.

However, as the extensive bibliography has shown, in this phase of the emergence of new processes that make it possible to envisage applications that were inconceivable less than 10 years ago, these trends can/should still evolve considerably. In fact, this is the whole point of the separation of intelligent and/or stimulable materials from the 3D or 4D printing that uses them in this chapter, which aims to estimate how we can respond through research to the legitimate expectations of the socio-economic world.

> Shoppers [...] form a mass [...]. Each of us is "gregarious" to the exact extent that our social attitude is, in any domain, determined by the influence of a mass. (Pech 2017)

1.4.1. *Creativity and 4D printing*

To express creativity, we can talk about intuition, attitudes, sensitivity, divergent and playful spirits, etc. Basically, this consists of putting names to complex concepts, without the hope that this is enough to describe them. Creativity is not easily "caught", but some of its manifestations can be understood, and there may even be methods of learning about creativity. To begin with, one of these seems easy to implement, and that is to make known phenomena unusual by calling them into question, by inverting them, by transforming causes into consequences, in short, to enter into the bissociation dear to Arthur Koestler (1976). If these processes of inversion do not systematically lead to new ideas, questioning can lead to a renewal of ways of thinking and acting (this can sometimes be observed in a focus group involving a "false" naive person or in certain role-playing games, which makes it possible to shift the objective by lateralization). This position may be held in particular by those who consider that experts are the least likely to discover a new idea, but there are other conservatisms at work and this assertion cannot be demonstrated. We must then move on to invention and its aftermath.

The word "creativity" is related to the word "creation" which implies that something new is put in place. According to the classical view of creativity based on Guilford's dichotomous divergence/convergence principle (1956, 1959), the creative process starts with the recognition of a problem. From there, a process of divergence begins ("*think different*"), and finally ends, through convergence, in a new solution to the problem. Creativity can be defined as the ability to achieve an original (new) and contextually appropriate result, with a performative dimension in the creative process (Ingold 2010). Creativity is based on personality, attitude, culture, development and motivation that can have an impact on the creative process, creative potential and creative ability. But the existence of these factors does not guarantee creativity (Eisenberger and Aselage 2009; Hennessy and Amabile 2010; Runco et al. 2010; Runco and Jaeger 2012; Simonton 2012; Runco 2015; Corazza 2016). Creativity can be seen not only as a quality coming from atypical individuals, but also as a useful skill in everyday life (use of imagination, original choices, resourcefulness, etc.). Moreover, the knowledge society that brings people, communities, creative processes and knowledge domains together to interact in new societal contexts imposes enhanced innovation and agility (associated idea of open innovation according to Trabucchi et al. (2018), Von Hippel et al. (2011) and Bogers et al. (2018)).

Box 1.1. *Creativity*

1.4.1.1. *Creativity and complexity*

> Either the researcher treats the object in a partial manner, ignoring or concealing all the links that inevitably exist between this partial view and the global biases that they cannot fail to have on the object. Or, they put their global biases at the heart of the scientific questioning: not as interesting elements to look at, but as the motors of a dynamic which then takes the inevitable desire for unity as a steppingstone towards what becomes a requirement for unity. (Alvarez-Pereyre 2003)

Complexity is a situation with a variety of elements of different forms and natures in dynamic interaction. Innovation moves the complexity of a situation by artificially modifying a relationship between interacting elements (it imposes an original path between elements in the form of an imposed convergence). When it is disruptive, with characteristics of interaction between multiple actors, uncertainty, unpredictability, co-evolution of the project and its context, etc., it is then part of a complexity issue. Some believe that it requires an evolving strategy to emerge (inventing while walking). It is this vision (André 2017c) that we wish to question, based on disruptive innovations involving the convergence of disjointed disciplines (see Figure 1.12).

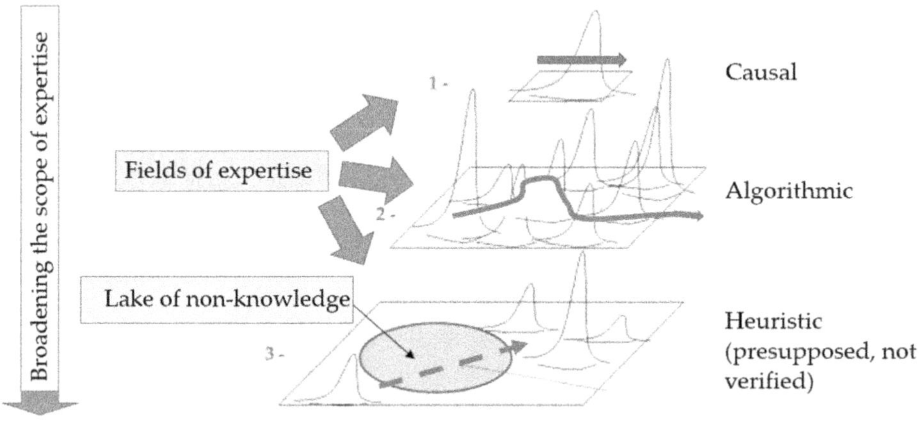

Figure 1.12. *Fields of expertise (the approximate Gaussian lines represent the scientific achievements from previous research work; depending on the form of expertise, the field is covered (causal, algorithmic) or partially covered (heuristic)). For a color version of this figure, see www.iste.co.uk/demoly/4Dprinting1.zip*

The heuristic approach can be defined as an empirical, approximate, intuitive, but fast and daring method to find a possible solution to a question for which all knowledge is not stabilized, and consequently to have a field of expertise, partially deductive, and perhaps to accelerate the resolution of the problem for which one (among others) possible solution is brought to light. No idea is born "good": it represents only a new hypothesis, which may, perhaps, become so.

1.4.1.2. *Estimating creativity*

In an in-depth peer review conducted for the principal actors, is the proposal from one researcher or a small group of researchers an appreciation of understood preferences or of the satisfaction of material constraints? Do these views provide conflicting or complementary perspectives? And so on. These are questions to be asked. But one must avoid a total "creativity" operation on the part of one's own evaluation and that of the responsible decision-maker, which would risk "nipping in the bud" any initiative. It would be better to do nothing, or to do as before!

To keep things simple, six criteria have been selected and presented in Figure 1.13, criteria of opportunity relative to the spirit of a disruptive project, and also of feasibility.

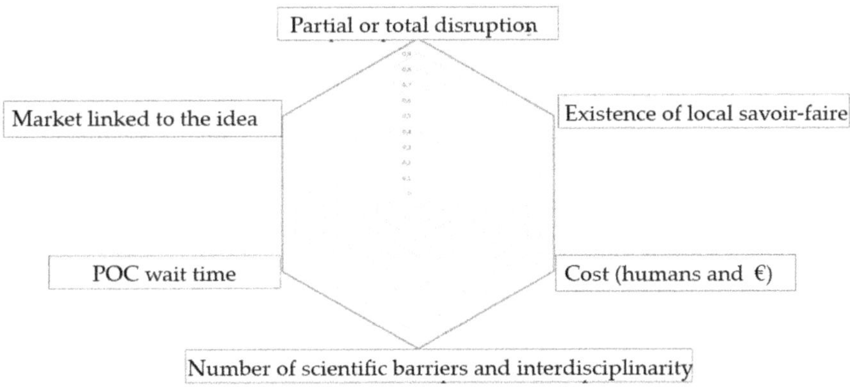

Figure 1.13. *Basis for "strategic" project evaluation. For a color version of this figure, see www.iste.co.uk/demoly/4Dprinting1.zip*

The idea is to represent as a polygon a (very approximate) digitization of each of the criteria presented in this figure and to examine whether the center of gravity of said polygon is located toward the top of the hexagon (favorable position) or, on the contrary, toward the bottom (questionable feasibility). To indicate the low degree of acumen retained, a duration is low (e.g. less than one month), reasonable (6 months–1 year) and high if it is more than several years. It would be up to the experts of a group of researchers and/or evaluators to adjust to each other based on a learning process that depends on the "means" implemented by the organization on which they depend. Depending on its financial strength and pioneering spirit, the criteria may naturally evolve.

To illustrate the applicability of this presentation, we have chosen the old example of the innovative 3D printing project, launched in 1984 (see Figure 1.14). At the time, it was a real breakthrough. With the contribution of adaptive materials (present on the "shelf"), the same graph can be proposed today for 4D printing.

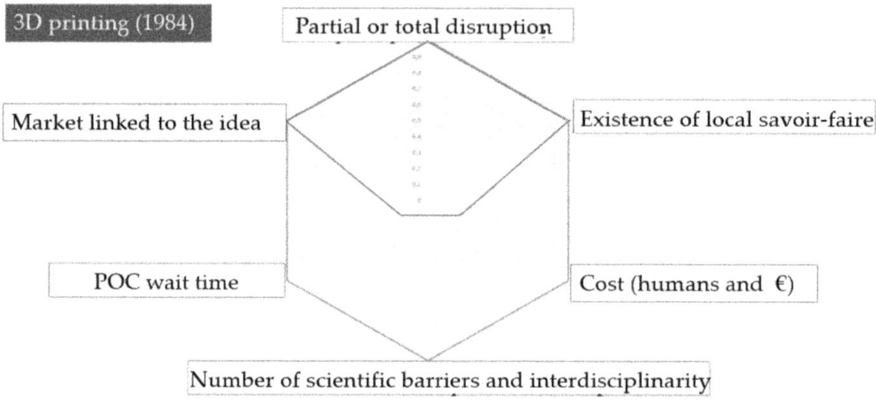

Figure 1.14. *Example of 3D printing (applicable to 4D printing). For a color version of this figure, see www.iste.co.uk/demoly/4Dprinting1.zip*

The local know-how was present in 1984 (processes, chemistry, polymers and automation) with a motivated person, the market was obvious (at the time it was primarily a question of making prototype parts with a 3D photocopier), the interdisciplinarity managed by a single person (!) which allowed a POC in a few days by borrowing resins from the neighboring macromolecular chemistry-physics laboratory. Clearly, the center of gravity of the polygon is well located upward and falls within the criteria presented above (André 2017c). Nevertheless, and less obviously, it is still possible to ask whether the proposal would be supported by scientific experts today (assuming that additive manufacturing had not been developed)?

Another example is bio-printing, a recently emerging field aiming to manufacture biological organs and/or tissues with obvious applications to replace limbs following accidents, organs following serious diseases, develop personalized medicine, etc. (André 2017c). Figure 1.15 (which could have been envisaged 8–10 years ago when the concept emerged) shows the authors' assessment of the proposed basis.

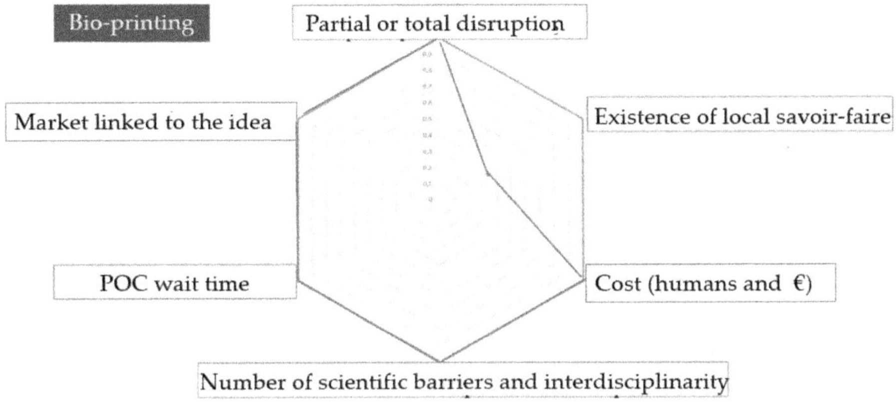

Figure 1.15. *Example of bio-printing. For a color version of this figure, see www.iste.co.uk/demoly/4Dprinting1.zip*

Bio-printing (4D printing of the living) represents a considerable market (probably several thousand billion €/year), but requires the convergence of many disciplines and the association of a very large number of researchers for medical applications. Obviously, the dimension of such a project goes far beyond the capacities of a single research team, which is generally mono-disciplinary. The conceptual development involves epistemological reflections, interdisciplinary convergence approaches and consequent risk-taking, which are generally poorly supported by the principal stakeholders. Indeed, it is necessary to move away from top-down trial-and-error methods that certainly complement the resumes, but do not allow viable solutions to emerge in the near future (Danzo and André 2020).

For Gozlan (2020), the evaluation of performance and the concentration of funds on the "best" ones are the key elements of the new system of governance of science. She writes: "This movement is characterized by an unprecedented formalization of scientific expertise practices. We are witnessing the creation of an arsenal of 'guidelines', 'charters', and other protocols that aim to frame the production of collegial judgement". What is important in the fields of research on borderline objects (as defined by Star and Griesemer (1989)) is the capacity to integrate disjointed scientific and technological knowledge, that is the possibility of leading interdisciplinary "project" groups, the capacity to produce, through heuristic approaches, programs involving a certain number of them, admittedly with obligations to question and take risks (Barrier 2011). However, not only are they difficult to evaluate quickly because of vertical disciplinary segregation (and especially quantitatively), but the risk-taking that can serve the long term can be seen as a possible use of the decision-makers' money. The short term is therefore strongly favored.

1.4.2. *Getting out of blindly following? Where to go?*

> What the engineer has gained in scientific spirit, they have lost in open-mindedness; is it not time, at the beginning of the 21st century, if they still want to play a role in technical progress, that the engineer rediscovers, from their illustrious predecessors, a universal concern and curiosity, in the face of the mutations that today affect all areas of human knowledge? (Jacomy 2015)

Once installed in the forms of a discipline, a methodology, a process, etc., groups of actors with common or shared visions are formed (learned societies, networks, etc.). From a historical point of view, the example of Thomas Edison can be recalled, imposing a view on the use of direct current (Corazza 2016). This situation, which may have met societal needs, was however undermined for scientific reasons by Tesla, imposing the use of alternating current. But it was necessary to go beyond corporatism, which is always a difficult process, especially if one does not have deep motivations.

1.4.2.1. *General context*

At the idea stage, almost nothing has been committed. To go a little further, it will be necessary to model and/or carry out POCs, which generally have the advantage of raising scientific, technical and epistemological questions, etc. An inherent dynamic of research (interdisciplinary or not) is to let oneself be carried away by one's curiosity or to commit oneself to spontaneous or stimulated requests for new questions or new objects, all the more so since in science, innovation should be a major criterion of success. This process sometimes leads to objects or questions that go beyond those defined by the cognitive and institutional structures that structure research, that is the disciplines. So, the innovator who has started with an idea, explored it and deepened it, can rely, at the outset, on a certain number of hypotheses on the elements of their perception of reality with a projection toward its possible use. It is from this stage that they may be able to carry out a POC, which may allow them to propose a draft of uses, if possible mirific. If a functional specification can be carried out, research and R&D work may allow, after many efforts, facing the complication of the system to arrive at a real demonstrator (algorithmic approach). But how can we go beyond the creative idea, the POC to reach concrete applications?

1.4.2.2. *Death valley*

A first "death valley" corresponding to this passage from POC to demonstrator exists and must be made explicit. The framework of possible development through external support (personnel, funding) is based on the importance of mutual understanding, sense-making and consensus-building between partners, both internally (especially if interdisciplinarity is involved in the action) and externally. Kijkuit and van den Ende (2007) consider breakthrough innovations to be focused on the structure and content of the different networks, and emphasize the need for strong linkages and related prior knowledge, the integration of decision-makers, and the need to move over time from a large, non-redundant and heterogeneous network structure to a smaller, more cohesive network structure capable of efficiency (see also Reiter-Palmon and Illies (2004)).

Innovation thus follows a complex process that leaves a lot of room for improvization and chance: idea, possibility of engaging in a deepening project, realization of a POC, market study, financing plans, business plan, initial financing, creation of an adapted organizational structure, industrial demonstrator, canvassing of new financial partners, complementary fund raising, prototyping and market tests, manufacturing, launching... An invention is not enough, because the road between the idea and its realization is long (Merle 2019). Figure 1.16 extracted in its principle from Harryson (2008) on the one hand, and Kotze and Purgathofer (2007), on the other hand, allows us to understand how one goes from a happy idea validated by a POC to an application.

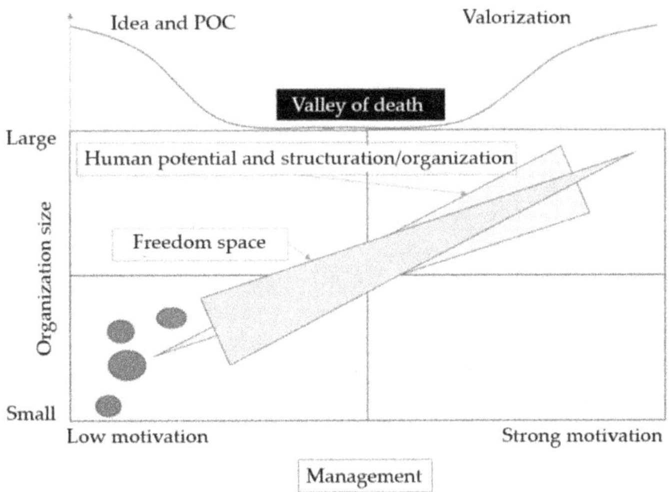

Figure 1.16. *From idea to application. For a color version of this figure, see www.iste.co.uk/demoly/4Dprinting1.zip*

But, on the frontiers of science and technology are placed ideas (brick red ovals in Figure 1.16) that increasingly require the mastery of complexity (changes in spatial and temporal scales (Dodet 2001; Puech 2016)) which are known to involve acts of interdisciplinary convergence (André 2017c). If interdisciplinarity is a means (which has its own implementation difficulties), can we propose approaches to tackle the complexity of situations from which alone the innovator must find scientific, technological and socio-economic alliances for the success of their idea? The role of foresight at this stage can be of great interest (as long as it is carried out in an appropriate manner and not, as is too often the case, to support certain lobbies…). The same is true of the strategy for supporting innovation, which must support creativity, the management of interdisciplinary projects, in short, risk-taking.

For this emergence, it is necessary to leave sole disciplinary research at least for a while to integrate spaces of convergence between normally disjointed disciplines which are by nature complex and which require, taking into account the associated risks, time and means, without being sure of the result. Enough to convince any politician to remain conservative (which they know is not satisfactory, but which is easily justifiable with regard to the electorate thanks to a low-risk management of public money)! And moreover, if it is too disruptive, would not the black veil of the precautionary principle make the happy and fragile idea disappear (even if the European Union tries to implement a principle of innovation…)? This means that it is necessary to consider taking risks in action, to manage lakes of uncertainty and lack of knowledge (heuristic approach), to find the paths (if they exist) of reasonable success…

Open innovation can remove traditional barriers by incorporating ideas, methods, technologies and solutions from external environments (Chesbrough and Crowther 2006; Enkel et al. 2009, West and Bogers 2017; Trabucchi et al. 2018). This situation where "customers" can bring their criticism, their proposals to industrialists and designers are not exactly the same in academia where paradigmatic approaches (Kuhn 1983) are generally maintained.

Traditional companies that develop new technologies have to reorganize to do so, which can lead to return on investment (ROI) difficulties and time delays. New entrants (startups), too small and with possibly prohibitive costs, cannot have reasonable margins, but they have nothing to lose when they enter the market and initially do not pose a threat to established companies. So, with disruptive, highly attractive technologies, their low inertia can allow them to operate sustainably and progress where incumbents could not. By penetrating more mature markets, they quickly compete with and disrupt incumbents.

Box 1.2. *Open innovation*

> No society can survive without a moral code based on values understood, accepted and respected by the majority of its members. We no longer have any of that. Can modern societies indefinitely master the fantastic powers that science has given them on the sole criterion of a vague humanism tinged with a kind of optimistic and materialistic hedonism? Can they resolve their intolerable tensions on this basis? Or will they collapse? (Monod 1970)

1.4.3. *Application to additive manufacturing*

To go further than the general view presented above, it may be interesting to examine the mapping that can be done in a case where we have real experience, that of additive manufacturing. The criteria used are the following:

1) potential market;

2) position of the work in relation to the current main axes (new 1, otherwise conformity 0);

3) risk taking;

4) interdisciplinarity;

5) solution to obstacles/challenges;

6) creation of new obstacles/challenges;

7) external funding (verifiable) and peer recognition (time between submission and acceptance);

8) ease of implementation.

Two extreme situations are targeted, the one corresponding to the start of the idea in 1984 and the present situation. They are shown in Figure 1.17. In 1984, the initial idea was, as already mentioned, to make a 3D photocopier. It was new, the risk-taking was high in terms of possible failure, interdisciplinarity was modest (one just had to know how to handle photo-crosslinkable resins with gloves), the obstacles were not known, funding was modest and did not require external support, recognition by peers was unthought of, and the simplicity of the process was high. But, in 2020, keeping the theme of stereolithography, things have changed with new markets, (conservative) principles acquired, modest risk-taking leading to incremental innovations, interdisciplinarity becoming modest (e.g. targeting the improvement of material properties), reduced barriers thanks to incremental innovation (with the possibility of creating new ones, such as finding new, more efficient materials), funding legitimized by industrial pressure and ease of implementation facilitated by the existence of 3D printers. The result is a clear evolution of mapping over time.

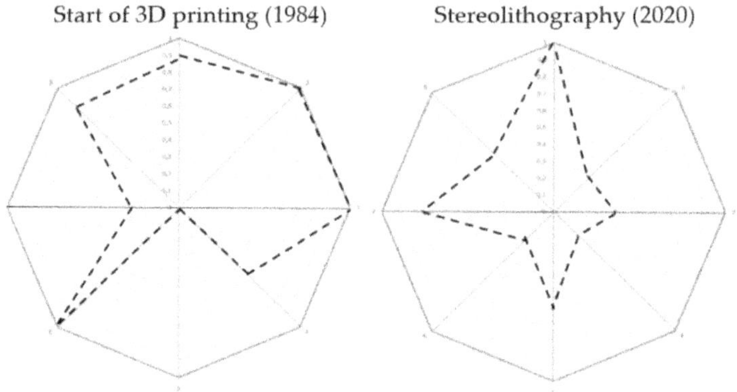

Figure 1.17. *Evolution of the mapping of scientific evolution. For a color version of this figure, see www.iste.co.uk/demoly/4Dprinting1.zip*

In the case of 4D printing, an element of complexity has been added corresponding to the exploitation of active materials allowing the exploration of space-time (with the stimulation not being restrictive in relation to the desired functionality).

1.4.4. *Application to 4D printing*

1.4.4.1. *General framework*

Figure 1.18 represents the authors' vision of 4D printing today with its limitations and its operation in supply logic. The potential market is questionable if the elements related to the specifications are not reached; for the second item, there is a need to operate differently in demand logic (top-down) with different partners, which defines specific risks and a possible loss of time to find the right partners and manage the scientific and technological project; these comments induce a low value attributed to items 3 and 4. In the same way, as it has been established, the coupling of 3D process – Materials + Stimulation – remains complex in general, which leads to the same assessment, with in item 6, the highlighting of difficulties of the same type with the probable appearance of new obstacles. Clearly, funding for such high-risk projects is unlikely to be plentiful, even if changes in risk-taking are beginning to emerge in some calls for tender. We have given an intermediate grade to item 8, because it will depend on the desired application activity.

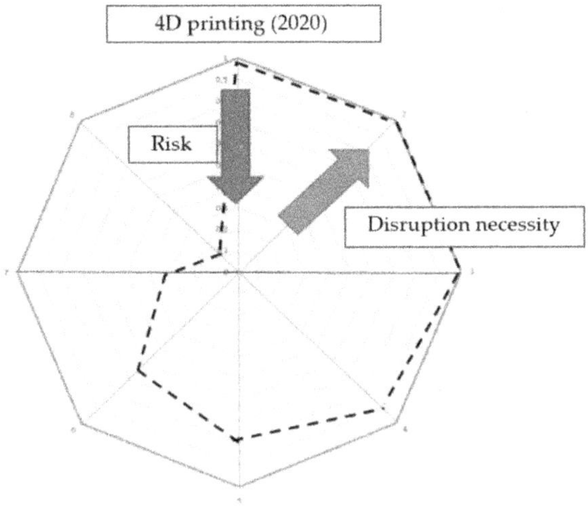

Figure 1.18. *Mapping of scientific evolution for 4D printing. For a color version of this figure, see www.iste.co.uk/demoly/4Dprinting1.zip*

1.4.4.2. *4D printing and structural electronics*

Two examples are discussed in this section: the integration of electronics into the core of objects (4D) and the realization of 4D acoustic wave absorption structures. Figure 1.19 illustrates the contrasting results between these two fields.

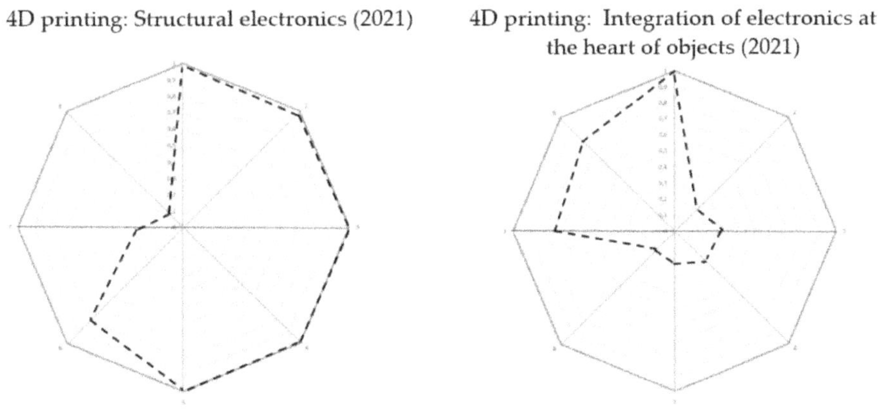

Figure 1.19. *Comparison between two situations. Left: electronics in the core of objects; right: absorption of acoustic waves. For a color version of this figure, see www.iste.co.uk/demoly/4Dprinting1.zip*

Some comments can be made to better understand these two figures. There is indeed a demand (more specialized in the second case than in the first one), but interdisciplinary management is very different since the case on the right corresponds to the realization of active origami/kirigami already dealt with elsewhere; consequently, the obstacles to be dealt with are more modest and should not raise new difficulties in the practical realization. Moreover, if the object is realized flat, in 2D, its transformation will only take place once thanks to a global stimulation (which cannot be the case of the inclusion of electronics in 4D objects). Finally, it is always easier to find funding when you are close to the goal (case on the right) than when you are far away with high risks.

These results therefore complement the previous proposals to operate in demand logic and to analyze the different criteria quickly presented to know if the risk is worth taking for a research activity to be carried out (independently of other criteria such as cost, ease of implementation, time needed before market introduction, environmental aspects, health and safety aspects, etc.).

1.5. Financial and organizational aspects

1.5.1. *Research funding and direction*

> Courage is a problem of organization. It remains to be seen who wants to be organized. (Malraux 1972)

> Innovation is essential to the formation and growth of firms and to the economic, social, and political health of nations. (Morton 1971)

From an innovation point of view, there is therefore a need to deal with the short term while having a vision of the long term, taking into account the major trends affecting the planet today, but also the various inertias that limit the possibilities of action. From this observation emerges the need for a general policy giving specific impulses to support 4D printing in its applicative possibilities and the creativity for innovations supporting the stated needs. On this basis, already in 1970 the Organization for Economic Co-operation and Development (OECD) Council (OCDE 1971) wrote regarding research: “It is important to pay more attention to its qualitative aspects [representing better living conditions] and to define the policies to be followed with regard to the major economic and social options implied by the allocation of increasing resources.”

Figure 1.20 reminds us of the obvious: the allocation of resources to research cannot go beyond a certain limit for various reasons. It is arithmetically impossible to exceed gross domestic product (GDP), there is a need to train researchers, research and R&D are only one element of the overall innovation process from primary innovation to market launch, etc. And then, for political reasons, there may be other priorities. However, in the 1960s, major impulses for the development of scientific research in France led to the belief that this age of support could continue for a long time to come, which induced, in the ambient conservatism of the constituted bodies, bitterness and violent reactions associated with financial limitations or attempts to optimize the credits allocated to public research.

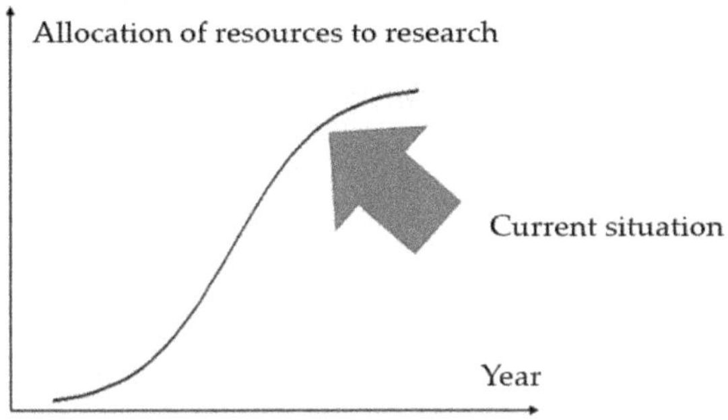

Figure 1.20. *Limitations on resource allocations to research*

> At least since the Renaissance, the market and democracy have been based on the organization of competition within a central core, a real critical mass of humans. These cities (Bruges, then Venice, Antwerp, Genoa, Amsterdam, London, Boston, New York, Los Angeles, etc.) have a local market (agricultural, industrial), a trading site (port in the city or nearby) and a power of resistance, if not aggression, against competitors acting according to the same methods. Their senescence is due to the existence of one or more new technologies possessed by the new competitors who then take over Western leadership (Malthus 1999; Attali 2011).
>
> These considerations imply a voluntarist attitude to sustaining progress through competition in order to maintain the supremacy of the structure, while at the same time proving necessary to have the warrior assets and information to compete successfully with other profit and power centers. Each structure of this type has lasted as long as it has been able to control its environment and has tended to disappear when the costs of maintaining the stability of its organization have become prohibitive (relative to the internal and especially to the resistance to various aggressions, hence the important place left to the improvement of war techniques). The capacity for technological innovation has enabled these competitive organizations to attain Western leadership, which was later replaced by that of the States, due to the obsolescence of their capacity to innovate (even if, as Japan did at the end of the last world war, it is possible to practice "second-hand" innovation for a short period). Similarly, there has been a shift from individual innovation to more collective processes relying on a body of researchers financed by the States.

It is therefore possible to consider maintaining one's place in international competition as long as it is possible to make strategic choices for research and R&D on the one hand, and to be able to finance it and recover the fruits on the other hand. In this context, the direction of technical progress is a condition of efficiency for the construction of the future (as long as we know where we want to go in this future). To achieve this objective, it is undoubtedly necessary to envisage reliable performance indicators that take into account the correct parameters of the social system and, above all, their complex interdependencies. Otherwise, by remaining in a rigid and conservative programming, not revisited (content and form), we risk leaving out new opportunities.

Box 1.3. *An obligation to innovate*

> Theory is when you know everything and nothing works; practice is when everything works and nobody knows why. Here we combine theory with practice: nothing works... and nobody knows why! (Einstein 2007)

For the citizen, what is visible are mainly commercialized objects, instruments used for medical applications, etc. While they have high technological values for the satisfaction of social needs, they only distantly represent the questions posed by unintentional science, an important mechanism for revitalizing knowledge (but which, in the 4D printing field, needs resourcing). Science and Technoscience are linked by a common fate. Figure 1.21, together with the results of the evolution of GDP as shown in Figure 1.22, represents the relationship between the importance of the social body of research (Research and R&D) relative to GDP per capita. Until about the 19th century, it was engineers (in the broadest sense) or relatively autonomous individuals who were the originators of innovations, then with the notable increase in purchasing power, research grew significantly exponentially to reach a considerable percentage of the GDP of Western States.

In Figure 1.21, technical progress is represented by a gray disk, with the resources allocated to scientific research in light beige. Between the beginning of the last century and the present, progress has mainly consisted of satisfying primary needs (food, shelter, access to electricity, etc.). With the increase in the standard of living, individual marginal utility decreases to be compensated by the satisfaction of collective goods, the number of which is increasing (infrastructures, health system, regional planning, transport, etc.). Several remarks can be associated with this figure:

– First of all, technical progress has had adverse effects in terms of pollution-related illnesses, stress, various accidents (at work, at home, etc.), even though people's life expectancy has increased for a long time.

– The environment has been deeply degraded (global warming, pollution, depletion of reserves).

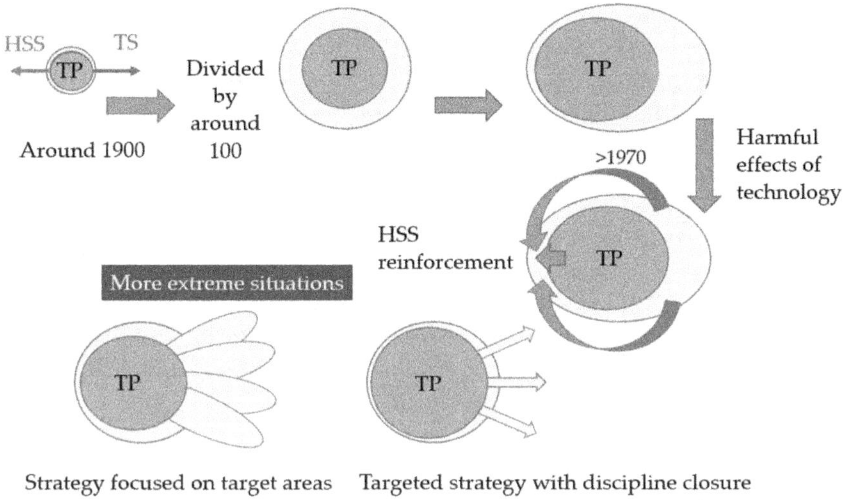

Figure 1.21. *Relationships between technological progress (TP) and strategic research policy (HSS: human and social sciences, associated with environmental sciences and hazard and risk management sciences; TS: technosciences and "hard" sciences). For a color version of this figure, see www.iste.co.uk/demoly/4Dprinting1.zip*

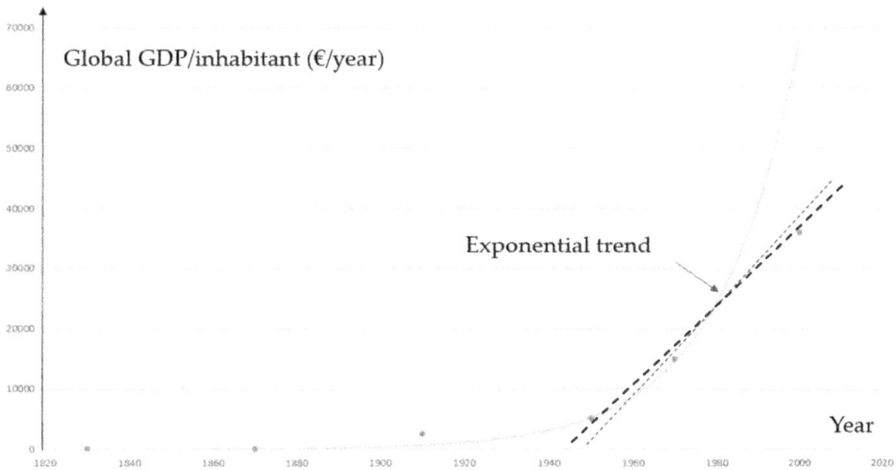

Figure 1.22. *Evolution of world GDP (black dotted line, linear evolution). For a color version of this figure, see www.iste.co.uk/demoly/4Dprinting1.zip*

These elements of social dissatisfaction linked to the risks involved have in return imposed support for research work in fields initially considered as low priority, on the HSS (human and social sciences) side. However, some developed countries may not take into consideration the need to find a balance between technological innovations and public acceptance (if only because the public is often a client). The two sub-figures at the bottom and left of Figure 1.21 characterize strategies where specific choices are privileged (the case of a nation like Israel, e.g).

However, if the share of national wealth devoted to research is important, as it reveals the willingness of a State to support its innovation (Saint-Etienne 2015) (e.g. a little less than 2.2% of GDP in France, more than 3% in Germany, 4.1% in Israel and around 5% in South Korea), the distribution of funds may be more or less oriented toward the development of major programs with a technological focus (see Figure 1.21). The internationalization of basic research then makes it possible to fill, at least in part, any gaps in certain disciplinary fields that are more or less neglected in a country's scientific strategy. This "cuckoo" method has its limits, however, because competence remains external.

> Rome being a city without commerce and almost without arts, plundering was the only means that individuals had to enrich themselves. (Montesquieu 1968)

What we observe today is, at least for France, with a certain stabilization of its GDP no longer evolving exponentially, an impossibility to respect its promise to reach the 3% of the national GDP devoted to research, which certainly leads to a scarcity of resources, but also to a certain obligation of choices, increasingly oriented toward other promises such as employment and a dynamic image with the concept of *Startup-Nation* (L'Usine Digitale 2018). In this desire for action, the President of the Republic (quoted in this article) defines four main indicators to achieve this goal:

– Scaling up: "The first challenge is to scale up and accelerate take-off".

– Talents: "There has never been a start-up that has worked, that has become a large group, a unicorn or that has failed just to bounce back, without talent".

– Simplified rules: "We need to simplify certain rules. We have started to do this and those who think we are going to stop are mistaken".

– Europe: "My final conviction is that all this will also be achieved with, in and through Europe".

This commitment illustrates the need expressed by the State to move toward a greater orientation with the satisfaction of economic needs, but is this possible in practice? If the amounts allocated to research and their orientations are part of a reasonable decision-making process in principle, it seems obvious that a long-term strategy (and not a short-term one that can be practiced by connecting points and tangents) is a political art in which science has little place. Castoriadis (1999) wrote the following on this subject: "Politics is not a matter of episteme [science], but of doxa [opinion] […]. All doxas are not equivalent, and there is a kind of knowledge in politics that is not 'science', but a matter of judgment, prudence and plausibility". Herbert Simon, who won the Nobel Prize in Economics in 1978 (quoted by Bruno Jarrosson (2017)), noted that, since everything cannot be known, decision-making is based on incomplete information about possible options and their consequences. Very often the decision-maker does not seek the optimal decision, but is satisfied with the first decision acceptable to them and their immediate environment.

Moreover, in the programming of research, one must take into account the immediate needs and those that will emerge over a longer period of time, which introduces an internal competition between short-term programmable needs and potentially useful speculations that will only produce their effects (if any) after the presence in office of politicians who have decided on behalf of the State to support this or that field relative to others… Of course, this uncertainty exists and is disturbing, because, unlike in the past when cities lost their influence relative to others that replaced them, we only have one planet!

1.5.2. *Constraints/opportunities related to research orientation*

> The more you want to speed up the progress of science, the sooner you will destroy science, just as a hen that is artificially forced to lay its eggs too quickly perishes. Science has made astonishingly rapid progress in the last ten years. Wonderfully so! But look at the scientists: exhausted hens. Really, they are not "harmonious natures"! They only know how to cluck more often than before, because they lay more eggs; it is true that these eggs are getting smaller and smaller. (Nietzche 1874)

It is therefore easy to understand the need to have the clearest possible vision of the whole system of interactions between technical, economic, financial, psychological, sociological and health factors associated with the transformation of society in a context of strong environmental constraints, if we are to try to optimize the direction of research ("toward a 'good' scientific policy"). At the same time, as Edouard Quinet (1984) commented on the French Revolution, everything is now moving very fast in terms of technological and organizational developments. He wrote: "Its genius is to suppress time. It puts nothing off until tomorrow, including the actions of years; it does not even give itself seven days to make a world". The purpose of this paragraph is to try to reveal some of the elements that limit the optimism of the statement with a proposal linked to the promotion of creativity as a grassroots element that brings out breakthrough innovations.

1.5.2.1. *Short term*

> Even if they are competent, honest and devoted to the common good, some technocrats [...] give me the impression that they are incomplete, boring, and stupid human beings [...]. They are so afraid of the fear of others and of their own fear, that they have never sought to understand it, nor to see it as an unsurpassable fact of the human condition. (Dupuy 2018)

Observation of the Brussels framework programs (H2020 and now Horizon Europe), the ANR (French National Research Agency), the National Research and Innovation Strategy (MESRI 2020a) and/or the Ministries in charge of research and industry (MESRI 2020b) reveal funding oriented toward the short term, as described synthetically in Figure 1.23 in terms of operation: the products produced are generally incremental innovations. In continuity with the past, previous actions are continued, with small changes, in a copy-and-paste logic that does not take into account potential events that could disrupt daily life. During their manufacture and use (as well as their recycling), various nuisances appear that need to be corrected thanks to new programs (e.g. "green deal", renewable energies and circular economy). There is thus a double increment, the second being temporally shifted, that linked to innovation and that linked to its effects, allowing in principle for a quasi-continuous improvement, whose temporal compatibility should (must) be set against the evolution of the major trends affecting the planet (social and environmental tension generated by temporal tectonics?). Most decision-makers are committed to supporting technological development, but there is nevertheless, and increasingly, a certain unease among the public because their response is considered insufficient for the survival of humanity.

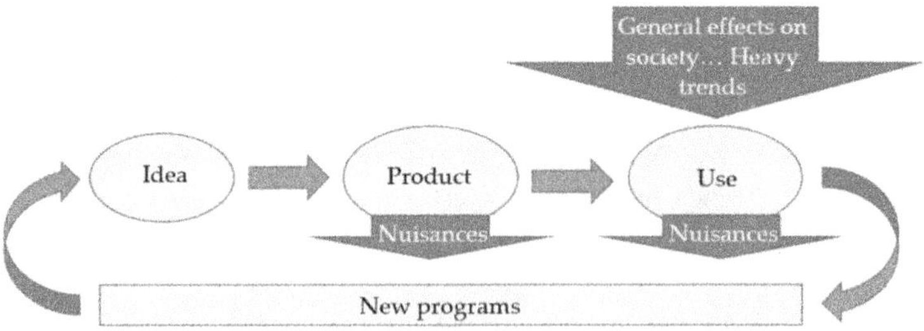

Figure 1.23. *Continuous improvement process associated with technological advancement. For a color version of this figure, see www.iste.co.uk/demoly/4Dprinting1.zip*

Moreover, the transformations of the economic system associated with the major research programs, a real instrumentalization of the future, do not only have positive effects on a society that is increasingly dubious about the effectiveness of these programs:

– Economic development is less and less shared (the "Gilets Jaunes" syndrome) with a gap between promises and facts.

– What is a new product for a few becomes an irrepressible need for the majority of society, reinforcing the nuisance; innovation creates new needs.

– Bouncing effects.

– The economic competition resulting from innovation leads to the decline of certain activities, which are partly compensated by less qualified jobs.

– The obsolescence of knowledge if only with the development of digital applications (see André (2019)) and progresses.

– Monopolistic companies impose their technologies.

– The effects of inertia (of dominant paradigms) leading by continuity to an intellectual *status quo* (Valaskakis 2014).

– And so on.

But, as Diemer (2007) reminds us: "The development of human activities has long led us to believe that the solution lay in improving techniques. However, these techniques are often reduced to a gain in productivity, and do not address the issue of the exponential growth in the consumption of natural resources". If, in addition, the cost of manufacture falls, the rebound effects will amplify the problem. We use more "low consumption" light bulbs, which may limit their interest for the environment.

For Viveret (2002), "Europe remains in the middle of the road, losing the last vestiges of its former dominating power without displaying the dynamism of a creative power".

In these contexts, in the current consumerist system, any radical attempt to assess and deal with dissatisfactions and problems expected in the long term, such as global warming, aiming at the possibility of formulating new objectives accepted by all States, goes against the policies established by those who have and continue to have a determining role in the appearance of nuisances. For France and for the EU, it is therefore necessary, in order to remain in the economic race, to take into account the others and to try to make proposals for improvements at the margin, without making a clean break. But is this enough?

1.5.2.2. *Inertia*

> If language is not available, trade cannot be carried out. (Confucius 2004)

> Like most who study history, Napoleon learned from past mistakes how to make new ones. (Taylor 1963)

If we accept the idea that it is possible to anticipate the demands that will emerge in the future (if only on the basis of a prospective approach), the crucial question will arise of how to reorient the body of academic research engaged in a permanent transition. Indeed, are we capable of organizing and directing with precision the progress resulting from science for an expected service to society? How, moreover, is this service to be defined? An excessive orientation of science could run counter to the objectives sought after because of the complexity of the questions to be dealt with; this consideration certainly does not mean that one should give up operational reflections, which are all the more important as the resources available cannot increase infinitely.

With the development of increasingly "specialized" scientific disciplines, will the vitality of research (already engaged in a search for funding that can affect its freedom and performance) be undermined? How will it be possible to develop training adapted to missions with changing disciplinary profiles? How long will it take to train these researchers? Is it possible to anticipate this training? Moreover, the French government has recently presented a new draft law on multi-year programming of research, called LPPR (*Loi de Programmation Pluriannuelle de la Recherche*). According to the Sud Education union (2020), this project has been prepared by three working groups, whose spokespeople are deputies of the presidential majority, presidents or vice-presidents of universities and research organizations and a former executive of multinational companies. "It is therefore not surprising that the three reports [for the LPPR] are in line with the government's policy by recommending an increase in the logic of competition, casualization and intensification of work".

This union, among others, considers that national research would be considered by these spokespeople as insufficiently competitive, requiring support for scientific excellence. This aspect is based in particular on a speech by the President of the CNRS, who is said to have said: "We need an ambitious, unequal law – yes, unequal, a virtuous and Darwinian law, which encourages the scientists, teams, laboratories, and establishments that perform best on an international scale, a law that mobilizes energies". At the risk of confusing Darwin with his brother-in-law Galton (the originator of the concept of eugenics), what Figure 1.3 shows with its disciplinary challenges does not support this desire to exclude the weak (unless one seeks to define the concept that is excellence along with all its baggage). It is true that Darwin (2008) wrote: "All organized beings struggle to fill the vacant places in the economy of Nature: consequently, if any species does not modify and perfect itself as quickly as its competitors, it must be exterminated". The message is clear, if a little violent. But, as Dufourmentelle (2013) points out: "To be gentle with things and beings is to understand them in their insufficiencies, their precariousness". Finally, it must be remembered that, in the current system (which may change), research is not (yet) an activity imposed at the finest level by programs to be "decanted" and specified as it descends to the rank-and-file performer. There is (still) a living research with an association between freedom and responsibility (Jonas 1998).

It is an all-encompassing concept defined in the Larousse dictionary as: "The eminent degree of quality, of value of someone, of something in its kind". But what is the status of this concept which can be applied to very different fields? Does it not raise the need for epistemological reflection on its use when we know that it can be applied in many fields: disciplinary relevance, originality, openness to others, management of interdisciplinary projects, etc.? The polysemy of this notion can lead to misuse in current use, in particular in judging a personal or team research activity.

Box 1.4. *Notion of excellence*

> But [people] seem to have observed that this newly-won power over space and time, this subjugation of the forces of nature, which is the fulfilment of a longing that goes back thousands of years, has not increased the amount of pleasurable satisfaction which they may expect from life and has not made them feel happier. (Freud 2010)

By provoking a body of researchers who have already been involved for a long time in the search for significant funding (with generally low success rates of around 15% for the ANR and for H2020, and with prohibitive preparation times for applications), it is difficult for all those who are concerned with their real activity as researchers to participate for the benefit of society. Indeed, in order to regain a minimum of serenity, it seems useful to have a little rationality, a weak partisan spirit and a real willingness to listen to different points of view. This is what the Minister in charge of research has been trying to do recently (Vidal 2020). Let us remember that the generation of ideologies based on reflections translates into simplifying incantatory formulas, but sometimes it is necessary for certain brutal positions to be taken in order to awaken collective responsibility. Who knows? In any case, by shifting the forms of thinking about national research toward two opposing ideologies, it is difficult to exchange through the sclerosis of discourse.

But, who to believe if we follow Roux (2011) who writes: "Political lucidity and maturity consist [...] in organizing the production of choice and action by giving voice to partisan interests and people of competence [...]. To orient political conduct according to a supposed general interest is therefore to sin by naivety." Is this a good way for informed and weakly state-supported research staff to manufacture consent (Bernays 2007)? With cognitive dissonance (Van Veen et al. 2009), the conservatism induced by belonging to structured networks where mimicry is the rule (Bohler 2019), each party runs the risk, in an attempt to alleviate them and recognize possible biases in their judgments, of reformulating their options in such a way as to justify their positions, without actually accepting the debate. The openness necessary in research, the scientific doubt, is likely to be undermined.

Indeed, while the method may be subject to criticism, it should be remembered that the French research and innovation system is not among the most efficient in comparable developed countries (Fernez-Walsh and Roman 2013). New ways must therefore be found to better operationalize the system of research at the service of society if we want to put to shame what Marc (1987) wrote: "How can we provide the necessary sustenance when anyone can shirk work, because no one is pressed by need and everyone can rest lazily on the zeal of others?"

1.5.2.3. *What about foresight?*

> Anticipation is not merely forecasting; it is much more than predicting what the next event will be. It is more than dealing with the next event, it is creating the next event. (Follett 1949)

In the preceding paragraphs, there has been much mention of defining the goals to be reached in the future with the means to reach them, and Follet's sentences that introduce this paragraph illustrate the importance of foresight in defining, on the basis of scenarios, the distant future and the goals to reach it. But for all that, the future that is taking shape must be considered in the same way as operations aimed at the short term, as long as we agree to get out of the cognitive biases, the theory of rational choice (Berthet 2018). In fact, the approach taken in prospective processes is based on complex heuristic approaches where not everything is rational. The reality that emerges is, let us remember, uncertain and determined by processes that cannot be isolated from each other (holistic reality).

"We are in the midst of a long and painful process leading to the emergence, in one form or another, of a global society whose probable structure it is not yet possible to imagine" (King 1999). Recent trends, which the reduction of a certain balance of terror during the Cold War has allowed, are reflected in the multiform phenomena of association on a European scale, of a globalization of exchanges on the one hand, but also of environmental concerns, of the re-emergence of forces of disenchantment, of irrationality, of xenophobia and other fundamentalisms, of a return to the self, of competition, etc., on the other hand. In short, the future outlined by the liberal system is no longer as certain as it was yesterday, which imposes a look at possible strategies that condition and are also conditioned by the digital world, which can no longer be ignored. Indeed, social structures are innervated by the new immediate digital information channels, suffer exclusion, no longer have an assured collective vision, see their values modified with their culture sometimes reaching a sterile uniformity, in a few decades many obvious things have become obsolete.

Three decades separate us from the horizon of 2050, and in all probability the world will be, in the middle of this century, quite different from what we are able to imagine. Figure 1.24 recalls some of the elements to be taken into consideration in order to estimate the needs of the future, based on new scientific paradigms, the field of economic possibilities, social, political, reserve and environmental aspects, etc. This figure defines possible objectives but does not aim to take into account all the means to achieve them.

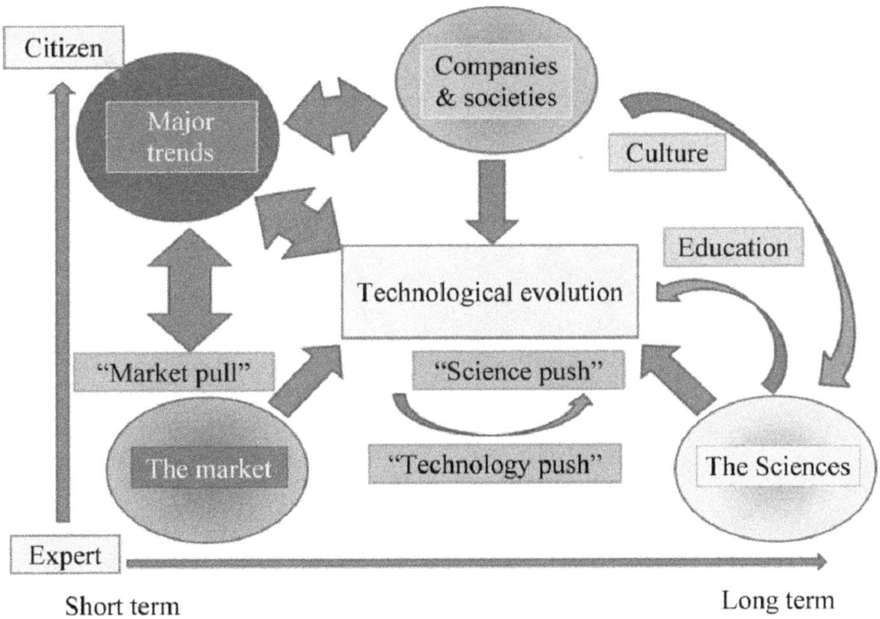

Figure 1.24. *A vision of linkages to be considered in techno-scientific foresight. For a color version of this figure, see www.iste.co.uk/demoly/4Dprinting1.zip*

In André (2019), major trends that will affect the planet and the production system were identified, leading to several possible scenarios. These data can be used to define a vision of the future, shared with society, coupled with the possibility of defining paths to reach it. In his inaugural speech, President Roosevelt (quoted by Pigasse (2014)) said: "Their efforts bore the stamp of an outmoded tradition. They came to exhortations, pleading with a tear in their eye for a return to trust. They knew only the rules of a selfish generation. They had no vision and without vision the people die." This observation of sharing induces a certain robustness to the projection on the future, robustness that research needs because time is an important element to take into consideration. Complexity is at work with paradoxical injunctions and uncertainties and, in a constrained framework, it is important not to make too many mistakes about one's choices (see Bauman (2006)).

Figure 1.25, inspired by Dieng (2014), highlights the different interacting elements that enable the definition of innovative activities relying on a probable future.

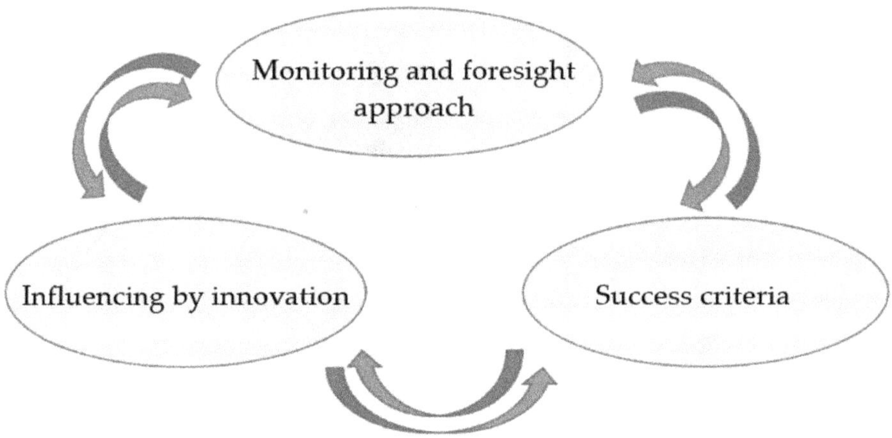

Figure 1.25. *Research-foresight association*

However, it will be necessary to take into account the various temporalities at work: that of technological and organizational innovations currently accelerating, the rhythm of life with the acceptance of the new and cultural changes. This asynchronous aspect discussed by Revault d'Allonnes (2012) results in losses of reference that can lead to misunderstandings, even rejections and crises, especially in a possible logic of decline (which no one takes into account in their forecasts). Perhaps then, in an interdisciplinary approach, incremental or disruptive innovators will need the insights that have become essential from specialists in the HSS. What will be done if they have not been considered in the LPPR?

1.5.2.4. *Supporting creativity as a solution*

> To accomplish anything worthwhile in research, one must necessarily go against the opinions of others. (Livio 2017)

> We must not mince words. This "engineering" is a modest tinkering that is much more akin to the empirical procedures of the amateur gardener than to the reasoned knowledge of the agronomist. But once again, this very discrepancy between the term and what it designates is instructive. On the one hand, we can recognize the prestige, supposedly legitimizing in our societies, of technology or its appearances, and on the other hand, the recurrent fantasy of control, extended in this case to the unforeseeable: "Even invention we manage to plan; producing innovation is our specialty; let's not rely on luck, let's force it". (Gordon 1961)

The European Union's response to meeting the needs of its citizens in its R&D framework programs is fairly simple in terms of innovation: it should rather be incremental with high TRLs, predictable, continuous with simply adaptations at the margin to keep the old principles that no longer work very well, but which are accepted by the countries of the Union (European Commission 2019). The case for disruption is a strong strategic element on how the EU intends to take global leadership on some key areas. Indeed, neither Amazon nor Apple were created in Europe, and the Union's strategic plan could have been written by slightly informed people who had watched a few TV shows about the future of the world. So, let us make a big place for digital (while forecasts for 2050 show that 50% of the energy produced by that date could be consumed by digital, as André reminded us in 2019), sustainable development, low-carbon economies, the circular economy, the "green deal", etc. This will certainly not be useless, but probably not enough.

Let us remember that digitization reinforces globalization by accentuating the movement of fluidity, as the German sociologist and philosopher Hartmut Rosa (2010) pointed out in his book *Acceleration*. He wrote: "Acceleration goes hand in hand with petrification, change accompanies inertia, and contemporary society is perceived as both fixed and frenetic." However, we can try to change the situation by moving from traditional programming to real time regarding what can be mobilized, from the short term to divergent ambitions for the future of Europe. But, in order to do this, will we be able to count on the numerous committees of experts in contact with decision-makers of all kinds who will give researchers the initiative and who will accept the risk of failure in their attempts to create something new?

To get out of the game, it is therefore necessary to go beyond the current increments that are already significant disruptors (as they occur faster than most citizens can accept, or the old frog syndrome applies…). Today, those who practice divergent thinking risk being eliminated from the group of researchers recognized by the evaluation structures. It is increasingly difficult for them to have their articles accepted and their applications for credit are generally rejected by their non-inclusion in the ambient paradigms. In fact, they do not have any influence on the research system. The system thus shows its unwillingness to take risks, a real form of conservatism that concerns the body of researchers as a whole (since as peers they participate in evaluations). So, the answers to the questions of innovation and support for wide-ranging creativity should be based on divergent minds (too little supported in the national academic world) and by the development of broadly interdisciplinary forms of action that assume radical innovations, but allow for possible failures, reorientations, agility and flexibility, and obtain trust on the part of the hierarchy, all of which are indispensable for exploring the complexity of the world. But this largely remains to be built (even if the cost of such operations is likely to be modest, because it is not a question of money, but of organization).

Alas, it is the human components among many policy makers, including those of the Union, which are difficult to quantify, that should be addressed. "The mistake in the face of disruption is to want to start from the old and adapt it to fit the new world" (Maillard 2018). It is therefore by inducing principles of economic and social responsibility that it might be possible to get out of the current technological slump from which the EU is suffering. But their caution in making strategic choices is probably due to the weakness of the theoretical resources available to analyze sudden and unexpected changes (Bessin et al. 2010). We should then trust those who wish to engage in this adventure of supporting divergence, those who accept without reservation to be responsible and answer to society for possible mistakes. But is it so easy to find today courageous decision-makers to support them in their desire to act for the satisfaction of society's future needs?

Since creativity leads to varying degrees of disruptive innovation – which is valued by the authors – the decision-maker must undoubtedly consider the relationship between continuity (incremental) and discontinuity (disruption). By disrupting the order of the company, or even of society, how can we support the disruptive moments of irruption that put the old continuities at risk? For a decision-maker who is not very committed, even in good faith, small day-to-day successes are likely to prevail over support for the breakers of the old order in order, perhaps, to satisfy society's needs in a different way. Indeed, it is hard to imagine candle manufacturers supporting research on electric lighting… except in situations where they are obliged to change.

Blumenberg (1993) writes in this regard: "Modernity does not so much resort to what is given to it before as it opposes it and responds to its challenge." The periods we are living through are conducive to change because of particular turning points in which the questions arrive in excess of the possible reasonable answers, favoring the emergence of ideologies and "fake news". Tensions in society correspond to over-solicitation, social, organizational or technological demands that the hierarchs cannot escape, challenges to previous certainties, etc. These strong reactions (the "gilets jaunes" syndrome, e.g.) either fall within nostalgia for the past or demand a future more in line with their innermost wishes by striking down the productions from previous periods as outdated. It is at this undoubtedly opportunistic stage, that creativity unleashes its full potential.

One might think that programs such as the "green deal" envisaged by the EU will follow the incremental logic that has proved successful in the past in reducing the environmental problems that threaten society. Reducing is good, but it only postpones the deadline. Giving imaginative people a chance to come up with radical innovations may be a way (though not the only one) to bring about original solutions. By trusting creative people, practicing divergent thinking, POCs could be realized and debated for applications satisfying the future demands of society, which is now forced to make hard choices if it wants to survive. As Marsan (2020) points out, "let us offer profusion, let us vitalize the diversity of productions and let natural selections take place through a sieve of our choice." To hope to survive in the near future (a few decades), humanity must accept to support risk-taking on the part of no doubt atypical researchers who offer POCs of their imagination, but supported by researchers more committed to scientific depth. These could contribute effectively to interdisciplinary project approaches.

Moreover with the Pact law voted in May 2019, a new impulse is given and extends the Allègre law of 1999, which gave the opportunity to researchers in the public domain to create startups in parallel to their public employment. According to Nguyen (2020), only 231 civil servant researchers have requested authorization to create their company since 2000, that is, less than 0.01% of the people working in public research. Today they no longer need to go before a deontology commission and can work part-time in their laboratory while working the rest of the time in the company they have created. They can also contribute their scientific expertise to the company developing the results of their research up to 50% of their time while retaining their full remuneration. They can also keep up to 49% of their company's shares for life (as opposed to the obligation to sell their shares within one year under the Allègre law). Everything is in place to help researchers' innovation, if only through the financial support of the State via the *Banque Publique d'Investissement* or BPI (Rolland 2019).

It is clear that the authors are taking the risk of making mistakes, but they wish to involve their readers in a reflection on ways of renewing innovation, which can no longer be satisfied with predictable increments in a world in open revolution, in a context where severe limits to technological development are clouding the atmosphere that was still serene a few decades ago. The ethics of the future therefore imply, through the changes necessary if human society is to survive, trusting responsible research, but promoting a break with the past in terms of imagination (while still using the rational knowledge of science).

1.6. A hopeful conclusion within an organization that learns

> My possibilities are naught, they make themselves possible. But probabilities do not make themselves probable; they are in themselves, as probabilities. (Sartre in *Being and Nothingness*, 1976, author's translation)
>
> The people always want the good, but of themselves they do not always see it. The general will is always right, but the judgment that guides it is not always enlightened. (Rousseau 2006)
>
> "[I]t is necessary for one to be certain, that to imagination be added something which is reasoning" (Spinoza 1954) or "it is necessary for one to change things, that to reasoning be added something called imagination" (Cargnelli 2020).

1.6.1. *General framework*

Hegel (1991) wrote: "It is not difficult to see that our epoch is an epoch of birth and passage into a new period; the spirit has broken with the world in which its existence and representation stood until then; it is about to bury this world in the past and is in the work of its own transformation." Everything is moving fast and, apart from elitist principles, what the author has tried to show is the difficulty of fine-tuned programming of research "from above", either because it is unrealistic, mere bookkeeping, or because it forgets that most radical innovations do not come from research in fields considered to be priorities carried out by "excellent" researchers, like the founder of Apple, Uber or Amazon.

In fact, for a general programming of science, it seems that it is necessary to have a better knowledge of the future in order to refine our vision of the means to be implemented to reach distant objectives of a few decades. It would be wise to master the system of interactions between different elements such as technological evolution, the economy, general politics, the psychology of citizens, access to reserves, current and future effects related to global warming, all elements of a complex system. On the basis of such an understanding, leading to a shared vision of the future to be allowed, we should try to identify sub-systems that are sufficiently disjointed for their study to be possible with the possibility of converging the knowledge acquired toward the final goal. We should try to make progress, if only by trial and error. Without this investment in multiple risks to be humbly shared with society and with the social body of research, it seems difficult to hope for credible results, except by chance, imposed by a too rigorous and too precise programming.

If we remain on a narrow panorama of foreseeable possibilities, we obliterate the many options necessary for the emergence of good options for tomorrow. To do this, we must undoubtedly change the paradigm of civilization, but for public research, this means re-founding our imaginations and developing a narrative on a revised basis that leaves more initiative to researchers. We must therefore encourage a craze for the truly new and exploit it as an opportunity to produce a responsible scientific narrative. The CNRS has partially undertaken this on the occasion of its 80th anniversary, even if there remain many continuities (CNRS 2020). Thus, in order to get out of a subjectivity sensitive to uncertain futures, perhaps we need to give researchers the chance to function differently, through a liberated creativity, accompanied at last by the hierarchy. Under these conditions, desire, passion and emotional intelligence must be supported in order to move forward. Perhaps, in a context where not everything can be decided, it is necessary to leave it to those who are the actors in the field to limit as much as possible the contingencies (but undoubtedly not in any old way). That is all the bad news we can wish them. Let them dare, trust them and better still, members of our hierarchy, help them! The risk is worth it! The future of our children depends on it.

1.6.2. *Organizing research in 4D printing*

> By placing enormous and very poorly defined costs and benefits on the same scale, normative evaluation is condemned to undecidability. (Dupuy and Roure 2004)

What has been shown throughout this chapter is a spectacular aspect which, admittedly, hides scientific and epistemological difficulties, but which can represent very interesting avenues in fields which are partly niches, but which can constitute at least partly autonomous fields, such as soft robotics or structural electronics, which correspond to obvious economic markets and for which France is no worse positioned than other industrialized countries. However, these two fields are aimed at more complex applications than those in medicine and drug delivery, which are essentially the result of work in homogeneous or hybrid 3D printing that has been pushed to its limit. If we want to satisfy the ecosystem defined in Table 1.1 (which is not exhaustive), the French position could be held by:

– a central core with a set of 3D printers which can, after some adaptations, cover the field of 4D printing and engage in generic research on the subject of 4D, in particular (but not exclusively) on heterogeneous 4D printing with the exploration of the obstacles defined at the beginning of this chapter (changes of scale, voxels of variable size, multi-materials);

– a local mirror research unit with knowledge in the fields of electronics and robotics allowing internal developments to the group constituted by the central 4D core and the "mirror" component;

– a very large external demand allowing upstream-downstream operations for the realization of 4D devices adapted to an external demand (see Appendices 1 and 2 in this chapter);

– scientific and technological advice to guide the choice of risky operations, taking into account the socio-economic impact of the work to be carried out;

– specialized research units located throughout France, selected for their expertise in fields such as materials, chemistry, mechanics, processes, etc.

> Rationalism can no more transcend knowledge by analogy than the famous Baron Münchhausen can pull himself out of quicksand by pulling on his own hair. (Stengers 2020)

The activities must allow on the one hand to continue to further the 4D theme while at the same time allowing the development of POCs in a first phase, then industrial demonstrators in a second phase. This means that it is imperative to associate a transfer structure with this type of proposal (which could advantageously work locally with external applicants who have expressed a specific need retained by the scientific opportunity board).

Faced with the increased demand for innovations that are of societal interest, we would like researchers to stop relying solely on their intuitions, to dare to create and to multiply exchanges with their application targets in order to develop solutions that respond to real difficulties and needs. To do this, we must abandon the classic "linear" processes for iterative, integrative, learning and shared approaches: brainstorming, openings, development, testing, evaluation of feedback, corrections, testing again… until the solution is finally ready to find its place in industrial appropriation.

This is a profound change of strategy in order to be a major competitor. By questioning future users, we exploit the full potential of co-creation. A product is no longer created by a person in research and R&D who has a hunch, but by a win-win alliance between designers and creative users. "This paradigm shift is transformative and innovative in companies" (Mayard 2021). Thinking about such co-construction is also based on a critique or deconstruction of previous technicist and rational approaches. Instead, we must seek to exploit principles for action-oriented reasoning, that is, a heuristic for defining and conducting co-constructivist devices. This is another form of intercultural (rather than interdisciplinary) project.

The approach cannot be based on a general method of optimizing the conduct of such upstream-downstream associations, which must be dealt with on a case-by-case basis, but on the postulate that a system can be designed collectively on the basis of reasoning and deliberation in confidence. Co-construction processes are obviously complex, contingent, situational and "are characterized by an irreducible unpredictability, because the actors always remain partially free and can never be totally instrumentalized in the service of an end that would be external to them" (Foudriat 2019). It is therefore, in a way, a question of achieving a 4D organization, flexible and stimulating, but coherent! In jest we have called this open group Min-Max 4D or 2M4D.

> The real world, unlike the sciences which carve out closed domains, is not divided into disciplines. It is the isolation of economics and the science devoted to it that is the problem, and it is also the need to build a renewed science of economics which, without ceasing to be an "autonomous" science and without reducing itself to a simple technique, recognizes that economic parthenogenesis is conditioned and submerged by the action of exogenous variables that define its environment and opens itself to multi-dimensionality. (Bartoli 1991)

> We must find the way back to a multidimensional way of thinking, which, of course, integrates and develops formalization and quantification, but does not get locked into it... The economic, the psychological, the demographic, which correspond to specialized disciplinary categories, are all sides of the same reality; they are aspects that must obviously be distinguished and treated as such, but they must not be isolated and made non-communicative. This is the call to multidimensional thinking. (Morin 1991)

But the transition that can be associated with 4D printing needs rough edges, a vision, topography to conquer in order to get out of the various solicitations that occupy all the available time (of which we are the consenting victims, but not always conscious). The co-constructions that are part of these concepts are not systematically systems of shared leadership aimed at developing the professional self-esteem of all the members participating in them. But for all that, in these operations where intellectuals are convened, the exchange can be horizontal.

Co-construction processes are complex and sometimes fraught with unpredictability because the applicant-designer couple remains, outside of the contract, partially free and can never be totally instrumentalized in the service of an end that is external to them (Foudriat 2019). Co-construction remains "a complex process, eminently contingent and situational, which is distinguished [...] by its high level of deliberation and involvement in the decision-making process" (Leyrie 2017). The interest is that all stakeholders can be involved, which leads them to get involved, to act with their diverse skills and to have an influence on the upstream-downstream project that concerns them.

In these operations, it is necessary to leave behind the belief in certain principles, first and foremost the principle of scientific determinism in the harmonic connection of things. It is necessary to construct new relationships between non-objectified desires and what is scientifically and technologically possible, based on the findings of experience in order to satisfy a requirement of positivity. The linkage of forms of determinism with laws that are first causal, then functional, and finally with the laws of probability based on statistics, can be defeated, which requires humility on the part of the researchers involved in these operations. Indeed, there is a rearrangement of previous knowledge and the emergence of new debates (with, for researchers in a hurry, a loss of time and a return to the previous paradigmatic focus).

In the minds of scientists, even when discussing interdisciplinarity, there is a concatenation of causes, each cause generating one or more effects normally known and controlled. However, when approaching the realization of an object in a co-construction, organization and individuality caught in a network of relationships, some of which are rigidified by habit, the existence of unevenly shared information, a memory organized in silos, uncertain anticipations and project management can have an effect on the desired result. This may be the result of interaction behaviors in the host structure (which will implement the translation of the idea) which is already organized, arranged and regulated, and in dependence on the operators' operating skills which have already been or are being constituted. The work of co-construction is the result of scientific research into "intention" and sharing, and is therefore a moving reality, rebellious to the idea of an absolute "law".

"Science is only illusorily definitive" (Granger 1960). Bartoli (1991) writes that because it is marred by errors and inadequacies, it corresponds to a risk "boldly" taken. In principle, he writes that it is "operative, hard-working, conquering, essentially unfinished, it never ceases to rectify, expand, complete, open up previous knowledge". But, in reality, the researcher's situation is more complex insofar as their evaluation takes into account their productions rather than divergent attempts (unless they are successful). Co-construction is therefore not a natural process. It is therefore necessary, as Gouil (2010) suggests, to relearn how to cooperate. This collective learning can, moreover, avoid exclusion, the incomprehension of a world that may seem forbidden, which prevents a will and a willingness to emulate in order to "be" in the adventure.

Riccardo Pucci (2021) in one of his emails wrote to us:

> Although we obviously cannot predict the kind of impact [4D printing] will have on the environment around us and the laws that dominate it, through creativity we can bring important elements in a two-way relationship between the creative and the engineer. Only time, patience and passion for 4D can show us the right methods to make this technology shine.

So let us co-create!

Box 1.5. *Co-construction*

> Knowledge is blind when it is reduced to its only quantitative dimension, and when the economy and the company are considered in a compartmentalized way. But the impermeable compartments have become established. The dominant logic being utilitarian and short-termist, we no longer explore fields, activities, specialties or ways of thinking other than our own, because a priori they do not directly and immediately serve the accomplishment of our tasks, although they could enrich it.
>
> Culture is not a luxury, it allows us to contextualize beyond the furrow that becomes rutted. The obligation to be technically ultra-performing in one's discipline results in a withdrawal into this discipline, the impoverishment of knowledge, and a growing in-culture. People believe that the only "valid" knowledge is that of their discipline, and that the notion of complexity, synonymous with interactions and feedbacks, is mere chatter. Is it any wonder then that the human and civilizational situation on the planet is so bad? To refuse the lucidities of complexity is to expose oneself to blindness in the face of reality. (Morin 2016)

1.7. Appendix 1: Processing an external file

> Our world has set itself a crazy mission: to "modernize" – for their own good – all humans on Earth. To eliminate all other worlds from the planet, such is its program. Is this not totalitarianism? This mad conquest of humanity began in the name of true faith, of civilization; it continued in the name of progress, of reason; it continues today in the name of freedom, development, democracy, Human Rights. Is the "major risk" not, more than ever, the gigantic violence that the totalitarian project of our world entails? (Solé 2004)

Figure 1.26 brings together the different elements that need to be taken into consideration in order to attempt to build a case (that could be supported later by 2M4D) from a creative idea that falls within the scope of 4D printing.

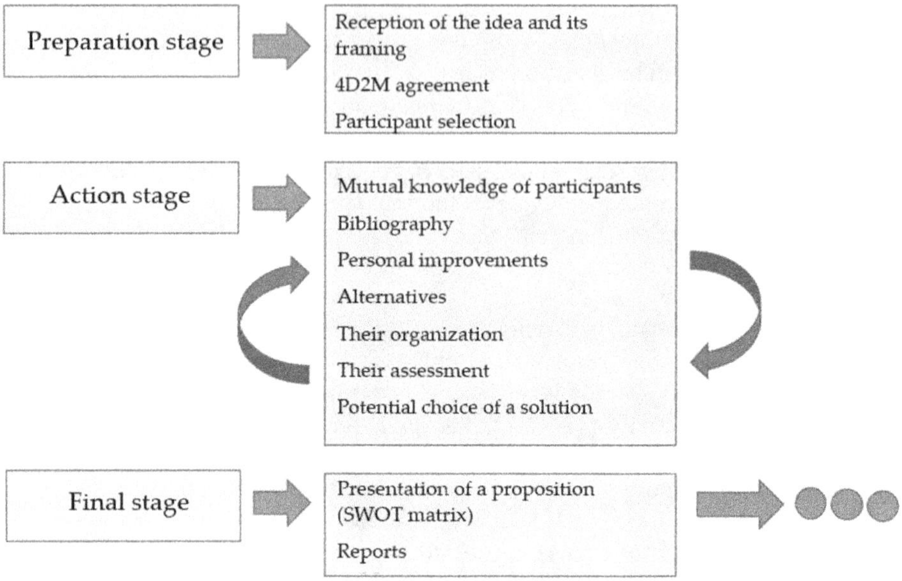

Figure 1.26. *Envisaged overall (flexible) procedure*

Between the choice of an action and its implementation, real feedback loops can occur over time, leading us to consider the decision as a dynamic process and not as an instantaneous act.

> Starting from an initial state of knowledge about the alternatives that the decision situation involves, the search for information is refined and then consists of inquiring only about the critical points of the problem on which the state of knowledge does not allow for a reliable definition of a decidable situation (indiscernibility of alternatives, uncertainty of an evaluation, unattained objectives, instability of evaluations, no consensus, etc.). To reach this situation, it is necessary to fill an information gap which, as long as it remains, induces epistemic uncertainty, in other words, a decision-making risk linked to the insufficient state of knowledge at a given moment in order to be able to decide. The further the quest for knowledge advances (when it is well managed), the more epistemic uncertainty is supposed to diminish and the situation that can be decided upon takes shape. The challenge of this dynamic cognitive process is to know how to conduct the search for information with insight during the deliberation: on which aspects of the problem will it be most judicious to acquire additional information to reduce this uncertainty as best as

possible, as quickly as possible, and thus accelerate the decision-making process? (Imoussaten 2011)

The choice to engage in new solutions from uncertain environments implies taking risks with a certain control of complexity, which is often the case in 4D printing. A confident and dynamic framework to deal with this type of situation is necessary to move forward in a "good" horizontal collective organization, which is very rarely practiced in the academic research world. It is necessary because of the need for mutualization of multiple forms of knowledge in the scientific, technical, social and organizational fields. There is therefore and a double responsibility in this operation, that of 2M4D, but also that of the participants who must really get involved with agility, motivation (Mauchand 2001) and confidence in these operations. For the Institute, the hopes of its managers must be confronted with operational constraints, the objectives with the means, and a dialog with the project group must be established to define achievable objectives. Facilitation by a 2M4D representative can be advantageously proposed to improve the organizational performance of the group. From the group's debates, a contradiction should emerge "between the social process of design, full of uncertainty and ambiguity, and the ideal image of the instrumental process in which the participants believe they are involved" (Darses 2009).

The purpose of this disruptive proposal is already to show that the conditions for involvement in forms of complexity that are on our doorstep with limits and/or constraints of various origins can be overcome and that we must invest in this new, henceforth inescapable challenge if we want to be recognized as credible actors capable of breaking out of the daily routine (training made reductive by the teaching of certainties, difficulty in recruiting "prepared" minds, effective management of interdisciplinarity, systemic approach, understanding of the issue in a world governed by the short term, involvement of the "best" in what is recognized as a promising field, etc.). In short, we must think outside the box… or else…

In an in-depth evaluation by 2M4D, is the "consensus" proposal from the project group one of preference modeling or one of meeting material constraints? Do these views provide antagonistic or complementary perspectives? And so on. These are questions to be asked. But the evaluation that comes out of the decision-maker must avoid a totally "creaticid" operation that would risk "nipping in the bud" any initiative. It would be better to do nothing!

But if 2M4D makes this proposal, it will only have credibility if its associates – anyone assigned to its research unit – feel concerned by this type of meaningful and hopeful action, otherwise Lewis Carroll will be right!

> In the Red Queen's kingdom, all subjects must run as fast as possible to stay put. As a puzzled and bewildered Alice remarks, this is a very tiresome way of getting nowhere! (Lewis Carroll 2010)

1.8. Appendix 2: Going a step further (working document)

> Individuals draw out their information and keep only that which is relevant to their system of thought: they seek to reinforce certainties already acquired in the past. The processes of anchoring and objectification account for this need. (Orfali 2005)

In this ambivalent setting, national research and university training systems have evolved in two modes: stable trajectories and a few bifurcations under dynamic conditions (the world is moving around us). They are continuously changing under the action of dynamic processes, both internal and external. If a structure emerges and is maintained in a sufficiently persistent manner to be recognizable, it is nonetheless the product of this evolution. This is why the engineering sciences (including computer science and materials) still constitute a respectable scientific force (even if it is a bit aging, on the side of research organizations). Finally, to the processes of self-organization, we must add the evolving specificity of regions wishing to redeploy their scientific and technical resources, which implies that the creative capacity of motivated university centers must also be taken into account, as the flow of innovations and the creation of new wealth are the only factors capable of explaining the continuation of an attractive dynamic. How then can science on an emerging subject be a real leavening agent of this design, which obviously concerns it only very imperfectly?

Any centralizing research and innovation system only really makes sense if it is thought through and challenged, at regular intervals, to maintain a level of excellence relative to some vision of the future. What we observe is that everyday mechanisms and procedures are put in place to control researchers and teacher-researchers in an increasingly sophisticated way, whose operations continue to be "business as usual". In the field of additive manufacturing, growth is 20% per year; in the particular case of 4D printing (adding a temporality to 3D printing, it is 40% per year, and for bio-printing the number of publications doubles every 18 months…). It is – at least in this example – not possible to remain in a wait-and-see logic.

The most successful systems today are open, pluralistic, living, agile, reactive… Moreover, we speak of "ecosystems", which increasingly imply integrating randomness, risk-taking, emerging actors in decisions, resourcing actions through science, etc. In fact, by making "objects" ever easier to acquire, by reducing their duration of use, by exploiting too quickly the reserves available at low cost, the consumer society knows without expressing it that it is running to its ruin (even if some people think that technology will remain the solution in the developed countries). Is there a desire on the part of the population and its elected representatives to be consumers of these goods or to participate in some aspects of a renewed economic activity that is always provisional and based on the knowledge of scientists?

Can they rely on a dynamic industrial environment? The answer is clearly negative, leading to local academic circles which are out of touch with the economy of their city. Can they then rely on the will of the so-called scholarly community? A recent survey conducted in a large laboratory in Nancy does not provide a long-term vision either. Defining an original societal need and the technical and organizational means of satisfying it does not, of course, systematically enter into the leadership of techno-scientific research, which must retain its scientific character, creativity and depth, while at the same time having specific approaches because of its position at the interface with the applicability of scientific concepts resulting from research. The legitimacy of science is in fact built by the dynamic and recursive mutualization of scientific knowledge with a view to an end whose origin comes from the scientific component or from that of the applicators or from a demand from decision-makers. But then, where are the motivated decision-makers at the local or national level?

> At a time when the space offered to individuals thanks to modern techniques of movement and communication is undergoing a process of enlargement, a promise of emancipation for individuals who are no longer riveted to a territory, the proliferation of misbehavior exerts the opposite pressure: it reduces the possible space. Beyond a certain level of disorder, a devastating doubt about the intentions of others takes hold of everyone. Incivilities, by expelling conviviality from the public domain, confine it to reserved niches. (Peyrat 2003)

Thus, in order to avoid confining researchers to their discipline, a discipline in which they exercise their freedom (their autonomy) and their activity, allowing them to go beyond the limits of knowledge, it is probably necessary to find and invent ways of "transgressing", of support for risk-taking, which must authorize and support new cultural couplings associating divergence (or disruption) and creativity. These must allow the promotion of adapted synergies, authorizing the creation of new concepts or new artifacts useful for society (notion of self-fulfilling competition). This is what is expressed by the (happy?) idea of coupling between science and technology and the social body. By supporting creativity at the interfaces, by developing creative hybridization, it is therefore a question of going beyond other frontiers of knowledge, but undoubtedly on the basis of new or revisited methods, by resituating research and researchers in the context of the life of the city rather than in national or European life. This exercise must be undertaken. 4D printing is there as a basis for experimentation.

With this aspect comes a singular element in research, that of rarity, of originality, which can induce exchange through the dynamics and attractiveness created. It is interesting to examine why and how exchange relationships (in the form of knowledge sharing, client-supplier relationships, bartering or money) are developed with difficulty internally or externally, how researchers in their scientific quest have financial and human support (this problematic raises the question of research management (or in other words, the researcher's autonomy) and possible cooperation to help some people stay on course in the "ocean liner" that a large laboratory can be). Indeed, how can the erratic funding experienced by researchers in the support of the agencies be smoothed out? What forces can act to ensure that separate researchers can "hold together" and form what the academic world calls a research unit whose mission is certainly to produce scientific knowledge, but also to be useful to society? The survey conducted (Horckmans 2016) attempted to uncover social mediations through which these individual-to-individual relationships are transformed and shaped to avoid anarchy, and also to fit into a meaningful global compatibility. What energy is supplied to reach this state of dynamic equilibrium, which makes it possible to fight against the second principle of thermodynamics where entropy dilutes efforts? In fact, is there not a clear conception of the world of research centered on itself, its promotion and a normative fetishism of the evaluation that is adapted to it? The relations between societal actors then appear in the background, almost in antagonism. Is this a diversion from the mission (as long as it is well defined)? So, what could be the link between the economy of the nation and an emerging subject like 4D printing?

One line of thought is to know whether we belong to an elite of scouts, of those who prevent us from thinking in circles, and how we can judge the societal impact of creativity, rational unreasonableness and the associated scientific depth of vision (foresight), provided we give it consistency and transparency; a second line of thought is that of the "virtuous followers" – wise, learned, useful, but compliant. Producers of stabilized knowledge represent the necessary backbone of a research unit that needs to rely on common and effectively shared values; a third, which is not opposed to the first, relies on the development of links, exchanges and sharing, both internally and externally, across disciplines and/or inter-disciplinarily.

> When a team makes arguments designed to appeal to funders or the public, but which members and their peers know are empty, the rule is silence. The citizens of the "Republic of Science" know that, in a self-interested society, such incentive mechanisms are "unfortunately necessary". (Stengers and Bensaude-Vincent 2003)

This represents another form of intellectual mobility. Between solidly established positions, the merit of scouts or openers can only develop if there is a call for air, if there is for a time, a space for action, human and financial support to explore the uncertain with its real risk-taking. There is therefore probably not a single merit, a single "excellence", but the diversity and harmony of skills (not only scientific) must constitute the wealth of a laboratory, as long as we get out of the symbolic balances, the non-aggressions, the opaque reality of certain local practices and that we calmly manage with empathy the tensions between these tendencies that sometimes come under the paradoxical injunction to be "readable", "credible", "driving", in short, "excellent" (a very vague notion…). It is this governance that must strengthen a collective imagination that is confident of having a space for freedom and joint action, freeing itself, as much as it can, from an administrative technocracy considered to be "superior", that thinks it is piloting research to meet the needs of society (as it unfortunately imagines them) and that continues to apply principles of "meritocracy" adapted to all and perhaps to nothing. But perhaps this is the price to pay for continuing to undertake, if one knows what one is and where one wants to go (and how).

Everything, a survey conducted a few years ago revealed, is about beliefs: belief in a future in continuity with the past because no major clouds obscure the sky, in leaders who assume scientific and administrative responsibility. Since then, Covid-19 has hit us. So, why try to practice activities that fall under the heading of divergent thinking (criteria that are generally unknown at the time of hiring), as long as there is no obligation to change? Are the temporal bases of resourcing adapted to the maintenance of the scientific and technological leadership of a research unit or to the personalities in this laboratory? That is the dilemma! Who should be credited for these sudden changes? This risk has apparently not been evaluated, and this in turn does not allow the urgency of the necessary changes and their anticipation to emerge. This work of social engineering is to be done in the academic world, for which the processes of integration and mutualization, of common or complementary values, have not been engaged in any notable way. The ball is in whose court?

These different "manifestations" of external tensions must therefore be shared in order to create a common sense of action and a shared collective feeling on the part of the members of the scientific community. Intelligible representations by the whole staff presuppose a minimum of trust between its members so that they can interact in a valid manner; one possible means could be a common or partially common scientific project. This material project would not be there to reveal the "mental" state with which the members are associated, but a means of bringing energies together to do things differently. "This solution implies an ability to identify the needs of the members and to find solutions to these needs. This solution implies an ability to identify reliable informants, who collegially test their ideas through demanding procedures" (Farina and Pasquinelli 2020). In order to win better together, it would consist of externalization in the movement enabling the creation of links, participation in the evolution of the image, strengthened because all the partners have contributed to its creation. This pooling would then have, in a rational anticipation in an ambient uncertainty, a renewed attractiveness and a strengthened power of integration, allowing shared risk-taking.

There is a need to explore the mobilization capacities of the "troops" in relation to the "incapacities of organized action", by situating it in the wake of reflections on the continuity between action "within" the organization and action "outside" – the latter being very often already organized and supervised (ANR, EU, CNRS, PIA, etc.). It is an in-depth exploration of the degrees of formalization that would give access to both the capacities of organized action and its incapacities and thus to an additional dimension of scientific anthropology. This work to be carried out should allow a conceptual clarification of the notion of incapacity to move the lines.

Torn between vocation and ambition, between emulation and competition, today's researcher is a somewhat "paradoxical" character, who only partially masters the logic of their actions (reflect on the interactions between scientific knowledge and actions toward/for society to better understand why and how it is possible to enter into a more responsible approach to research, due to very important developments). The quest for a scientific reputation now resembles a sport competition; a race for "refereed" publications, under the umbrella of a new research management culture that is more reminiscent of the old Soviet central planning than of the free market of ideas (which is modestly allowed to blossom). The researcher finds themselves more than ever a slave to a system that they helped create, somewhat in spite of themselves, and from which they find it increasingly difficult to distance themselves.

The scientific world in Nancy would do well to critically examine these new reputational mechanisms, if only to enable researchers to better master the rules so that they can be more concerned with their humanist mission, which is open to the needs of society and the dynamics of the city, instead of with their "simple" evaluation based on criteria that are partially removed from the scientific target and their mission. Evaluation mechanisms have consequences on the way personal and collective reputations are built. The tangle of different tools, norms and customs reveals a mixture of practices of recognition of esteem and prestige (the classic mechanism of peer review, the citation system, the economy of greatness indexed on distinctions, prizes, and national and international rankings). How does individual performance contribute to the recognition of the scientific activity of a research unit (and vice versa)? How can this performance, when it exists, validly participate in the entrepreneurial dynamics of the city? Obviously, this point of view does not answer the question, even if it raises more of them.

> Leaders seem to have been trained, even formatted, in a world of realist thinking that leads them to perceive the world as something external to them: this intellectual heritage [...] seems, in fact, to prevent or at least slow down the apprehension of this complexity, constituting a sort of realist filter. (De Geuser and Fiol 2003)

But several new things are happening today. Apart from a decent average quality of national research (see Shanghai ranking), without any locally useful nuggets of gold, research is totally globalized. The companies cover the same world and we must give value to these companies if we want to participate in a public utility contribution from (and turned toward) the interest of our country.

Moreover, in the present situation, an essential element must be taken into account, that of the reduction of temporalities, something that stems first of all from an authentic technical "leap" that has taken place in just a few years. Ten years ago, the development of smartphones was only in the latent state; the autonomous car was still a distant project; the prospect of a computer expressing itself as fluidly as a human or composing musical works "in the manner of" resembling an authentic work was a matter of the imagination. This technical leap is generating cultural evolutions, a change in lifestyles, such as an addiction to the instantaneous fulfillment of desire, to the individualization of performance. The interaction between the digital world and the real world is increasingly close: the exchanges born in the former are increasingly translated into the physical world and, conversely, our activities continuously generate data that enlarge and feed the virtual world. This sudden irruption of a new reality disturbs, upsets and creates tensions. But it also raises expectations and offers tremendous prospects for development. Who, on the research side, is actively participating in this revolution?

While there were 138 in 2015, the number of digital platforms is close to 250 in 2019 with a business model based on matchmaking. But these platforms, which are at the heart of the ongoing economic revolution, do not constitute a technological breakthrough: they use existing digital processes (albeit particularly innovative ones). The effects of such an ecosystem, particularly on productivity gains, are clearly not the same as those of the widespread use of "digital" in the industry. Who in the country has participated in this organizational revolution, in this change of economic and social paradigm, linked to the use and dissemination of new technologies, when it calls into question the very foundations and structures of the industrial economy? Even Paris is not in the top 25 in the world on this subject!

In addition to these external dynamics, there is another, cross-cutting one, which stems from the intrinsic characteristics of the economy and the systems that are generating the current upheavals. It has to do with the scale on which we can act, which can no longer be that of the national territory. The platform economy is a global phenomenon, whose instruments and levers for action are almost inherently a-territorial. Acting on too small a scale – that is, in this case, on the scale of a single research unit – means running the risk of losing innovation without gaining anything in return.

> The daily life of individuals and organizations is made up of poor decisions, because it is difficult to agree and work together, human intelligence has its limits and hazards are inevitable. Poor decisions are the result of action that revolves around the goal. (Morel 2002)

1.8.1. *Can we break the deadlock?*

It is therefore appropriate to ask what the ideal of local government can be, that which is at the extreme of programming without resourcing, and that at the extreme of resourcing without any intentionality. Both are incapable of bringing about an "acceptable" harmony between the freedom/autonomy of the researcher and the general local interest. However, can we not consider that responsibility does not have only one face, that of the researcher, but that it also involves society as a whole?

> We are all "guilty" and all "responsible" for the situations we live in, and this in spite of ourselves, because we did not choose them or build them. We are responsible because this structural "equality" is not an elite that is called upon to make decisions at the critical points of the situation, it is all of us. (Benasayag 1994).

What this little reflection tries to show is the absence of a clear steering from those in charge and the presence of the conservative "tertiary deposit" that constitutes the academic world committed to goals that do not sufficiently understand the future of their country. There are several directions that should be explored in an attempt to get out of the sad trends presented above:

What future exists for France?

It is up to citizens and their representatives to reflect on the dynamic image they wish for their city. Various well-framed organizations (or certain personalities) can make a scientific contribution on the factual aspects and on prospective approaches, but it is up to the social body to determine itself (ends, means, communication, openings, etc.). Among the reflections, it would be useful to examine the multiple consequences of inequalities: they prevent those with less income from reaching their full potential in terms of education and inventiveness, which results in a decrease in the spirit of entrepreneurship. Furthermore, if wealth is concentrated in a small segment of the population, the demand for imported luxury goods increases, but creates little or no industrial development.

> In general, only rough assertions and estimates are possible regarding the concrete nature of these distant effects. (Rapp 2000)

Introducing disruptions in the mechanics of research

The present situation leads to a "Darwinian" life for researchers, linked to the survival of the most able to manage in the emerging normative universe linked to the modifications of their institutional research environment, thanks to a good perception (or even anticipation) of the new rules, explicit and especially implicit, of the functioning of the academic system. By introducing changes, the conservative body (at least in their discipline) of many researchers is led to modify their ways of acting, which can be beneficial. Indeed, in the absence of stimulation, an instinct develops tending toward the resurrection of the old order (the Freudian automatisms of repetition). Specialization is probably a necessary evil in the sciences, but how can we avoid premature compartmentalization and unsuitable programs which risk "ossifying" specialists, walled up in their system of thought and incapable of rejuvenation (Elgozy 1966)?

Fostering interdisciplinarity

The disciplinary approach is too scattered, hence the key idea of percolating through disciplinary boundaries so that the complexity paradigm can truly unfold, particularly because the re-composition of categories of thought is no longer based on disciplinary boundaries and objects, but on border objects. This context requires the convergence of disciplines. When working on border subjects aimed at the creation of instruments, machines, processes and software, the scientific aspects must integrate broader forms of convergence concerning uses, human and economic aspects, and involve integrated work with technologists, making it possible to integrate their skills as far upstream as possible (or even to develop them, as well as those of researchers) in order to satisfy the application needs (both technological research and societal) in a robust manner.

Supporting creativity

In systems where unknown areas exist, it is necessary to enter into a heuristic approach to try to find "profitable" avenues of action (rather than, as is often the case, waiting for others to do it for you). How to create interdisciplinary groups practicing "divergent" thinking and having expertise in the chosen subject, likely to define, with external input, what is possible and therefore known, as well as partial or total areas of unknown knowledge. Depending on the goals defined, it must be possible to define scientific and technological scenarios adapted to the achievement of the scientific goal (originality) and why not industrial goals as well.

etc.

Other ways are possible, we must think about them together, but they must avoid the "ship continuing to sink slowly".

> We are on the threshold of a new cultural compromise in which traditional humanism will have to settle its affairs with the third culture of which it has never taken note. And the third culture is neither the hard sciences nor the humanities, but all the knowledge of engineers. (Sloterdijk 2000)

1.8.2. *So what?*

> In a civilization saturated by technology, there is no more adventure, only the risk of being late. (Sloterdijk 2006)

We propose just one example to illustrate how we can support the search for disruptive ideas that could create economic development (depending on possible feedback, other ideas for action could be expressed, including a complementary action of prospective reflection). The principles of the action initiated by and for 2M4D are based on the general document presented above:

Project summary

A candidate, with about 5 years of higher education (without this being a strong selection criterion), develops, with the help of 2M4D and research structures, an original, even revolutionary or disruptive idea (some would say "crazy" because it comes out of a "reassuring" continuity), but likely to trigger a creative cascade leading to the subsequent development of the job. The selected candidate is supported financially, but above all by researchers and teacher-researchers, company managers and their engineers, 2M4D's technical services and the local media, in particular during an initial 3-month stay with the host scientific and/or technological structures. The support to the project can be concretized by the attribution of a prize allowing the continuation of the work.

Call for proposals

In the current internationalization of science, it is possible to provide the widest possible information, beyond the hexagon. Support by the scientific advisers of the embassies could allow this externalization (to stabilize). Normally, the call concerns all citizens, but it is hoped to reach young people who are likely to be mobile and who are not yet too formatted by their educational and/or professional activities. Essentially, it is expected to use knowledge or devices in an original way to take up challenges that should be able to solve concrete problems from an economic activity focused on 2M4D.

Response to the call

As already mentioned, apart from the classical administrative aspects, the candidate will be asked to make a proposal for an original scientific challenge that excites them in one line (illustrative example: being able to read an (old) book without turning the pages corresponding to an achievement of the MIT in the USA; a disruptive proposal that has applications for the reading of very old documents). A few lines of explanation on how the candidate envisages achieving their objective. It is indeed originality in the 4D domain that is targeted with a framing by the candidate's knowledge of the real and the possible.

Initial assessment (E1)

A committee (C1) composed of representatives of 2M4D and independent scientists (to be defined) sorts through the applications received and, if necessary, directs the pre-reviewed applications to external experts. According to the answers, the candidates are informed via 2M4D of the committee's decision.

Interview with the candidate

On the basis of the selection (also taking into account the financial means envisaged by 2M4D), the candidates will be called for an interview with the C1 committee and experts designated by this committee. During the interview, this enlarged committee (C2) measures the candidate's willingness to show initiative (if possible leadership) to address an original issue. Based on this first phase, the committee may intervene to refine the project with the candidate.

Consecutive expertise

Applicants selected by the enlarged committee are asked to revise their proposal, taking into account the opinions and advice of the C2 committee. The file is examined for possible commitment (interest of the proposal and possibility of funding) and the results are communicated to the applicants. At the same time, C2 will propose to the selected candidates the possibility of a grant, the amount of which should not exceed €10,000, distributed as follows: €5,000 for the candidate (accommodation and travel expenses); €5,000 for the host laboratory (by host laboratory, we mean academic structures and industrial centers, with, in both cases, the need to clarify the industrial property aspects, which must be used in the event of success for an activity under the local control of 2M4D). In this context, C2 will seek, in consultation with 2M4D, to find host laboratories for a period of 3 months.

After the course

After the internship, the candidate will prepare a confidential report (taking into account the possible added value in terms of innovation, which must be at least partially owned by 2M4D), in consultation with their tutors, which will be sent to C1. The examination of the file will have to show the implementation of the ideas of the file validated by C2 and their positive impact in terms of potential valorization. This file should also showcase a measure of their ability to promote their ideas, of their competence (analysis of the aspects of inspiration, creativity, spirit of cooperation, risk taking, technological opportunism, etc.).

Toward a creativity award

2M4D and C2 will seek to set up a consortium with partners from the State, the EU and the business community to form the 2M4DF Foundation, which should be used to award a prize to the best candidates (and with appropriate information to the public).

Depending on the available funding, it could be envisaged to award research grants for a period of more than 3 months (up to 3 years), bearing in mind that the results are (at least in part) the property of 2M4D. If the recipient develops an activity in France, 2M4D could transfer its property rights within a framework to be defined. This operation could be carried out in an adapted festive framework with a large-scale public information (after Covid-19!).

> Indecision-making is reflected in a whole series of measures – in this case the re-launching of research and the postponement of the final decision – that allow the range of possibilities to be expanded [...], it is a mode of exercising power. (Barthe 2006)

> By maintaining an upstream object-centered division of science, some finite and reactionary researchers maintain a comfortable illusion that they are not really being evaluated by their peers. (Lecointre 2005)

1.9. References

Abdel-Rahman, A. and Tafrihi, E. (2018). Heat-actuated auxetic facades [Online]. Available at: https://www.researchgate.net/profile/Amira-Abdel-Rahman-2/publication/326633575_Heat-Actuated_Auxetic_Facades/links/5db9edf4a6fdcc2128f0c509/Heat-Actuated-Auxetic-Facades.pdf [Accessed 25 March 2022].

Abdelmohsen, L.K.E.A., Nijemeisland, M., Pawar, G.M., Janssen, G.J.A., Nolte, R.J.M., van Hest, J.C.M., Wilson, D.A. (2016). Dynamic loading and unloading of proteins in polymeric stomatocytes: Formation of an enzyme-loaded supramolecular nano-motor. *ACS Nano*, 10, 2652–2660.

Ackoff, R.L. (1973). *Méthodes de planification de l'entreprise*. Editions de l'organisation, Paris.

Agarwala, S., Goh, G.L., Goh, G.D., Dikshit, V., Yeong, W.Y. (2020). 3D and 4D printing of polymer/CNTs-based conductive composites. In *3D and 4D Printing of Polymer Nanocomposite Materials – Processes, Applications and Challenges*, Sadasivumi, K.K., Deshmukh, K., Almaadeed, M.A. (eds). Elsevier, Amsterdam.

Akbari, S., Sakhaei, A.H., Panjwani, S., Kowsari, K., Serjouei, A., Ge, Q. (2019). Multi-material 3D printed soft actuators powered by shape memory alloy wires. *Sensors and Actuators A: Physical*, 290, 177–189.

Alcimed (2021). PIA 4 : quelles perspectives pour les acteurs français de l'innovation ? [Online]. Available at: https://www.alcimed.com/fr/les-articles-d-alcim/pia-4-programme-investissements-avenir-perspectives-acteurs-francais-innovation/.

Alderete, N.A., Medina, L., Lamberti, L., Sciammarella, C., Espinosa, H.D. (2021). Programmable 3D structures via Kirigami engineering and controlled stretching. *Extreme Mechanics Letters*, 43, 101146.

Algoud, J.P. (2014). *Systémique – vie et mort de la civilisation occidentale 1150–2050*. L'indiscipline, Lyon.

Alvarez-Pereyre, F. (2003). *L'exigence interdisciplinaire*. Les éditions de la Maison des sciences de l'homme, Paris.

Amra, C., André, J.C., Gallais, L. (2017). Procédé pour la réalisation d'un objet tridimensionnel par un processus de photo-polymérisation multi-photonique et dispositif associé. French patent 16-59211 from 28-09-2016, Dépôt d'une demande internationale de brevet (PCT) concernant l'invention citée en référence : numéro PCT/FR2017/052622 du 27-09-2017.

André, J.C. (2017a). *From Additive Manufacturing to 3D/4D Printing 1: From Concepts to Achievements*. ISTE Ltd, London, and John Wiley & Sons, New York.

André, J.C. (2017b). *From Additive Manufacturing to 3D/4D Printing 2: Current Techniques, Improvements and their Limitations*. ISTE Ltd, London, and John Wiley & Sons, New York.

André, J.C. (2017c). *From Additive Manufacturing to 3D/4D Printing 3: Breakthrough Innovations: Programmable Material, 4D Printing and Bio-printing*. ISTE Ltd, London, and John Wiley & Sons, New York.

André, J.C. (2019). *Industry 4.0 – Paradoxes and Conflicts*. ISTE Ltd, London, and John Wiley & Sons, New York.

André, J.C., Le Méhauté, A., De Witte, O. (1984). Dispositif pour réaliser un modèle de pièce industrielle. French patent no. 84 11 241 from 16/07/1984.

Apsite, I., Biswas, A., Li, Y., Ionov, L. (2020). Microfabrication using shape-transforming soft materials. *Advanced Functional Materials*, 30, 1908028.

Arenas, L.F., Ponce de León, C., Walsh, F.C. (2017). 3D-printed porous electrodes for advanced electrochemical flow reactors: A Ni/stainless steel-electrode and its mass transport characteristics. *Electrochemistry Communications*, 77, 133–137.

Attali, J. (2011). *Une brève histoire de l'avenir*. Poche, Paris.

Bachelard, G. (1943). *L'air et les songes*. José Corti Editions, Paris.

Badillo, P.Y. (2013). Les théories de l'innovation revisitées : une lecture communicationnelle et interdisciplinaire de l'innovation ? Du modèle "Émetteur" au modèle communicationnel. *Les Enjeux de l'information et de la communication*, 14, 19–34.

Bajpai, A., Baigent, A., Raghav, S., Brádaigh, C.O., Koutsos, V., Radacsi, N. (2020). 4D printing: Materials, technologies, and future applications in the biomedical field. *Sustainability*, 12, 10628.

Barnes, M., Sajadi, S.M., Parekh, S., Rahman, M.M., Ajayan, P.M., Verduzco, R. (2020). Reactive 3D printing of shape-programmable liquid crystal elastomer actuators. *ACS Applied Materials Interfaces*, 12, 28692–28699.

Barthe, Y. (2006). *Le pouvoir d'indécision : la mise en politique des déchets nucléaires*. Editions Economica, Paris.

Bartoli, H. (1991). *L'économie multidimensionnelle*. Editions Economica, Paris.

Bauman, Z. (2006). *La vie liquide*. Le Rouergue, Arles.

Beites, S. (2013). Morphological behavior of shape memory polymers toward a deployable, adaptive architecture [Online]. Available at: http://papers.cumincad.org/data/works/att/acadia13_121.content.pdf [Accessed 25 March 2022].

Bell, D. (1999). *Vers la société postindustrielle*. Robert Laffont, Paris.

Benasayag, M. (1994). *Penser la liberté, la décision, le hasard et la situation*. La Découverte, Paris.

Benyus, J. (2017). *Bio-mimétisme*. Livre de poche, Paris.

Bernays, E. (2007). *Propaganda – Comment manipuler l'opinion en démocratie*. La Découverte, Paris.

Berthet, V. (2018). *L'erreur est humaine*. CNRS Editions, Paris.

Bessen, J.E., Denk, E., Kim, J., Righi, C. (2020). Declining industrial disruption. Research paper 20–28, Boston University, School of Law, Law and Economics [Online]. Available at: https://ssrn.com/abstract=3682745; http://dx.doi.org/10.2139/ssrn.3682745 [Accessed 25 March 2022].

Bessin, M., Bidart, C., Grossetti, M. (2010). *Bifurcations – les sciences sociales face aux ruptures et à l'événement*. La Découverte, Paris.

Biswas, M.C., Chakraborty, S., Bhattacharjee, A., Mohammed, Z. (2021). 4D printing of shape memory materials for textiles: Mechanism, mathematical modeling, and challenges. *Advanced Functional Materials*, 2100257.

Blackiston, D., Lederer, E., Kriegman, S., Garnier, S., Bongard, J., Levin, M. (2021). A cellular platform for the development of synthetic living machines. *Science Robotics*, 6, eabf1571.

Blumenberg, H. (1993). *La légitimité des temps modernes*. Gallimard, Paris.

Bogers, M., Chesbrough, H., Moedas, C. (2018). Open innovation: Research, practices, and policies. *California Management Review*, 60, 5–16.

Bohler, S. (2019). *Le bug humain*. Robert Laffont, Paris.

Boni, C. and Royer-Carfagni, G. (2021). Equilibrium of bi-stable flexural-tensegrity segmental beams. *Journal of Mechanics and Physics of Solids*, 152, 104411.

Boudon, R. (2003). *Raisons, bonnes raisons*. PUF, Paris.

Bouët, C. and Duréault-Thoméré, M.C. (2011). Créativité et intelligence collective [Online]. Available at: http://reseauculture21.fr/labo/2011/11/05/creativite-et-intelligence-collective/.

Bradberry, S.J., Byrne, J.P., McCoy, C.P., Gunnlaugsson, T. (2015). Lanthanide luminescent logic gate mimics in soft matter: [H+] and [F−] dual-input device in a polymer gel with potential for selective component release. *Chemical Communication*, 51, 16565.50.

Braudel, F. (1993). *La Prospective – pratiques et méthodes*. Economica, Paris.

Bumjin, J. (2018). Nanobot propulsion methods: Theory and experiments. PhD, ETH Zürich [Online]. Available at: https://www.research-collection.ethz.ch/handle/20.500.11850/265997.

Burela, R.G., Kamineni, J.N., Harursampath, D. (2020). Multifunctional polymer composites for 3D and 4D printing. In *3D and 4D Printing of Polymer Nanocomposite Materials – Processes, Applications and Challenges*, Sadasivumi, K.K., Deshmukh, K., Almaadeed, M.A. (eds). Elsevier, Amsterdam.

Camacho-Hübner, E. (2007). De l'interdisciplinarité comme paradigme de recherche. Espaces temps.net [Online]. Available at: http://espacetemps.net/document3842.html.

Cardon, D. (2005). La trajectoire des innovations ascendantes : inventivité, coproduction et collectifs sur Internet. Actes du Colloque Innovations, Usages, Réseaux [Online]. Available at: http://halshs.archives-ouvertes.fr/docs/00/13/49/04/PDF/Cardon_IUR2006.pdf.

Cargnelli, K. (2020). Private communication.

Carroll, L. (2010). *Au-delà du miroir*. Poche, Paris.

Castoriadis, C. (1999). *Figures du pensable*. Le Seuil, Paris.

Cesewski, E., Haring, A.P., Tong, Y., Singh, M., Thakur, R., Laheri, S., Read, K.A., Powell, M.D., Oestreich, K.J., Johnson, B.N. (2018). Additive manufacturing of three-dimensional (3D) microfluidic-based microelectromechanical systems (MEMS) for acousto-fluidic applications. *Lab-on-Chip*, 18, 2087.

Ceylan, H., Yasa, I.C., Yasa, O., Tabak, A.F., Giltinan, J., Sitti, M. (2019). 3D-printed biodegradable micro-swimmer for theranostic cargo delivery and release. *ACS Nano*, 13, 3353–3362 [Online]. Available at: https://pubs.acs.org/doi/full/10.1021/acsnano.8b09233 [Accessed 25 March 2022].

Chan Vili, Y.Y.F. (2007). Investigating smart textiles based on shape-memory materials. *Textiles Research Journal*, 77, 290–300.

Charaudeau, P. (2021). Polémique sur l'"islamo-gauchisme" : revenir à l'esprit de la recherche [Online]. Available at: https://theconversation.com/polemique-sur-l-islamo-gauchisme-revenir-a-lesprit-de-la-recherche-155889?utm_medium=email&utm_campaign=La%20lettre%20de%20The%20Conversation%20France%20du%2025%20fvrier%202021%20-%201872618260&utm_content=La%20lettre%20de%20The%20Conversation%20France%20du%2025%20fvrier%202021%20-%201872618260+CID_6f09a49970f3d258343e1cbe52898ac4&utm_source=campaign_monitor_fr&utm_term=Polmique%20sur%20l%20islamo-gauchisme%20%20%20revenir%20%20lesprit%20de%20la%20recherche.

Chen, T. and Shea, K. (2018). An autonomous programmable actuator and shape reconfigurable structures using bistability and shape memory polymers. *3D Printing and Additive Manufacturing*, 5, 91–101.

Chen, C., Chang, X., Teymourian, H., Ramírez-Herrera, D.E., Esteban-Fernández de Ávila, B., Lu, X., Li, J., He, S., Fang, C., Liang, Y., Mou, F., Guan, J., Wang, J. (2018). Bioinspired chemical communication between synthetic nano-motors. *Angewandte Chemie*, 130, 247.

Chen, H.T., Song, Y., Cheng, X.L., Zhang, H.X. (2019a). Self-powered electronic skin based on the triboelectric generator. *Nano Energy*, 56, 252.

Chen, X., Liu, X., Ouyang, M., Chen, J., Taiwo, O., Xia, Y., Childs, P.R.N., Brandon, N.P., Wu, B. (2019b). Multi-metal 4D printing with a desktop electrochemical 3D printer. *Scientific Reports*, 9, 3973.

Chen, C., Wang, X., Wang, Y., Yang, D., Yao, F., Zhang, W., Wang, B., Sewvandi, G.A., Yang, D., Hu, D. (2020). Additive manufacturing of piezoelectric materials. *Advanced Functional Materials*, 2005141.

Chesbrough, H. (2003). *Open Innovation: The New Imperative for Creating and Profiting from Technology*. Harvard Business School Press, Boston.

Chesbrough, H. and Bogers, M. (2014). Explicating open innovation: Clarifying an emerging paradigm for understanding innovation. In *New Frontiers in Open Innovation*, Chesbrough, H., Vanhaverbeke, W., West, J. (eds). Oxford University Press, Oxford.

Chesbrough, H., Vanhaverbeke, W., West, J. (2006). *Open Innovation: Researching a New Paradigm*. Oxford University Press, New York.

Chillara, V.S.C. and Tapino, M.J. (2020). Review of morphing of laminated composites. *Applied Mechanics Reviews*, 72, 010801 [Online]. Available at: https://par.nsf.gov/servlets/purl/10159251 [Accessed 25 March 2022].

Cholleti, E.R. (2018). A review on 3D printing of piezo-electric-materials. *IOP Conferences Series: Materials Science and Engineering*, 455, 012046.

Clough, E.C., Plaisted, T.A., Eckel, Z.C., Cante, K., Hundley, J.M., Schaedler, T.A. (2019). Elastomeric microlattice impact attenuators. *Matter*, 1, 1519–1531.

CNRS (2020). 80 Prime : les projets lauréats 2020 dévoilés [Online]. Available at: http://www.cnrs.fr/fr/cnrsinfo/80-prime-les-projets-laureats-2020-devoiles.

Cohn, D., Widlan, F., Zarek, M., Bloom, A. (2020). Engineering in situ weldable vascular devices [Online]. Available at: https://ssrn.com/abstract=3674176; http://dx.doi.org/10.2139/ssrn.3674176.

Confucius (2004). *Un langage pour l'organisation – l'approche OSSAD*. Presses Polytechniques et Universitaires Romandes, Lausanne.

Corazza, G.E. (2016). Potential originality and effectiveness: The dynamic definition of creativity. *Creativity Research Journal*, 28, 258–267.

Cowen, T. (2011). *The Great Stagnation: How America All the Low-Hanging Fruit of Modern History, Got Sick and Will (Eventually) Feel Better*. Dutton, New York.

Crozier, M. and Tillette, B. (2000). *Quand la France s'ouvrira*. Fayard, Paris.

Dagognet, F. (1989). *Rematérialiser*. Vrin, Paris.

Danzo, A. and André, J.C. (2020). Que peut apporter une modélisation mathématique à la maitrise du bio-printing ? *Entropie*, 20-1, 32 [Online]. Available at: http://openscience.fr/Que-peut-apporter-une-modelisation-mathematique-a-la-maitrise-du-bio-printing.

Darbelay, F. (2020). L'érudition : faire du neuf avec du vieux ? Interdisciplinarité, polymathie, indiscipline Frédéric Darbellay. *Hermès – La revue*, 87, 205–212.

Darses, F. (2009). Résolution collective des problèmes de conception. *Le travail humain*, 72, 43–59.

Darwin, C. (2008). *L'origine des espèces*. Livre de Poche, Paris.

Das, O. and Dujana, W.A. (2020). Review based proposal on the use of 4D printing smart materials for automobile light cover through process structure performance linkage with a sustainable high throughput material development technique. *International Conference on Mechanical, Industrial and Energy Engineering*, 19–21 December [Online]. Available at: https://www.kuet.ac.bd/icmiee2020/Technical_and_Keynote_papers/All_Paper/ICMIEE20-183.pdf [Accessed 25 March 2022].

Daynes, S. and Weaver, P.M. (2013). Review of shape morphing automobile structures: Concepts and outlook. *Proceedings of the Institution of Mechanical Engineers, Part D: Journal of Automobile Engineering*, 227, 11 [Online]. Available at: https://www.researchgate.net/profile/Stephen-Daynes/publication/283874241_Review_of_shape-morphing_automobile_structures_Concepts_and_outlook/links/5a221e0fa6fdcc8e86652450/Review-of-shape-morphing-automobile-structures-Concepts-and-outlook.pdf [Accessed 25 March 2022].

De Angelis, M. (2019). Microsoft archived "Superman" on its Project Silica glass storage medium [Online]. Available at: https://www.engadget.com/2019-11-04-microsoft-archived-superman-project-silica.html?guccounter=1 [Accessed 25 March 2022].

De Beer, M.P., van der Laan, H.L., Cole, M.A., Whelan, R.J., Burns, M.A., Scott, T.F. (2019). Rapid, continuous additive manufacturing by volumetric polymerization inhibition patterning. *Science Advances*, 5, eaau8723.

De Brabandère, L. (2019). Gaston Trouvetou, recrue de choix pour DRH. In *Gaston, un philosophe au travail*, Latour, B. (ed.). Philosophie Magazine, Paris.

De Gennes, P.G. (2020). *Détournement de science*. Ecosociété, Québec, Montreal.

De Geuser, F. and Fiol, M. (2003). Faire face aux situations complexes : la blessure narcissique des managers. In *Peut-on former les dirigeants ? L'apport de la recherche*, Moingeon, B. (ed.). L'Harmattan, Paris.

Decroly, G., Toncheva, A., Blanc, L., Raquez, J.M., Lessinnes, T., Delchambre, A., Lambert, P. (2020). Programmable stimuli-responsive actuators for complex motions in soft robotics: Concept, design and challenges. *Actuators*, 9, 131.

Deng, D., Yang, Y., Chen, Y., Lan, X., Tice, J. (2017). Accurately controlled sequential self-folding structures by polystyrene film. *Smart Materials and Structures*, 26, 085040.

Di Giacomo, F., Fakharuddin, A., Jose, R., Brown, T.M. (2016). Progress, challenges and perspectives in flexible perovskite solar cells. *Energy & Environmental Science*, 9, 3007.

Diemer, A. (2007). La technologie au cœur du développement durable. In *Culture de la provocation, réflexions indisciplinaires*, Francfort, D. (ed.). Presses Universitaires de Nancy, Nancy.

Dieng, D. (2014). *Intelligence stratégique – guide pour le recherche et l'innovation.* Presses Universitaires de Namur, Namur.

Dodet, M. (2001). Innovation et société. 22–23 Passages 106/107.

Donovan, B.R., Matavulj, V.M., Ahn, S.K., Guin, T., White, T.J. (2019). All-optical control of shape. *Advanced Materials*, 31, 1805750.

Doualle, T., André, J.C., Gallais, L. (2021). 3D-printing of silica glass through multiphoton polymerization process. *Optical Letters*, 46, 364–367.

Drucker, P.F. (1985). The discipline of innovation. *Harvard Business Review*, 63, 67–72.

Drucker, P.F. (2006). *Innovation and Entrepreneurship*. Harper, New York.

Drucker, P.F. (2007). *L'avenir du management*. Pearson, Paris.

Du Sautoy, M. (2020). *Le code de la créativité – comment l'IA apprend à écrire, peindre et penser*. Editions Héloïse d'Ormesson, Paris.

Duerig, T., Pelton, A., Stockel, D. (1999). An overview of nitinol medical applications. *Materials Science and Engineering*, A273–A275, 149–160.

Dufourmentelle, A. (2013). *Puissance de la douceur*. Manuels Payot, Paris.

Dupuy, J.P. (2018). Nous n'avons pas peur de ce qui devrait nous terrifier. In *De quoi avons-nous peur ?*, Birnbaum, J. (ed.). Folios-Essais, Paris.

Dupuy, J.P. and Roure, F. (2004). *Les nanotechnologies : éthique et prospective industrielle*. Conseil Général des Mines, Paris.

Duriez, C. (2020). C'est dur d'inventer des robots souples [Online]. Available at: https://theconversation.com/cest-dur-dinventer-des-robots-souples-150207.

Eid, A., Hester, J.G.D., Tentzeris, M.M. (2021). 5G as a wireless power grid. *Scientific Reports*, 11, 636.

Einstein, A. (2007). *L'espoir économique : vers la révolution du quaternaire*. Bourin, Paris.

Eisenberger, R. and Aselage, J. (2009). Incremental effects of reward on experienced performance pressure: Positive outcomes for intrinsic interest and creativity. *Journal of Organizational Behavior*, 30, 95–117.

Elahinia, M.H., Hashemi, M., Tabesh, M., Bhaduri, S.B. (2012). Manufacturing and processing of NiTi implants: A review. *Progress in Materials Science*, 57, 911–946.

Elgozy, G. (1966). *Le paradoxe des technocrates*. Denoël, Paris.

Enkel, E., Gassmann, O., Chesbrough, H. (2009). Open R&D and open innovation: Exploring the phenomenon. *R&D Management*, 36, 311–316.

Ergene, B. and Yalçın, B. (2019). A review on 4D printing technology and applications. *5th International Conference on Advances in Mechanical Engineering*, Istanbul, 17–19 December [Online]. Available at: https://www.researchgate.net/publication/338037057_A_REVIEW_ON_4D_PRINTING_TECHNOLOGY_AND_APPLICATIONS [Accessed 25 March 2022].

Espalin, D., Muse, D.W., McDonald, E., Wicker, R.B. (2014). 3D printing multi-functionality: Structures with electronics. *The International Journal of Advanced Manufacturing Technology*, 72, 963–978.

Espera Jr., A.H., Dizon, J.R.C., Chen, Q., Advincula, R.C. (2019). 3D-printing and advanced manufacturing for electronics. *Progress in Additive Manufacturing*, 4, 1–23.

European Commission (2019). Orientations towards the first strategic plan implementing the research and innovation framework programme Horizon Europe [Online]. Available at: https://ec.europa.eu/research/pdf/horizon-europe/ec_rtd_orientations-towards-the-strategic-planning.pdf [Accessed 25 March 2022].

Farina, M. and Pasquinelli, E. (2020). *L'art de faire confiance*. Odile Jacob, Paris.

Fernandes, D.F., Majidi, C., Tavakoli, M. (2019). Digitally printed stretchable electronics: A review. *Journal of Material Chemistry*, C7, 14035.

Fernez-Walsh, S. and Roman, F. (2013). *Management de l'innovation – de la stratégie aux projets*. Vuibert, Paris.

Foudriat, M. (2019). *La co-construction – une alternative managériale*. Presses de l'EHESP, Paris.

Franquin, A. (1973). *Les petits formats*. Dupuis, Marcinelle-Charleroi.

Fraternali, F., Daraio, C., Rimoli, J. (2019). Multiscale lattices and composite materials: Optimal design, modeling and characterization. *Frontiers in Materials*, 6, 199.

Fresneau, T. (2020). Etude de transistors organiques innovants pour des applications en électronique structurelle et biocapteurs [Online]. Available at: http://www.theses.fr/s233034

Freud S. (2010). *Civilization and Its Discontents*, translated by James Strachey. W.W. Norton & Company, New York, USA.

Frochot, C., Schaer, E., André, J.C. (2020). Il ne faut pas avoir peur des "nano", c'est juste une nano-crise mal anticipée. *Environnement risques et santé*, 19, 446–466.

Fuller, E.J., Gabaly, F.E., Léonard, F., Agarwal, S., Plimpton, S.J., Jacobs-Gedrim, R.B., James, C.D., Marinella, M.J., Talin, A. (2017). Li-ion synaptic transistor for low power analog computing. *Advanced Materials*, 29, 1604310.

Future Bridge (2020). 4D printing – The technology of the future [Online]. Available at: https://www.futurebridge.com/industry/perspectives-mobility/4d-printing-the-technology-of-the-future/ [Accessed 25 March 2022].

Ganet, F. (2018). Fabrication additive de matériaux électro-actifs pour applications à la mécatronique [Online]. Available at: http://theses.insa-lyon.fr/publication/2018LYSEI011/these.pdf.

Gao, W., Zhang, Y., Ramanujan, D., Ramani, K., Chen, Y., Williams, C.B., Wang, C.C.L., Shin, Y.C., Shang, S., Zavattieri, P.D. (2015). The status, challenges, and future of additive manufacturing in engineering. *Computer-Aided Design*, 69, 65–89.

Garces, I.T. and Ayranci, C. (2020). Active control of 4D prints: Towards 4D printed reliable actuators and sensors. *Sensors and Actuators*, A301, 111717.

Garcia, I. (2020). 4D printing: An overview into a disruptive technology [Online]. Available at: https://chemindustryrevolution.com/4d-printing-an-overview-into-a-disruptive-technology/ [Accessed 25 March 2022].

Ge, Q., Dunn, C.K., Qi, H.J., Dunn, M.L. (2014). Active origami by 4D printing. *Smart Material Structures*, 23, 094007.

Georgantzinos, S.K., Giannopoulos, G.I., Bakalis, P.A. (2021). Additive manufacturing for effective smart structures: The idea of 6D printing. *Journal of Composites Science*, 5, 119.

Gordon, W.J.J. (1961). *Synectics*. Harper & Row, New York.

Gordon, W.J.J. (1965). *Stimulation des facultés créatrices dans les groupes de recherche par la méthode synectique*. Hommes et Techniques, Paris.

Gordon, R. (2016). *The Rise and Fall of American Growth*. Princeton University Press, Princeton.

Gouil, H. (2010). *Réapprendre à coopérer*. Yves Michel, Gap.

Gramaccia, G. (2008). *Démocratie participative et communication territoriale. Vers la micro-représentativité*. L'Harmattan, Paris.

Granger, G.G. (1960). *Pensée formelle et sciences de l'homme*. Aubier-Montaigne, Paris.

Grassi, G., Sparrman, B.E., Tibbits, S. (2020). Material agency and 4D printing. In *Material Balance – A Design Equation*, Paolleti, I. and Nasri, M. (eds). Springer, Cham [Online]. Available at: https://re.public.polimi.it/retrieve/handle/11311/1159548/580239/NASTRI%20-%20Material%20Balance%20-%20Chapter%208%20-%20Springer.pdf [Accessed 25 March 2022].

Grojean, O. (2015). Sensibilisation et coercition ; à propos de quelques questions relatives au façonnage institutionnel des individus. In *Emotions et expertises ; les modes de coordination des actions collectives*, Taïni, C. (ed.). Presses Universitaires de Rennes, Rennes.

Gu, Y., Alt, E.A., Wang, H., Li, X., Willard, A.P., Johnson, J.A. (2018). Photo-switching topology in polymer networks with metal-organic cages as crosslinks. *Nature*, 560, 65–80.

Gu, G., Xu, H., Peng, S., Li, L., Chen, S., Lu, T., Guo, X. (2019a). Integrated soft ionotronic skin with stretchable and transparent hydrogel-elastomer ionic sensors for hand-motion monitoring. *Soft Robotics*, 6, 368–376.

Gu, M.S., Hong, J., Kim, S.Y., Shin, T.J., Kotov, N.A., Park, S., Kim, B.S. (2019b). Stretchable batteries with gradient multilayer conductors. *Science Advances*, 5, eaaw1879.

Guilford, J.P. (1956). The structure of intellect. *Psychological Bulletin*, 53, 267–293.

Guilford, J.P. (1959). Traits of creativity. In *Creativity and its Cultivation*, Anderson, H.H. (ed.). Harper, New York.

Guillaud, H. (2021). L'illusion de l'innovation… et l'illusion de sa critique [Online]. Available at: https://www.internetactu.net/2021/03/04/lillusion-de-linnovation-et-lillusion-de-sa-critique/.

Gul, J.Z., Sajid, M., Rehman, M.M., Siddiqui, G.U., Shah, I., Kim, K.H., Lee, J.W., Choi, K.H. (2018). 3D printing for soft robotics – A review. *Science and Technology of Advanced Materials*, 19, 243–262.

Hagaman, D.E., Leist, S., Zhou, J., Ji, H.F. (2018). Photo-activated polymeric bilayer actuators fabricated via 3D printing. *ACS Applied Materials and Interfaces*, 10, 27308–27315.

Hamza, H. (2018). Dental 4D printing: An innovative approach. *Innovation*, 1, e17.

Hasani, R., Lechner M., Amini A., Rus D., Grosu R. (2020). *Liquid Time-constant Networks*. arXiv:2006.04439v4.

Hegel, G.W.T. (1991). *Préface de la phénomènologie de l'esprit*. Aubier, Paris.

Hennessey, B. and Amabile, T. (2010). Creativity. *Annual Review of Psychology*, 61, 569–598.

Henry, C. (1998). Le "je" intellectuel et le "jeu" interdisciplinaire. *Le Genre humain/Le Seuil*, 33, 155–170.

Herbert, R., Kim, J.H., Kim, Y.S., Lee, H.M., Yeo, W.H. (2018). Soft material-enabled, flexible hybrid electronics for medicine, healthcare, and human-machine interfaces. *Materials*, 11, 187.

Hoa, S.V. and Rosca, D.I. (2020). Formation of letters in the alphabet using 4D printing of composites. *Materials Today Communications*, 25, 101115.

Horckmans, M. (2016). Prospective en génie des procédés. Research and Development Project, ENSIC – Université de Lorraine, Lorraine.

Hu, J. and Lu, J. (2014). Smart polymers for textile applications. In *Smart Polymers and their Applications*, Aguilar, M.R. and San Roman, J. (eds). Woodhead Publishing, Cambridge.

Huang, H.Y., Ge, C., Zhang, Q.H., Liu, C.X., Du, J.Y., Li, J.K., Wang, C., Gu, L., Yang, G.Z., Jin, K.J. (2019). Electrolyte-gated synaptic transistor with oxygen ions. *Advanced Functional Materials*, 29, 1902702.

Huber, C., Abert, C., Bruckner, F., Groenefeld, M., Muthsam, O., Schuschnigg, S., Sirak, K., Thanhoffer, R., Teliban I., Vogler, C. et al. (2016). 3D print of polymer bonded rare-earth magnets, and 3D magnetic field scanning with an end-user 3D printer. *Applied Physical Letters*, 109, 162401.

Hubert, B. (2005). L'interdisciplinarité en sciences sociales/sciences de la nature dans les recherches sur problème. In *Entre connaissance et organisation : l'activité collective*, Lorino, P. and Teulier, R. (eds). La Découverte, Paris.

IBM (2012). A boy and his atom: The world's smallest movie. IBM Company [Online]. Available at: http://www.research.ibm.com/articles/madewithatoms.shtml [Accessed 25 March 2022].

IBM (2017). IBM somehow crammed data into a single atom [Online]. Available at: https://www.cnet.com/news/ibm-storage-atom-breakthrough-quantum-computing-research/ [Accessed 25 March 2022].

Idumah, C.I., Nwuzor, I., Odera, S.R. (2020). Recent advancements in self-healing polymeric hydrogels, shape memory, and stretchable materials. *International Journal of Polymeric Materials and Polymeric Biomaterials*, 1–26.

Imoussaten, A. (2011). Modélisation et pilotage de la phase de délibération dans une décision collective : vers le management d'activités à risques. Thesis, École Nationale Supérieure des Mines de Paris, Paris.

Inphonlek, S., Sunintaboon, P., Léonard, M., Durand, A. (2020). Chitosan/carboxymethylcellulose-stabilized poly(lactide-co-glycolide) particles as bio-based drug delivery carriers. *Carbohydrates Polymers*, 242, 116417.

Ingold, T. (2010). Bringing things to life: Creative entanglements in a world of materials. Working Paper No. 15. NCRM Working Paper Series [Online]. Available at: http://eprints.ncrm.ac.uk/1306/1/0510_creative_entanglements.pdf [Accessed 25 March 2022].

Irimia-Vladu, M., Troshin, P.A., Reisinger, M., Shmygleva, L., Kanbur, Y., Schwabegger, G., Bodea, M., Schwödiauer, R., Mumyatov, A., Fergus, J.W. et al. (2010). Biocompatible and biodegradable materials for organic field-effect transistors. *Advances in Functional Materials*, 20, 4069.

Islam, M.K., Hazell, P.J., Escobedo, J.P., Wang, H. (2021). Biomimetic armour design strategies for additive manufacturing: A review. *Materials & Design*, 205, 109730.

Jaakkola, H., Henno, J., Mäkelä, J., Thalheim, B. (2017). Today is the future of yesterday; What is the future of today? In *40th International Convention on Information and Communication Technology, Electronics and Microelectronics (MIPRO)*, 22–26 May, Opatija.

Jacomy, B. (2015). *Une histoire des techniques*. Points Science, Paris.

Jarroson, B. (2017). *Oser la confiance : propos sur l'engagement des dirigeants*. Eyrolles, Paris.

Jasiuk, I., Abueidda, D.W., Kozuch, C., Pang, S., Su, F.Y., Mckittrick, J. (2018). An overview on additive manufacturing of polymers. *JOM – Journal of the Minerals, Metals & Materials Society*, 70, 275–283.

Jeong, H.Y., An, S.C., Seo, I.C., Lee, E., Ha, S., Kim, N., Jun, Y.C. (2019). 3D printing of twisting and rotational bistable structures with tuning elements. *Scientific Reports*, 9, 324.

Jeong, Y.S., Lee, L., An, S.C., Lim, Y., Jun, Y.C. (2020). 3D and 4D printing for optics and meta-photonics. *Nano-photonics*, 9, 1139–1160.

Jia, W., Wang, Y., Liu, R., Yu, X., Gao, H. (2021). Shape transformable strategies for drug delivery. *Advanced Functional Materials*, 2009765.

Jiang, Y., Korpas, L.M., Raney, J.R. (2019). Bifurcation-based embodied logic and autonomous actuation. *Nature Communications*, 10, 128.

Jonas, H. (1998). *Le principe responsabilité*. Flammarion, Paris.

Joshi, A., Goh, J.K., Goh, K.E.J. (2020). Polymer based conductive composites for 3D and 4D printing of electrical circuits. In *3D and 4D Printing of Polymer Nanocomposite Materials – Processes, Applications and Challenges*, Sadasivumi, K.K., Deshmukh, K., Almaadeed, M.A. (eds). Elsevier, Amsterdam.

Jung, Y.H., Park, B., Kim, J.U., Kim, T.I. (2019). Bioinspired electronics for artificial sensory systems. *Advanced Materials*, 31, 1803637.

Kamat, A.M., Pei, Y., Kottapalli, A.G.P. (2019). Bioinspired cilia sensors with graphene sensing elements fabricated using 3D printing and casting. *Nanomaterials*, 9, 954.

Kamyshny, A. and Magdassi, S. (2019). Conductive nanomaterials for 2D and 3D printed flexible electronics. *Chemical Society Review*, 48, 1712.

Kalyan, M.V.D.S., Kumar, H., Nagdeve, L. (2021). Latest trends in additive manufacturing. *IOP Conferences Series: Materials Science and Engineering*, 1104, 012020.

Kijkuit, B. and van den Ende, J. (2007). The organizational life of an idea: Integrating social network, creativity and decision-making perspectives. *Journal of Management Studies*, 44, 863–882.

Kim, H., Lee, J.A., Ambulo, C.P., Lee, H.B., Kim, S.K., Naik, V.V., Haines, C.S., Aliev, A.E., Ovalle-Robles, R., Baughman, R.H. et al. (2019). Intelligently actuating liquid crystal elastomer-carbon nanotube composites. *Advanced Functional Materials*, 29, 1905063.

King, A. (1999). *Géopolitique du chaos*. Folio Actuel, Paris.

Kirillova, A. and Ionov, L. (2019). Shape-changing polymers for biomedical applications. *Journal of Material Chemistry*, B7, 1597.

Koch, H.C., Schmelzeisen, D., Gries, T. (2021). 4D textiles made by additive manufacturing on pre-stressed textiles – An overview. *Actuators*, 10, 31.

Koestler, A. (1976). *The Ghost in the Machine*. Random House, New York.

Konaković, M., Crane, K., Deng, B., Bouaziz, S., Piker, D., Pauly, M. (2016). Beyond developable: Computational design and fabrication with auxetic materials. *ACM Transactions on Graphics*, 35, 89.

Kotikian, A., Truby, R.L., Boley, J.W., White, T.J., Lewis, J.A. (2018). 3D printing of liquid crystal elastomeric actuators with spatially programmed nematic order. *Advanced Materials*, 30, 1706164.

Kotze, P. and Purgathofer, P. (2007). A framework for design exercises – From creativity to real-world use [Online]. Available at: https://publik.tuwien.ac.at/files/pub-inf_5251.pdf [Accessed 25 March 2022].

Kuang, X., Chen, K., Dunn, C.K., Wu, J., Li, V.C.L., Qi, H.J. (2018). 3D printing of highly stretchable, shape-memory, and self-healing elastomer toward novel 4D printing. *ACS Applied Materials & Interfaces*, 10, 7381–7388.

Kuang, X., Roach, D.J., Wu, J., Hamel, C.M., Ding, Z., Wang, T., Dunn, M.L., Qi, H.J. (2019). Advances in 4D printing: Materials and applications. *Advances in Functional Materials*, 29, 1805290.

Kuhn, T. (1983). *La structure des révolutions scientifiques*. Flammarion, Paris.

Kumar, K.P.A. and Pumera, M. (2021). 4D-printing to mitigate COVID-19 pandemic. *Advanced Functional Materials*, 2100450.

Kwon, K.S., Rahman, M.K., Phung, T.H., Hoath, S.D., Jeong, S., Kim, J.S. (2020). Review of digital printing technologies for electronic materials. *Flexible and Printed Electronics*, 5, 043003.

Lamichhane, S., Bashyal, S., Keum, T., Noh, G., Eun, S.J., Bastola, R., Jaewoong, C., Hwan, S.D., Sangkil, L. (2019). Complex formulations, simple techniques: Can 3D printing technology be the Midas touch in pharmaceutical industry? *Asian Journal of Pharmaceutical Sciences*, 14, 465–479.

Le, D.H., Xu, Y., Tentzeris, M.M., Lim, S. (2020). Transformation from 2D meta-pixel to 3D meta-pixel using auxetic kirigami for programmable multifunctional electromagnetic response. *Extreme Mechanics Letters*, 36, 100670.

Le Blanc, D. (2017). *Smart World – comment de simples idées deviennent-elles de grandes innovations*? Edilivres, Saint-Denis.

Le Good, G., Resnick, M., Boardman, P., Clough, B. (2021). An investigation of aerodynamic effects of body morphing for passenger cars in close-proximity. *Fluids*, 6, 64.

Lee, J., Kim, H.C., Choi, J.W., Lee, I.H. (2017a). A review on 3D printed smart devices for 4D printing. *International Journal of Precision Engineering and Manufacturing–Green Technology*, 4, 373–383.

Lee, J.Y., An, J., Chua, C.K. (2017b). Fundamentals and applications of 3D printing for novel materials. *Applied Materials Today*, 7, 120–133.

Lee, A.Y., Zhou, A., An, J., Chua, C.K., Zhang, Y. (2020). Contactless reversible 4D-printing for 3D-to-3D shape morphing. *Virtual and Physical Prototyping*, 15, 481–495.

Lee, M., Oh, J., Lim, H., Lee, J. (2021). Enhanced liquid transport on a highly scalable, cost-effective, and flexible 3D topological liquid capillary diode. *Advanced Functional Materials*, 2011288.

Lecointre, G. (2005). Ce qui peut diviser les chercheurs. In *Charlie ramène sa science*, Fischetti, A. and Lecointre, G. (eds). Vuibert & Charlie Hebdo, Paris.

Lefranc, J. (2016). Philosophie et interdisciplinarité et Sur une proposition de Programme européen. *L'Enseignement philosophique*, 66, 73–76.

Lemouchi, C. (2010). Moteurs moléculaires cristallins photo-pilotés chimie organique. Thesis, Université d'Angers, Angers [Online]. Available at: https://tel.archives-ouvertes.fr/tel-00801226/document.

Leyrie, C. and Boivin, S. (2017). Gestion de projet et co-construction : utopie ou voie du succès ? Une réflexion exploratoire. *Revue organisation et territoires*, 26, 117–129.

Li, G. (2016). Polymères et hydrogels à mémoire de forme ultrason-répondants. PhD, Université de Sherbrooke, Quebec [Online]. Available at: https://savoirs.usherbrooke.ca/bitstream/handle/11143/8359/Li_Guo_PhD_2016.pdf?sequence=9&isAllowed=y.

Li, X. and Su, X. (2018). Multi-functional smart hydrogels: Potential in tissue engineering and cancer therapy. *Journal of Material Chemistry*, B6, 4714–4730.

Lin, W.C., Fan, F.Y., Cheng, H.C., Lin, Y., Shen, Y.K., Lai, J.S., Wang, L., Ruslin, M. (2021). Optimization shape-memory situations of a stimulus responsive composite material. *Polymers*, 13, 697.

Liten-CEA (2019). Électronique structurelle – l'électronique structurelle place l'intelligence au cœur des objets [Online]. Available at: https://liten.cea.fr/cea-tech/liten/Pages/Axes-de-recherche/Economie-circulaire/Electronique-structurelle.aspx.

Liu, R., Kuang, X., Deng, J., Wang, Y.C., Wang, A.C., Ding, W., Lai, Y.C., Chen, J., Wang, P., Lin, Z. et al. (2018a). Shape memory polymers for body motion energy harvesting and self-powered mechano-sensing. *Advanced Materials*, 30, 1–8.

Liu, Y., Zhang, W., Zhang, F., Lan, X., Leng, J., Liu, S., Jia, X., Cotton, C., Sun, B., Gua, B. et al. (2018b). Shape memory behavior and recovery force of 4D printed laminated Miura-origami structures subjected to compressive loading. *Composites*, Part B, 153, 233–242.

Liu, J., Gao, Y., Wang, H., Poling-Skutvik, R., Osuji, C.O., Yang, S. (2020a). Shaping and locomotion of soft robots using filament actuators made from liquid crystal elastomer–carbon nanotube composites. *Advanced Intelligent Systems*, 2, 1900163.

Liu, K., Hand, L., Hua, W., Ji, L., Zhu, S., Wan, Z., Yang, X., Wei, Y., Dai, Z., Zhao, Z. et al. (2020b). 4D printed zero Poisson's ratio metamaterial with switching function of mechanical and vibration isolation performance. *Materials and Design*, 196, 109153.

Liu, Z., Hong, P., Huang, Z., Zhang, T., Xu, R., Chen, L., Xiang, H., Liu, X. (2020c). Self-healing, reprocessing and 3D printing of transparent and hydrolysis-resistant silicone elastomers. *Chemical Engineering Journal*, 387, 124142.

Liu, X., Li, Y., Sun, X., Yu, R., Dai, L., Shui, J. (2021). Off/on switchable smart electromagnetic interference shielding aerogel. *Matter*, 4, 1735–1747.

Livio, M. (2017). *Fabuleuses erreurs – de Darwin à Einstein*. CNRS Editions, Paris.

Lopez-Valdeolivas, M., Liu, D., Broer, D.J., Sánchez-Somolinos, C. (2018). 4D printed actuators with soft-robotic functions. *Macromolecular Rapid Communications*, 39, 3–9.

Loterie, D., Delrot, P., Moser, C. (2020). High-resolution tomographic volumetric additive manufacturing. *Nature Communications*, 11, 852.

Lou, Z., Li, L., Wang, L., Shen, G. (2017). Recent progress of self-powered sensing systems for wearable electronics. *Small*, 13, 1–27.

Lu, B., Lan, H., Liu, H. (2018b). Additive manufacturing frontier: 3D printing electronics. *Opto-Electronic Advances*, 1, 170004 [Online]. Available at: http://www.oejournal.org/j/PDFFull/A180209000001.pdf [Accessed 25 March 2022].

Ludwig, T., Seckendorff, J., Troll, C., Fischer, T., Tonigold, M., Rieger, B., Hinrichsen, O. (2018). Additive manufacturing of Al_2O_3-based carriers for heterogeneous catalyst. *Chemie Ingenieur Tecknik*, 90, 703–709.

Lui, Y.S., Sow, W.T., Tan, L.P., Wu, Y., Lai, Y., Li, H. (2019). 4D printing and stimuli-responsive materials in biomedical applications. *Acta Biomaterialia*, 92, 19–36.

L'Usine Digitale (2018). A Station F, Emmanuel Macron édicte ses 4 grandes convictions pour faire de la France une start-up Nation [Online]. Available at: https://www.usine-digitale.fr/editorial/a-station-f-emmanuel-macron-edicte-ses-4-grandes-convictions-pour-faire-de-la-france-une-start-up-nation.N753194.

Ma, Z., Yang, Z., Gao, Q., Bao, G., Valiei, A., Yang, F., Huo, R., Wang, C., Song, G., Ma, D. et al. (2021). Bioinspired tough gel sheath for robust and versatile surface functionalization. *Science Advances*, 7, eabc3012.

Magistrelli, G. (2021). Comment la robotique aide-t-elle à construire une usine de fabrication additive smart et efficace ? [Online]. Available at: https://www.a3dm-magazine.fr/magazine/toutes-industries/robotique-fabrication-additive-usine-futur.

Mahy, I. (2014). L'art de l'inter à travers l'œil du cœur. In *L'interdisciplinarité racontée ; chercher hors frontières, vivre l'interdisciplinarité*, Lemay, V. and Darbellay, F. (eds). P. Lang, Bern.

Maillard, S. (2018). *Disruption ; préparez-vous à changer le monde*. Dunod, Paris.

Malraux, A. (1972). *L'espoir*. Livre de Poche, Paris.

Malthus, R. (1999). *Essai sur le principe de population*. Poche, Paris.

Marasso, E.L., Cocuzza M., Bertora V., Perruci F., Tommasi A., Ferrero D., Scaltrito O., Piri F. (2018). PLA conductive filament for 3D printed smart sensing applications. *Rapid Prototyping Journal*, 24, 739–743.

Marc, T. (1987). *L'utopie*. Flammarion, Paris.

Market Analysis Report (2017). 4D printing market size, share & trends analysis report by material (programmable carbon fiber, programmable wood, programmable textiles), by end-use, by region, and segment forecasts, 2018–2025 [Online]. Available at: https://www.grandviewresearch.com/industry-analysis/4d-printing-market [Accessed 25 March 2022].

Market Intellica (2019). 4D printing market research: Global status & forecast by geography, type & application (2015, 2025) [Online]. Available at: https://www.marketintellica.com/report/MI74639-4d-printing-market-research-global-status [Accessed 25 March 2022].

Marsan, C. (2020). Risquerions-nous un déficit d'imaginaire ? [Online]. Available at: https://up-magazine.info/index.php/imaginaire-sf-et-prospective/32940-un-deficit-dimaginaire/?utm_source=ActiveCampaign&utm_medium=email&utm_content=La+quotidienne+de+UP++Magazine+-++TODAY&utm_campaign=ACTION%21+Quotidienne+Mardi+11++f%C3%A9vrier.

Mathijsen, D. (2020). The rise of compliant mechanisms: Composites' role in enabling an engineering revolution. *Reinforced Plastics*, 64, 254–260.

Mauchand, P. (2001). Motivation sous incertitude : étude de l'influence de l'heuristique d'ancrage et d'ajustement sur les cognitions, le comportement et la performance. Thesis, Université de Lyon, Lumière, Lyon.

Maurel, A., Courtry, M., Fleutot, B., Tortajada, H., Prasantha, K., Armand, M., Grugeon, S., Panier, S., Dupont, L. (2018). Highly loaded graphite-polylactic acid composite based filament for lithium-ion battery three dimensional printing. *Chemical Materials*, 30, 7484–7493.

Mayard A. (2021). Innovation : comment expérimenter ? [Online]. Available at: https://www.ladn.eu/entreprises-innovantes/innovation-comment-experimenter/.

Mayer, F., Ryklin, D., Wacker, I., Curticean, R., Čalkovský, M., Niemeyer, A., Dong, Z., Levkin, P.A., Gerthsen, D., Schröder, R.R. et al. (2020). 3D two-photon microprinting of nanoporous architectures. *Advanced Materials*, 32, 2002044.

McDonald, E., Salas, R., Espalin, D., Perez, M., Aguilera, E., Muse, D., Wicker, R.B. (2014). 3D printing for the rapid prototyping of structural electronics. *IEEE Access*, 2, 234–242 [Online]. Available at: https://ieeexplore.ieee.org/stamp/stamp.jsp?arnumber=6766751 [Accessed 25 March 2022].

McKinsey (2013). Disruptive technologies: Advances that will transform life, business, and the global economy [Online]. Available at: https://www.mckinsey.com/~/media/McKinsey/Business%20Functions/McKinsey%20Digital/Our%20Insights/Disruptive%20technologies/MGI_Disruptive_technologies_Full_report_May2013.ashx [Accessed 25 March 2022].

Medina-Sanchez, M., Schwarz, L., Meyer, A.K., Hebenstreit, F., Schmidt, O.G. (2016). Cellular cargo delivery: Toward assisted fertilization by sperm-carrying micro-motors. *Nano Letters*, 16, 555–561.

Medina-Sanchez, M., Xu, H., Schmidt, O.G. (2018). Micro- and nano-motors: The new generation of drug carriers. *Therapeutic Delivery*, 9, 303–316.

Melly, S.K., Liu, L., Liu, Y., Leng, J. (2020). On 4D printing as a revolutionary fabrication technique for smart structures smart materials structures . In press [Online]. Available at: https://doi.org/10.1088/1361-665X/ab9989.

Melocchi, A., Inverardi, N., Uboldi, M., Baldi, F., Maroni, A., Pandini, S., Briatico-Vangosa, F., Zema, L., Gazzaniga, A. (2019). Retentive device for intravesical drugo delivery based on water-induced shape memory response of poly(vinyl alcohol): Design conception and 4D printing feasibility. *International Journal of Pharmaceutics*, 559, 299–311.

Merle, B. (2019). Fonder une start-up, une aventure à risques mais qui peut rapporter gros ! [Online]. Available at: https://www.futura-sciences.com/tech/questions-reponses/tech-fonder-start-up-aventure-risques-mais-peut-rapporter-gros-12444/.

MESRI (2020a). France Europe 2020 [Online]. Available at: https://www.enseignementsup-recherche.gouv.fr/pid29645/france-europe-2020.html.

MESRI (2020b). Faire de la France une économie de rupture technologique – soutenir les marchés émergents à forts enjeux de compétitivité [Online]. Available at: https://cache.media.enseignementsup-recherche.gouv.fr/file/Mediatheque/41/1/Rapport_college_experts_06_02-2_1242411.pdf.

Miao, S., Zhu, W., Castro, N.J., Nowicki, M., Zhou, X., Cui, H., Fisher, J.P., Zhang, L.G. (2016). 4D printing smart biomedical scaffolds with novel soybean oil epoxidized acrylate. *Scientific Reports*, 6, 27226.

Miao, D., Wang, X., Yu, J., Ding, B. (2021). A biomimetic transpiration textile for highly efficient personal drying and cooling. *Advanced Functional Materials*, 2008705.

Michelin (2016). Nous demain [Online]. Available at: http://www.michelin.com/fre/groupe-michelin/strategie/Nous-demain.

Milana, E., Bellotti, M., Gorissen, B., Qian, J., De Volder, M., Reynaerts, D. (2020). Shaping soft robotic micro-actuators by wire electrical discharge grinding. *Micro-machines*, 11, 661.

Miriyev, A., Stack, K., Lipson, H. (2017). Soft material for soft actuators. *Nature Communications*, 8, 596.

Mirzaali, M.J., Herranz de la Nava, A., Gunashekar, D., Nouri-Goushki, M., Doubrovski, E.L., Zadpoor, A.A. (2019). Fracture behavior of bio-inspired functionally graded soft–hard composites made by multi-material 3D printing: The case of co-linear cracks. *Materials*, 12, 2735.

Mirzanejad, H. and Agheli, M. (2019). Soft force sensor made of magnetic powder blended with silicone rubber. *Sensors and Actuators*, A293, 108–118.

Mitchell, A., Lafont, U., Hołyńska, M., Semprimoschnig, C. (2018). Additive manufacturing – A review of 4D printing and future applications. *Additive Manufacturing*, 24, 606–626.

Modes, C. and Warner, W. (2016). Shape-programmable materials. *Physics Today*, 69, 32–38.

Moigne, A. (2020). Des matériaux changeant de forme à volonté, conçus par impression 4D réactive [Online]. Available at: https://www.techniques-ingenieur.fr/actualite/articles/des-materiaux-changeant-de-forme-a-volonte-concus-par-impression-4d-reactive-82270/.

Momeni, F. and Ni, J. (2018). Nature-inspired smart solar concentrators by 4D printing. *Renewable Energy*, 122, 35–44.

Monino, A. (2017). Tweel, the 3D printed airless wheel that will change the automobile world [Online]. Available at: https://www.ennomotive.com/tweel-3d-printed-airless-wheel/ [Accessed 25 March 2022].

Monot, J. (ed.) (1970). La science et les valeurs. In *Pour une éthique de la connaissance*. La Découverte, Paris.

de Montaigne, M. (2020). *Dictionnaire amoureux de Montaigne*. Plon, Paris.

Montesqieu, de Secondat C.L. (1968). *Considérations sur les causes de la grandeur des romains et de leur décadence*. Garnier-Flammarion, Paris.

Moonmoon, F.T.Z. (2019). Possibilities of 3D & 4D printing technologies for personalized medicine. Pharmacy Thesis, Brac University, Dhaka [Online]. Available at: http://dspace.bracu.ac.bd/xmlui/bitstream/handle/10361/13766/15146021_PHR.pdf?sequence=1 [Accessed 25 March 2022].

Morel, C. (2002). *Les décisions absurdes*. Gallimard, Paris.

Morin, E. (1991). De la complexité : complexus. In *Les théories de la complexité – autour de l'œuvre d'Henri Atlan*, Fogelman-Soulié, F. (ed.). Le Seuil, Paris.

Morin, E. (2016). Eduquer à la paix pour résister à l'esprit de guerre [Online]. Available at: https://www.lemonde.fr/idees/article/2016/02/07/peut-on-prevenir-la-formation-du-fanatisme_4860871_3232.html.

Morteza Hosseini, S., Mohammadi, M., Guerra-Santin, O. (2019). Interactive kinetic façade: Improving visual comfort based on dynamic daylight and occupant's positions by 2D and 3D shape changes. *Building and Environment*, 165, 106396 [Online]. Available at: https://www.sciencedirect.com/science/article/pii/S0360132319306067 [Accessed 25 March 2022].

Morton, J. (1971). *Organizing for Innovation: A System Approach to Technical Management*. McGraw Hill, New York.

Moskvitch, K. (2015). Programmable matter: Shape-shifting micro-bots get it together [Online]. Available at: https://eandt.theiet.org/content/articles/2015/05/programmable-matter-shape-shifting-microbots-get-it-together/ [Accessed 25 March 2022].

Mouzakis, D.E. (2018). Advanced technologies in manufacturing 3D-layered structures for defense and aerospace. In *Lamination – Theory and Application*, Osheku, C.A. (ed.). Intech, Zagreb [Online]. Available at: https://www.intechopen.com/books/lamination-theory-and-application/advanced-technologies-in-manufacturing-3d-layered-structures-for-defense-and-aerospace [Accessed 25 March 2022].

Municoy, S., Álvarez Echazú, M.I., Antezana, P.E., Galdopórpora, J.M., Olivetti, C., Mebert, A.M., Foglia, M.L., Tuttolomondo, M.V., Alvarez, G.S., Hardy, J.G. et al. (2020). Stimuli-responsive materials for tissue engineering and drug delivery. *International Journal of Molecular Science*, 21, 4724.

Muzzafor, A., Ahamed, M.B., Deskmukh, K., Kovarik, T., Krenek, T., Pasha, S.K.K. (2020). 3D and 4D printing of pH-responsive and functional polymers and their composites. In *3D and 4D Printing of Polymer Nanocomposite Materials – Processes, Applications and Challenges*, Sadasivumi, K.K., Deshmukh, K., Almaadeed, M.A. (eds). Elsevier, Amsterdam.

Nguyen, D. (2020). Chercheurs, sortez de votre zone de confort : entreprenez ! [Online]. Available at: https://www.lesechos.fr/idees-debats/cercle/opinion-chercheurs-sortez-de-votre-zone-de-confort-entreprenez-1171877.

Ni, Y., Ji, R., Long, K., Bu, T., Chen, K., Zhuang, S. (2017). A review of 3D-printed sensors. *Applied Spectroscopy Reviews*, 52, 623–652.

Nicolas, G., Geiger, D., Oddou, C., Hamonet, C. (2000). Des fils d'alliages à mémoire de formes utilisés comme actionneurs d'orthèses dynamiques programmables. In *Matériaux et structures des sciences de l'ingénieur et du vivant*, Ribreau (ed.). Ted & Doc, Paris.

Nietzche, F. (1874). De l'utilité et l'inconvénient des études historiques pour la vie [Online]. Available at: https://fr.wikisource.org/wiki/De_l%E2%80%99utilit%C3%A9_et_de_l%E2%80%99inconv%C3%A9nient_des_%C3%A9tudes_historiques_pour_la_vie.

Ning, H., Zhang, Y., Zhu, H., Ingham, A., Huang, G., Mei, Y., Solovev, A.A. (2018). Geometry design, principles and assembly of micro-motors. *Micro-machines*, 9, 75.

Novakovic, K., Matcham, S., Scott, A. (2018). Intelligent hydrogels as drug delivery systems. *Hydrogels: Recent Advances*, 1–28.

Novel, A.S. and Riot, S. (2013). *Vive la co-révolution ; pour une société collaborative*. Manifestô-Gallimard, Paris.

Novotný, P., Wang, H., Pumera, M. (2020). Nano-robots: Machines squeezed between molecular motors and micro-motors. *Chem*, 6, 867–884.

Ntouanoglou, K., Stavropoulos, P., Mourtzis, D. (2018). 4D printing prospects for the aerospace industry: A critical review. *Procedia Manufacturing*, 18, 120–129.

OCDE (1971). *Science, croissance et société*. OCDE, Paris.

Oh, H., Ta, T.D., Suzuki, R., Gross, M.D., Kawahara, Y., Yao, L. (2018). PEP (3D Printed Electronic Papercrafts): An integrated approach for 3D sculpting paper-based electronic devices. *CHI*, 1–12.

Ollivier, B. (2002). Enjeux de l'interdisciplinarité. *L'année sociologique*, 51, 337–354.

Orfali, B. (2005). *La société face aux événements extraordinaires*. Zagros, Paris.

Ould Moussa, M. (2011). Modélisation du comportement dynamique non-linéaire des structures en matériaux à mémoire de forme. Thesis, Ecole Polytechnique, Palaiseau [Online]. Available at: https://pastel.archives-ouvertes.fr/pastel-00675919/document.

Park, Y., Kim, M.K., Lee, J.S. (2020). Emerging memory devices for artificial synapses. *Journal of Material Chemistry*, C8, 9163.

Papetti, V., Dimopoulos-Eigenschwiler, P., Della Tora, A., Lucci, F., Ortana, A., Montnegro, G. (2018). Additive manufactured open cell polyhedral structures as substrate for automotive catalysts. *International Journal of Heat and Mass Transfer*, 126, 1035–1047.

Papoutsis-Kiachagias, E.M., Asouti, V.G., Giannakoglou, K.C., Gkagkas, K., Shimokawa, S., Itakura, E. (2019). Multi-point aerodynamic shape optimization of cars based on continuous adjoint. *Structural and Multidisciplinary Optimization*, 59, 675–694.

Patel, S.K., Cao, J., Lippert, A.R. (2017). A volumetric three-dimensional digital light photoactivatable dye display. *Nature Communications*, 8, 15239.

Poppinga, S., Schenck, P., Speck, O., Speck, T., Bruchmann, B., Masselter, T. (2021). Self-actuated paper and wood models: Low-cost handcrafted biomimetic compliant systems for research and teaching. *Biomimetics*, 6, 42.

Pech, T. (2017). *Insoumission – Portrait de la France qui vient.* Le Seuil, Paris.

Peng, B., Yang, Y., Cavicchi, C.A. (2020). Sequential shapeshifting 4D printing: Programming the pathway of multi-shape transformation by 3D printing stimuli-responsive polymers. *Multifunctional Materials*, 3, 042002.

Pestre, D. (2014). Du gouvernement du progrès technique et de ses effets. In *Le gouvernement des technosciences ; gouverner le progrès et ses dégâts depuis 1945*, Pestre, D. (ed.). La Découverte, Paris.

Petrella, R. (1992). Le printemps du "Technology Assessment" en Europe ; faits et questionnements. FAST report (Forecasting and Assessment in the Field of Science and Technology), Brussels.

Peyrat, D. (2003). *Éloge de la sécurité.* Gallimard, Paris.

Pigasse, M. (2014). *Éloge de l'anormalité.* Plon, Paris.

Pinho, A.C., Buga, C.S., Piedade, A.P. (2020). The chemistry behind 4D printing. *Applied Materials Today*, 19, 100611.

Poty, M., Lumay, G., Vandewalle, N. (2017). Customizing mesoscale self-assembly with 3D printing [Online]. Available at: https://arxiv.org/pdf/1310.4847.pdf [Accessed 25 March 2022].

Prigogine, I. and Stengers, I. (1979). *La Nouvelle Alliance – Métamorphose de la science.* Gallimard, Paris.

Przyczyna, D., Zawal, P., Mazur, T., Gentili, P.L., Szaciłowski, K. (2020). In materio neuromimetic devices: Dynamics, information processing and pattern recognition. *Japanese Journal of Applied Physics*, 59, 050504.

Puech, M. (2016). *Homo Spiens Technologicus.* Le Pommier, Paris.

Puren, V. (2013). Vers l'impression 4D ou l'ère des matériaux "intelligents" [Online]. Available at: https://www.presse-citron.net/vers-impression-4d-ou-ere-materiaux-intelligents/.

Quinet, E. (1984). *Le Christianisme et la révolution française*, Fayard, Paris.

Rafiee, M., Farahani, R.D., Therriault, D. (2020). Multi-material 3D and 4D printing: A survey. *Advanced Science*, 7, 1902307.

Rai, V., Tiwari, N., Rajput, M., Joshi, S.M., Nguyen, A.C., Mathews, N. (2017). Reversible electrochemical silver deposition over large areas for smart windows and information display. *Electrochimica Acta*, 255, 63–71.

Rander (2020). 4D printing: A revolution across industries [Online]. Available at: https://www.machinedesign.com/3d-printing-cad/article/21837454/4d-printing-a-revolution-across-industries [Accessed 25 March 2022].

Rapp, F. (2000). L'évaluation des choix technologiques : remarques sur la discussion en Allemagne. In *Peut-on encore croire au progrès ?*, Bourg, D. and Besnier, J.M. (eds). PUF, Paris.

Rastogi, P. and Kandasubramanian, B. (2019). Breakthrough in the printing tactics for stimuli-responsive materials: 4D printing. *Chemical Engineering Journal*, 366, 264–304.

Rayate, A. and Jain, P.K. (2018). A review on 4D printing material composites and their applications. *Materials Today: Proceedings*, 5, 20474–20484.

Regehly, M., Garmshausen, Y., Reuter, M., König, N.F., Israel, E., Kelly, D.P., Chou, C.Y., Koch, K., Asfari, B., Hecht, S. (2020). Xolography for linear volumetric 3D printing. *Nature*, 588, 620–624.

Reiter-Palmon, R. and Illies, J.J. (2004). Leadership and creativity: Understanding leadership from a creative problem solving perspective. *The Leadership Quarterly*, 15, 55–77.

Ren, L., Li, B., Song, Z., Liu, Q., Ren, L., Zhou, X. (2019). Bioinspired fiber-regulated composite with tunable permanent shape and shape memory properties via 3D magnetic printing. *Composites. Part B, Engineering*, 164, 458–466.

Ren, T., Chen, P., Gu, L., Ogut, M.G., Demirci, U. (2020). Soft ring-shaped cellu-robots with simultaneous locomotion in batches. *Advanced Materials*, 32, 1905713.

Revault d'Allonnes, M. (2012). *La crise sans fin – essai sur l'expérience moderne du temps*. Le Seuil, Paris.

Reyes, C., Samogyi, R., Nin, S., Cruz, M.A., Yang, F., Catenacci, M.J., Rhodes, C.P., Wiley, B.J. (2018). Three dimensional printing of a complete lithium-ion battery with filament fabrication. *ACS Applied Energy Materials*, 1, 5268–5279.

Robertson, J., Rodriguez, R., Holmes Jr., L., Mather, P., Wetzel, E. (2016). Thermally driven microfluidic pumping via reversible shape memory polymers. *Smart Materials Structures*, 25, 085043.

Rodriguez, R.D., Shchadenko, S., Murastov, G., Lipovka, A., Fatkullin, M., Petrov, I., Tran, T.H., Khalelov, A., Saqib, M., Villa, N.E. et al. (2021). Ultra-robust flexible electronics by laser-driven polymer-nanomaterials integration. *Advanced Functional Materials*, 2008818.

Rolland, S. (2019). Plan Deep-Tech : 800 millions d'euros d'aides nouvelles pour l'innovation de rupture [Online]. Available at: https://www.latribune.fr/technos-medias/plan-deep-tech-800-millions-d-euros-d-aides-nouvelles-pour-l-innovation-de-rupture-805594.html.

Rosa, H. (2010). *Accélération*. La Découverte, Paris.

Rösch, A.G., Gall, A., Aslan, S., Hecht, M., Franke, L., Mallick, M., Penth, L., Bahro, D., Friderich, D., Lemmer, U. (2021). Générateurs thermoélectriques en origami entièrement imprimés pour la récupération d'énergie. NPI Flex Electron 5, 1 [Online]. Available at: https://doi.org/10.1038/s41528-020-00098-1.

Rostaing, H. (2004). Conception, modélisation et fabrication d'un micro-actionneur bistable, hors plan et magnétique. Thesis, Engineering Sciences, Institut National Polytechnique de Grenoble, Grenoble [Online]. Available at: https://tel.archives-ouvertes.fr/tel-00408339/document.

Rothwell, R. and Zegveld, W. (1982). *Innovation and the Small and Medium Sized Firm: Their Role in Employment and in Economic Change*. Frances Pinter, London.

Roubelat, P. (2015). Le futur vu de l'action. Le State of the Future 2015–2016. *Prospective et stratégie*, 6, 157–159.

Roudbarian, N., Baghani, M., Baniasadi, M., George, D., Mohammadi, A. (2019). An experimental investigation on the energy storage in a shape-memory-polymer system. *Energy Equipment Systems*, 7, 309–316.

Rousseau, J.J. (2006). *Du contrat social*. Flammarion, Paris.

Roux, M. (2011). *J'ai demandé un rapport – la politique est-elle une affaire d'experts ?* Flammarion, Paris.

Royal Society (2021). Animate materials – Perspective [Online]. Available at: https://royalsociety. org/-/media/policy/projects/animate-materials/animate-materials-report.pdf [Accessed 25 March 2022].

Rubio, L.D., Potomkin, M., Baker, R.D., Sen, A., Berlyand, L., Aranson, I.S. (2021). Self-propulsion and shear flow align active particles in nozzles and channels. *Advanced Intelligent Systems*, 3, 2000178.

Runco, M.A. (2015). Meta-creativity: Being creative about creativity. *Creativity Research Journal*, 27, 295–298.

Runco, M.A. and Jaeger, G. (2012). The standard definition of creativity. *Creativity Research Journal*, 24, 92–96.

Runco, M.A., Kaufman J.C., Halladay L.R., Cole, J.C. (2010). Change in reputation as index of genius and eminence. *Historical Methods*, 43, 91–96.

Rus, D. and Tolley, M.T. (2015). Design, fabrication and control of soft-robots. *Nature*, 521, 467–475.

Saint-Etienne, C. (2015). *La France 3.0 – agir, espérer, réinventer*. Odile Jacob, Paris.

Saleh, E. (2020). 3D and 4D printed polymer composites for electronic applications. In *3D and 4D Printing of Polymer Nanocomposite Materials – Processes, Applications and Challenges*, Sadasivumi, K.K., Deshmukh, K., Almaadeed, M.A. (eds). Elsevier, Amsterdam.

Salmon, A.R., Cormier, S., Wang, W., Abell, C., Baumberg, J.J. (2019). Motile artificial chromatophores: Light-triggered nanoparticles for micro-droplet locomotion and color change. *Advanced Optical Materials*, 7, 1900951.

Sartre, J.P. (1976). *L'être et le néant.* Gallimard, Paris.

Sauvage, J.P. (2003). Les nano-machines moléculaires : de la biologie aux systèmes artificiels et aux dispositifs [Online]. Available at: https://culturesciences.chimie.ens.fr/sites/default/files/2020-08/Nanomachines_JP_Sauvage.pdf.

Sauvons l'Université (2019). Antoine Petit se lâche à l'occasion des 80 ans du CNRS – Tribune, Les Echos, 26 November [Online]. Available at: http://www.sauvonsluniversite.com/spip.php?article8551.

Scalet, G. (2020). Two-way and multiple-way shape memory polymers for soft robotics: An overview. *Actuators*, 9, 10.

Schaffner, M., Faber, J.A., Pianegonda, L., Rühs, P.A., Coulter, F., Studart, A.R. (2018). 3D printing of robotic soft actuators with programmable bioinspired architectures. *Nature Communications*, 9, 878.

Schara, S., Blau, R., Church, D.C., Pokorski, J.K., Lipomi, D.J. (2021). Polymer chemistry for haptics, soft robotics, and human–machine interfaces. *Advanced Functional Materials*, 2008375.

Schmookler, J. (1966). *Invention and Economic Growth.* Harvard University Press, Cambridge, MA.

Schreiber, F., Manns, M., Morales, J. (2019). Design of an additively manufactured soft ring-gripper. *Procedia Manufacturing*, 28, 142–147.

Schumpeter, J. (1942). *Capitalisme, socialisme et démocratie*. Payot, Paris.

Schumpeter, J. (1952). *The Theory of Capitalist Development.* Transaction Books, Harvard.

Sculpteo (2020). 4D printing: A technology coming from the future [Online]. Available at: https://www.sculpteo.com/en/3d-learning-hub/best-articles-about-3d-printing/4d-printing-technology/ [Accessed 25 March 2022].

Sellitto, A. and Riccio, A. (2019). Overview and future advanced engineering applications for morphing surfaces by shape memory alloy materials. *Materials (Basel)*, 12, 708.

Semba, J.A., Mieloch, A.A., Rybka, J.D. (2020). Introduction to the state-of-the-art 3D bio-printing methods, design, and applications in orthopedics. *Bio-printing*, 18, e00070.

Serres, M. (1985). *Les cinq sens*. Grasset, Paris.

Shao, L.H., Zhao, B., Zhang, Q., Xing, Y., Zhang, K. (2020). 4D printing composite with electrically controlled local deformation. *Extreme Mechanics Letters*, 39, 100793.

Sharbati, M.T., Du, Y., Torres, J., Ardolino, N.D., Yun, M., Xiong, F. (2018). Low-power, electrochemically tunable graphene synapses for neuromorphic computing. *Advanced Materials*, 30, 1802353.

Shen, Z.C., Xia, Y., Ding, Y.G., Zhao, C.Q., Yang, Y.B. (2019). 4D printing and its key technologies. *Journal of Materials Engineering*, 47, 11–18.

Shen, B., Erol, O., Fang, L., Kang, S.H. (2020a). Programming the time into 3D printing: Current advances and future directions in 4D printing. *Multi-functional Materials*, 3, 012001.

Shen, Z., Chen, F., Zhu, X., Yong, K.T., Gu, G. (2020b). Stimuli-responsive functional materials for soft robotics. *Journal of Material Chemistry*, B8, 8972–8991.

Shi, Q. and Lee, C. (2019). Self-powered bio-inspired spider-net-coding interface using single-electrode triboelectric nano-generator. *Advanced Science*, 6, 1900617.

Shi, W., Guo, Y., Liu, Y. (2020). When flexible organic field-effect transistors meet bio-mimetics: A prospective view of the internet of things. *Advanced Materials*, 2019, e1901493.

Shie, M.Y., Shen, Y.F., Astuti, S.D., Lee, A.K.X., Lin, S.H., Dwijaksara, N.L.B., Chen, Y.W. (2019). Review of polymeric materials in 4D printing biomedical applications. *Polymers*, 11, 1864.

Siddharth, J., Krishna, R., Karunakaran, C., Vasudevan, R., Arun, T.M., Koziolc, K., Thakur, V.K., Balan, A.S.S. (2020). 4D printing of materials for the future: Opportunities and challenges. *Applied Materials Today*, 18, 100490 [Online]. Available at: https://dspace.lib.cranfield.ac.uk/bitstream/handle/1826/14652/4D_printing_of_materials_for_the_future-2019.pdf.pdf?sequence=4&isAllowed=y [Accessed 25 March 2022].

Silverzahn, P. (2021). Un enjeu français : le pari de Pascal de l'innovation [Online]. Available at: https://www.contrepoints.org/2021/02/24/391745-un-enjeu-francais-le-pari-de-pascal-de-linnovation.

Simeral, J.D., Hosman, T., Saab, J., Flesher, S.N., Vilela, M., Franco, B., Kelemen, J., Brandman, D.M., Ciancibello, J.G., Rezaii, P.C. et al. (2021). Home use of a percutaneous wireless intra-cortical brain-computer interface by individuals with tetraplegia. *Transactions on Biomedical Engineering*, 00114-2020.R1 [Online]. Available at: https://ieeexplore.ieee.org/stamp/stamp.jsp?tp=&arnumber=9390339 [Accessed 25 March 2022].

Simoneau, C. (2013). Modélisation de matériaux composites adaptatifs munis d'actionneurs en alliage à mémoire de forme. Mechanical Engineering Thesis, École de technologie supérieure, Université de Québec, Montreal [Online]. Available at: https://espace.etsmtl.ca/id/eprint/1177/1/SIMONEAU_Charles.pdf.

Simonton, D.K. (2012). Taking the U.S. Patent Office criteria seriously: A quantitative three-criterion creativity definition and its implications. *Creativity Research Journal*, 24, 97–106.

Sloterdijk, P. (2000). *Critique de la raison cynique*. Christian Bourgeois, Paris.

Sloterdijk, P. (2006). *Le palais de cristal : à l'intérieur du capitalisme planétaire*. Moren Sell, Paris.

Solé, A. (2004). Critique de la complexité [Online]. Available at: http://forum.aceboard.net/63455-2145-3752-0.htm.

Song, W.J., Lee, S., Song, G., Park, S., (2018). Stretchable aqueous batteries: Progress and prospects. *ACS Energy Letters*, 4, 177.

Spinoza, B. (1954) *Œuvres complètes*. Gallimard, Paris.

Staedter, T. (2019). How 4D printing will transform the future [Online]. Available at: https://now.northropgrumman.com/how-4d-printing-will-transform-the-future/ [Accessed 25 March 2022].

Stano, G., Di Nisio, A., Lanzolla, A.M., Ragolia, M., Percoco, G. (2020). Fused filament fabrication of commercial conductive filaments: Experimental study on the process parameters aimed at the minimization, repeatability and thermal characterization of electrical resistance. *The International Journal of Advanced Manufacturing Technology*, 111, 2971–2986.

Star, S.L. and Griesemer, J. (1989). Institutional ecology, "Translations", and boundary objects: Amateurs and professionals on Berkeley's Museum of Vertrebate zoology. *Social Studies of Science*, 19, 387–420.

Stengers, I. (2020). *Réactiver le sens commun – lecture de Whitehead en temps de débâcle*. Les empêcheurs de penser en rond, Paris.

Stengers, I. and Bensaude-Vincent, B. (2003). *100 mots pour commencer à penser les sciences*. Les empêcheurs de penser en rond, Paris.

Stevenson, K. (2020). Xolo reveals some secrets of volumetric 3D printing in new research paper [Online]. Available at: https://www.fabbaloo.com/news/xolo-reveals-some-secrets-of-volumetric-3d-printing-in-new-research-paper?utm_source=newsletter&utm_medium=email &utm_campaign=fabbaloo_daily_3d_printing_news&utm_term=2020-12-23 [Accessed 25 March 2022].

Su, J.W., Wang, J., Zheng, Y., Jiang, S., Jiang, S., Lin, J. (2018a). Mechanically guided assembly of monolithic three-dimensional structures from elastomer composites. *ACS Applied Materials and Interfaces*, 10, 44716–44721.

Su, Y., Wang, X., Wu, B., Wang, F., Wang, J., Xing, B. (2018b). Application potential of 4D printing technology in development of aircraft. *Hang kong cai liao xue bao*, 38, 59–69.

Su, I., Jung, G.S., Narayanan, N., Buehler, M.J. (2020). Perspectives on three-dimensional printing of self-assembling materials and structures. *Current Opinion in Biomedical Engineering*, 15, 59–67.

Sud Education (2020). Ce qui nous attend dans le projet de Loi de programmation pluriannuelle de la recherche (LPPR) [Online]. Available at: https://www.sudeducation.org/Ce-qui-nous-attend-dans-le-projet-de-Loi-de-programmation-pluriannuelle-de-la.html.

Sun, Y., Evans, J.S., Lee, T., Senyuk, B., Keller, P., He, S., Smalyukh, I.I. (2012). Optical manipulation of shape-morphing elastomeric liquid crystal micro-particles doped with gold nanocrystals. *Applied Physics Letters*, 100, 241901 [Online]. Available at: https://arxiv.org/ftp/arxiv/papers/1701/1701.04849.pdf [Accessed 25 March 2022].

Sun, W., Starly, B., Daly, A.C., Burdick, J.A., Groll, J., Skeldon, G., Shu, W., Sakai, Y., Shinohara, M., Nishikawa, M. et al. (2020). The bio-printing roadmap. *Bio-fabrication*, 12, 022002.

Surjadi, J.U., Gao, L., Du, H., Li, X., Xiong, X., Fang, N.X., Lu, Y. (2019). Mechanical metamaterials and their engineering applications. *Advanced Engineering Materials*, 21, 1800864.

Svetozarevic, B., Nagy, Z., Hofer, J., Jacob, D., Begle, M., Chatzi, E., Schlueter, A. (2016). SoRo-Track: A two-axis soft robotic platform for solar tracking and building-integrated photovoltaic applications. *Proceedings of the 2016 IEEE International Conference on Robotics and Automation (ICRA)*, 16–21 May, Stockholm.

Tabrizi, S., Cao, Y.Y., Lin, H., Jia, B.H. (2017). Two-photon reduction: A cost-effective method for fabrication of functional metallic nanostructures. *Science China-Physics, Mechanics, Astronomy*, 60, 034201.

Tahouni, Y., Krüger, F., Poppinga, S., Wood, D., Pfaff, M., Rühe, J., Speck, T., Menges, A. (2021). *Programming Sequential Motion Steps in 4D-Printed Hygromorphs by Architected Mesostructure and Differential Hygro-Responsiveness*. IOP Publishing, 16, 055002. In press.

Tao, Y., Lee, Y.C., Liu, H., Zhang, X., Cui, J., Mondoa, C., Babaei, M., Santillan, J., Wang, G., Luo, D. et al. (2021). Morphing pasta and beyond. *Science Advances*, 7, eabf4098.

Taylor, A.J.P. (1963). *The First World War*. Berkley Publishing Group, New York.

The Chinese University of Hong Kong (2019). New technique increases 3-D printing speed by 1,000 to 10,000 times [Online]. Available at: https://phys.org/news/2019-12-technique-d.html.

The Hague Centre for Strategic Studies (2021). Capstone report: Robotic and autonomous systems in a military context [Online]. Available at: https://hcss.nl/sites/default/files/files/reports/RAS%20Capstone%20%28Final%29.pdf [Accessed 25 March 2022].

Thomas, C., Tibbits, S., Banning, G. (2014). The programmable world. *Scientific American*, 311, 60–65.

Tibbits, S. (2013). The emergence of 4D printing [Online]. Available at: https://www.ted.com/talks/skylar_tibbits_the_emergence_of_4d_printing [Accessed 25 March 2022].

Tibbits, S. (2014). 4D printing: Multi-material shape change. *Architectural Design*, 84, 116–121.

Tofail, S.A.M., Koumoulos, E.P., Bandyopadhyay, A., Bose, S., O'Donoghue, L., Charitidis, C. (2018). Additive manufacturing: Scientific and technological challenges, market uptake and opportunities. *Materials Today*, 21, 22–37.

Toffler, A. (1991). *Les nouveaux pouvoirs*. Fayard, Paris.

Trabucchi, D., Buganza, T., Dell'Era, C., Pellizzoni, E. (2018). Exploring the inbound and outbound strategies enabled by user generated Big Data: Evidence from leading smartphone applications. *Creativity and Innovation Management*, 27, 42–55.

Trott, P. (2008). *Innovation Management and New Product Development.* Financial Times/Prentice Hall, New York.

Tyagi, M., Spinks, G.M., Jager, E.W.H. (2021). 3D printing micro-actuators for soft micro-robots. *Soft Robotics*, 8, 19–27.

Utela, B., Storti, D., Anderson, R., Gautir, M. (2008). A review of process development steps for new material system in three dimensional printing. *Journal of Manufacturing Processes*, 10, 96–104.

Valaskakis, K. (2014). La prospective en trois leçons ; mon expérience personnelle. In *La prospective scientifique en action*, Durance, P. (ed.). Odile Jacob, Paris.

Van Veen, V., Krug, M.K., Schooler, J.W., Carter, C.S. (2009). Neural activity predicts attitude change in cognitive dissonance. *Nature Neurosciences*, 12, 1469–1474.

Vatankhah-Varnosfaderani, M., Keith, A.N., Cong, Y., Liang, Y., Rosenthal, M., Sztucki, M., Clair, C., Magonov, S., Ivanov, D.A., Dobrynin, A.V. et al. (2018). Chameleon-like elastomers with molecularly encoded strain-adaptive stiffening and coloration. *Science*, 359, 1509–1513.

Verpaalen, R.C.P., Varghese, S., Froyen, A., Pilz da Cunha, M., Pouderoijen, M.J., Severn, J.R., Bhatti, M.R., Peijs, T., Bastiaansen, C.W.M., Debije, M.G. et al. (2020). Fast, light-responsive, metal-like polymer actuators generating high stresses at low strain. *Matter*, 2, 1522–1534.

Vidal, F. (2020). A mes collègues scientifiques, je veux dire que le gouvernement a entendu leur appel à réinvestir massivement dans la recherche [Online]. Available at: https://www.lemonde.fr/idees/article/2020/02/10/frederique-vidal-a-mes-collegues-scientifiques-je-veux-dire-que-le-gouvernement-a-entendu-leur-appel-a-reinvestir-massivement-dans-la-recherche_6029000_3232.html.

Vinck, D. (2000). *Pratiques de l'interdisciplinarité.* Presses Universitaires de Grenoble, Grenoble.

Vinsel, L. (2020). Seven theses on technology and the US economy [Online]. Available at: https://medium.com/swlh/seven-theses-on-technology-and-the-us-economy-c02ba562587a [Accessed 25 March 2022].

Vinsel, L. and Russel, A.L. (2020). *The Innovation Delusion.* Penguin Random House, New York.

Viveret, P. (2012). *La cause humaine – du bon usage de la fin du monde*. Les liens qui libèrent, Paris.

Von Hippel, E. (2013). User innovation. In *Leading Open Innovation*, Huff, A.S., Moslein, K.M., Reichwald, R. (eds). MIT Press, Cambridge, MA.

Von Hippel, E., Ogawa, S., De Jong, J.P.J. (2011). The age of the consumer-innovator. *MIT Sloan Management Review* [Online]. Available at: https://dspace.mit.edu/bitstream/handle/1721.1/77937/Von%20Hippel_The%20age.pdf?sequence=2&isAllowed=y [Accessed 25 March 2022].

Wagg, D., Bond, I., Weaver, P., Grisweel, M. (2007). *Adaptive Structures – Engineering Applications*. John Wiley & Sons Ltd, New York.

Wan, Z., Zhang, P., Liu, Y., Lv, L., Zhou, Y. (2020). Four-dimensional bio-printing: Current developments and applications in bone tissue engineering. *Acta Biomaterialia*, 101, 26–42.

Wang, H. and Pumera, M. (2020). Coordinated behaviors of artificial micro/nano-machines: From mutual interactions to interactions with the environment. *Chemical Society Review*, 49, 3211.

Wang, F. and Xu, Z. (2020). Graphene and graphene-oxide – Reinforced 3D and 4D printable composites. In *3D and 4D Printing of Polymer Nanocomposite Materials – Processes, Applications and Challenges*, Sadasivumi, K.K., Deshmukh, K., Almaadeed, M.A. (eds). Elsevier, Amsterdam.

Wang, C., Wang, C., Huang, Z., Xu, S. (2018a). Materials and structures toward soft electronics. *Advanced Materials*, 30, 1801368.

Wang, H., Wang, C., Jian, M., Wang, Q., Xia, K., Yin, Z., Zhang, M., Liang, X., Zhang, Y. (2018b). Super-elastic wire-shaped supercapacitor sustaining 850% tensile strain based on carbon nanotube@ graphene fiber. *Nano Research*, 11, 2347.

Wang, Q., Tian, X., Huang, L., Li, D., Malakhov, A.V., Polilovb, A.N. (2018c). Programmable morphing composites with embedded continuous fibers by 4D printing. *Materials and Design*, 155, 404–413.

Wang, J., Wang, Z., Song, Z., Ren, L., Liu, Q., Ren, L. (2019a). Biomimetic shape–color double-responsive 4D printing. *Advanced Materials Technology*, 4, 1900293.

Wang, Q., Wang, H., Du, P., Liu, J., Liu, D., Liu, P. (2019b). Porous poly-lactic acid/carbon nanotubes/polyaniline composite film as flexible free-standing electrode for supercapacitors. *Electrochimica Acta*, 294, 312.

Wang, A.J., Maharjan, S., Liao, K.S., McElhenny, B.P., Wright, K.D., Dillon, E.P., Neupane, R., Zhu, Z., Chen, S., Barron, A.R. et al. (2020a). Poly(octadecylacrylate)-grafted multi-walled carbon nanotube composites for wearable temperature sensors. *ACS Applied Nano Materials*, 3, 2288–2301.

Wang, D., Gao, C., Zhou, C., Lin, Z., He, Q. (2020b). Leukocyte membrane-coated liquid metal nano-swimmers for actively targeted delivery and synergistic chemo-photo-thermal therapy. *AAAS Research*, 3676954 [Online]. Available at: https://downloads.spj.sciencemag.org/research/2020/3676954.pdf [Accessed 25 March 2022].

Wang, L., An, M.R., Liu, H.T. (2020c). Compression spin bio-inspired arm: A conceptual model based on compression–torsion cubic mechanical metamaterials with variable cross-section. *Extreme Mechanics Letters*, 41, 101069.

Wang, P., Hu, M., Wang, H., Chen, Z., Feng, Y., Wang, J., Ling, W., Huang, Y. (2020d). The evolution of flexible electronics: From nature, beyond nature, and to nature. *Advanced Science*, 7, 2001116 [Online]. Available at: https://www.ncbi.nlm.nih.gov/pmc/articles/PMC7578875/ [Accessed 25 March 2022].

Wei, H., Zhang, Q., Yao, Y., Liu, L., Liu, Y., Leng, J. (2017). Direct-write fabrication of 4D active shape-changing structures based on a shape memory polymer and its nanocomposite. *ACS Applied Materials Interfaces*, 9, 876–883.

Weil, S. (2021). *L'enracinement – prélude à une déclaration des devoirs envers l'être humain*. Payot, Paris.

West, J. and Bogers, M. (2017). Open innovation: Current status and research opportunities. *Innovation – Organization & Management*, 19, 43–50.

Whitehead, A.N. (1994). *La science et le monde moderne*. Editions du Rocher, Monaco.

Wiese, M., Thiede, S., Herrmann, C. (2020). Rapid manufacturing of automotive polymer series parts: A systematic review of processes, materials and challenges. *Additive Manufacturing*, 36, 101582 [Online]. Available at: https://www.researchgate.net/profile/Mathias-Wiese-2/publication/344226316_Rapid_manufacturing_of_automotive_polymer_series_parts_A_systematic_review_of_processes_materials_and_challenges/links/5f5f45b0a6fdcc116410b58e/Rapid-manufacturing-of-automotive-polymer-series-parts-A-systematic-review-of-processes-materials-and-challenges.pdf [Accessed 25 March 2022].

Willis, K.D.D., Brockmeyer, E., Hudson, S.E., Poupyrev, I. (2012). Printed optics: 3D printing of embedded optical elements for interactive devices [Online]. Available at: http://www.ivanpoupyrev.com/wp-content/uploads/2016/10/Printed-Optics-UIST2012.pdf [Accessed 25 March 2022].

Wu, J., Jang, B., Harduf, Y., Chapnik, Z., Avci, O.B., Chen, X., Puigmartí-Luis, J., Ergeneman, O., Nelson, B.J., Or, Y. et al. (2021). Helical klinotactic locomotion of two-link nano-swimmers with dual-function drug-loaded soft polysaccharide hinges. *Advanced Science*, 2004458 [Online]. Available at: https://onlinelibrary.wiley.com/doi/pdf/10.1002/advs.202004458 [Accessed 25 March 2022].

Xiong, G., He, P., Lyu, Z., Chen, T., Huang, B., Chen, L., Fisher, T.S. (2018). Bioinspired leaves-on-branchlet hybrid carbon nanostructure for supercapacitors. *Nature Communication*, 9, 790.

Xue, W., Wei, H., Zhang, F., Liu, Y., Leng, J. (2019). 3D printing of shape memory poly(d,l-lactide-co-trimethylene carbonate) by direct ink writing for shape-changing structures. *Journal of Applied Polymer Science*, 136, 48177.

Xue, R., Cui, X., Zhang, P., Liu, K., Li, Y., Wu, W., Liao, H. (2020). Mechanical design and energy absorption performances of novel dual scale hybrid plate-lattice mechanical metamaterials. *Extreme Mechanics Letters*, 40, 100918.

Yang, X., Fang, Y., Yu, Z., Wang, Z., Zhang, T., Yin, M., Lin, M., Yang, Y., Cai, Y., Huang, R. (2016). Non-associative learning implementation by a single memristor-based multi-terminal synaptic device. *Nanoscale*, 8, 18897–18904.

Yang, J.C., Mun, J., Kwon, S.Y., Park, S., Bao, Z., Park, S. (2019). Electronic skin: Recent progress and future prospects for skin-attachable devices for health monitoring, robotics, and prosthetics. *Advanced Materials*, 31, e1904765.

Yang, Q., Gao, B., Xu, F. (2020). Recent advances in 4D bio-printing. *Biotechnology Journal*, 15, 1900086.

Yao, T., Wang, Y., Zhu, B., Wei, D., Yang, Y., Han, X. (2021). 4D printing and collaborative design of highly flexible shape memory alloy structures: A case study for a metallic robot prototype. *Smart Materials and Structures*, 30, 015018.

You, I., Choi, S.E., Hwang, H., Han, S.W., Kim, J.W., Jeong, U. (2018). E-skin tactile sensor matrix pixelated by position-registered conductive micro-particles creating pressure-sensitive selectors. *Advances in Functional Materials*, 28, 1801858.

Yu, S. (2018). Neuro-inspired computing with emerging nonvolatile memories. *Proceedings of the IEEE*, 106, 260–285.

Yu, J., Seipel, S., Nierstrasz, V.A. (2018). Digital inkjet functionalization of water-repellent textile for smart textile application. *Journal of Materials Science*, 53, 13216–13229.

Yu, Y., Kim, G., Mathur, K. (2020). A critical review of additive manufacturing: An innovation of mass customization. *Journal of Textiles and Apparel, Technology and Management*, 11, 1–16 [Online]. Available at: https://ojs.cnr.ncsu.edu/index.php/JTATM/article/viewFile/16727/7796 [Accessed 25 March 2022].

Yuan, H., Chapelle, F., Fauroux, J.C., Balandraud, X. (2018). Concept for a 3D-printed soft rotary actuator driven by a shape-memory alloy. *Smart Materials and Structures*, 27, 55005.

Zamiatine, E. (2020). *Nous autres*. Galimard, Paris.

Zarek, M., Layani, M., Cooperstein, I., Sachyani, E., Cohn, D., Magdassi, S. (2016a). 3D printing: 3D printing of shape memory polymers for flexible electronic device. *Advanced Materials (Weinheim)*, 28, 4166–4166.

Zarek, M., Layani, M., Eliazar, S., Mansour, N., Cooperstein, I., Shukrun, E., Szlar, A., Cohn, D., Magdassi, S. (2016b). 4D printing shape memory polymers for dynamic jewelry and fashionwear. *Virtual and Physical Prototyping*, 11, 263–270.

Zarek, M., Mansour, N., Shapira, S., Cohn, D. (2017). 4D printing of shape memory-based personalized endoluminal medical devices. *Macromolecular Rapid Communications*, 38, 1600628.

Ze, Q., Kuang, X., Wu, S., Wong, J., Montgomery, S.M., Zhang, R., Kovitz, J.M., Yang, F., Qi, H.J., Zhao, R. (2020). Magnetic shape memory polymers with integrated multifunctional shape manipulation. *Advanced Materials*, 32, 1906657.

Zhang, J. and Xiao, P. (2018). 3D printing of photopolymers. *Polymer Chemistry*, 9, 1530–1540.

Zhang, I.M., Ji, Q., Liu, Y., Huang, J., Duan, H. (2018a). An integrated micro-milli-fluidic processing system. *Lab-on-Chip*, 18, 3393–3404.

Zhang, W., Zhang, F., Lan, X., Leng, J., Wu, A.S., Bryson, T.M., Cotton, C., Gu, B., Sun, B., Chou, T.W. (2018b). Shape memory behavior and recovery force of 4D printed textile functional composites. *Composites Science and Technology*, 160, 224–230.

Zhang, X., Chen, L., Lim, K.H., Gonuguntla, S., Lim, K.W., Pranantyo, D., Yong, W.P., Tyler Yam, W.J., Low, Z., Teo, W.J. et al. (2019). The pathway to intelligence: Using stimuli-responsive materials as building blocks for constructing smart and functional systems. *Advanced Materials*, 31, 1804540.

Zhang, D., Ren, B., Zhang, Y., Xu, L., Huang, Q., He, Y., Li, X., Wu, J., Yang, J., Chen, Q. et al. (2020). From design to applications of stimuli-responsive hydrogel strain sensors. *Journal of Material Chemistry*, B8, 3171–3191.

Zhu, J. (2020). Carbon black – Reinforced 3D and 4D printable conductive polymer composites. In *3D and 4D Printing of Polymer Nanocomposite Materials – Processes, Applications and Challenges*, Sadasivumi, K.K., Deshmukh, K., Almaadeed, M.A. (eds). Elsevier, Amsterdam.

Zhu, M., Xie, M., Lu, X., Okada, S., Kawamura, S. (2020). A soft robotic finger with self-powered triboelectric curvature sensor based on multi-material 3D printing. *Nano Energy*, 73, 104772.

Zolfagharian, A., Kaynak, A., Khoo, S.Y., Zhang, J., Nahavandi, S., Kouzani, A. (2019). Control-oriented modelling of a 3D-printed soft actuator. *Materials*, 12, 71.

Zolfagharian, A., Kaynak, A., Bodaghi, M., Kouzani, A.Z., Gharaie, S., Nahavandi, S. (2020). Control-based 4D printing: Adaptive 4D-printed systems. *Applied Science*, 10, 3020.

2

Is There External Creativity to Support 4D Printing?

> Matter has neither bag nor skin. Only the cell has a membrane [...]. It is by providing itself with an insulating layer, whose role is not to prohibit but to regulate the exchange between the inside and the outside, that a living being can form and grow. (Debray 2020)

Figure 2.1. *Spectacular example of a 4D object made with heat sensitive polymers, validation for additive manufacturing according to Ge et al. (2016). For a color version of this figure, see www.iste.co.uk/demoly/4Dprinting1.zip*

2.1. Introduction

As recalled in the introduction, according to Tibbits (2013):

> An unprecedented revolution is underway. It is the ability to program biological and physical materials to change shape, change properties [...]. The idea behind 4D printing is to do 3D printing with multiple materials, so you can put multiple materials in, and add a new capability: transformation. As soon as they come out, the parts can transform from one shape to another, directly, by themselves. It's like robotics without wires or motors. (Tibbits 2016)

You print this part, and it can turn into something else. Who hasn't dreamed of this after seeing the movie "Terminator"? With this visionary speech, Tibbits (2016) thus defines a new objective to 3D printing with, in anticipation, its inclusion in the context of Industry 4.0 (Jardim-Goncalves, Romero and Grilo, 2017; André, 2019), with the expression of a new need, because a degree of freedom is added to 3D manufacturing. That said, Tibbits opens up a considerable field for creativity and imagination of distant goals. This chapter examines how the public perceives this field of possibilities and, with the presence of scientific and technical barriers as presented in the previous chapters, possible applications for the short term.

2.2. A survey for the general public

> There is an amusing game that consists of going to the texts of science fiction authors of the past to find the announcements of what has actually come true, to rave about the quality of their predictions, and thus claim a place for science fiction in the field of foresight, or on the contrary, what has not happened in order to mock and thus try to prove the indigence of the genre, and its uselessness. (Gévart 2018)

With the coronavirus pandemic, it has become common to hear that reality has caught up with fiction. Among the visions of the future of medicine are classic phenomena in which the wealthiest enhance themselves (particularly through bio-printing, a form of 4D printing of the living (André 2017c)) and replace their defective organs through cybernetics. It is something of this kind of disruptive vision that is sought in a survey aimed at the general public. However, the results show that this could not be achieved. We expected unrealistic or unattainable proposals, "not serious" (see Figure 2.2). What seemed interesting to us was to imagine a receptive and proactive public that would have suggested the real disruptive innovation, the one that no one saw coming. "As futile as it may be, all-ground coffee, suitcases with wheels, smartphones, the Internet... The disruption" (Gévart 2018).

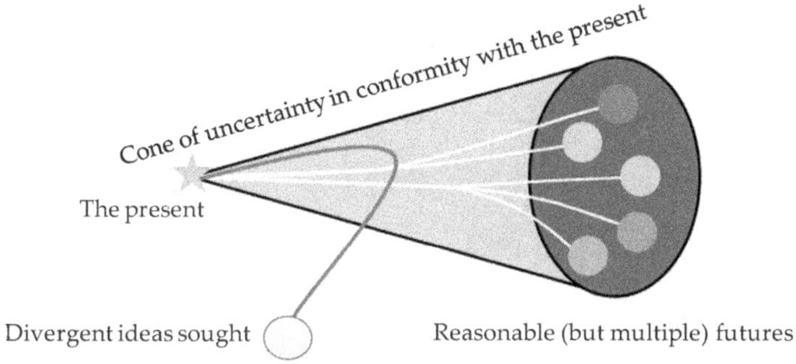

Figure 2.2. *Between compliance and divergence in the survey. For a color version of this figure, see www.iste.co.uk/demoly/4Dprinting1.zip*

In a world where technology is omnipresent and faced with the public's appetite, whether identified or not, these fictions can be rich in situations and their reading appears useful, even if the gaps between projected applications and current possibilities can be very large. Guéniot (2020) considers that

> on the one hand, the overproduction associated with the project-based mode of organization accelerates environmental catastrophe; on the other hand, the accumulation of institutionalized projects leads to a saturation of the imagination, and it becomes difficult to conceive other modes of action and production.

If she is right, then the proposed exercise has its own limits. This is what we will examine below by attempting to assess whether the proposals "obey a structure of investment in the future that consists of exchanging present goods for future goods, which are considered superior" (Guéniot 2020). Whether these risky proposals seemed to be means to a future, or tended to become an end in themselves, we were comfortable with all this open information. In short, "the expression of a desire, that of overcoming uncertainty" (Weil-Dubuc 2014) was expected.

2.2.1. *The survey*

> This rhetoric, omnipresent, intimates that we should never stop inventing, creating, imagining new horizons for the future. (Guéniot 2020)

This survey relayed by Factuel, one of the sites of the University of Lorraine, and carried out by one of last year's students (Azran Mostapha) from one of the Grandes Ecoles of Engineering (ENSIC), was presented in the following way.

2.2.1.1. *4D printing*

The Reactions and Process Engineering Laboratory (LRGP CNRS, University of Lorraine) is launching a survey on the perception of 4D printing and our vision of the future. Interview with Jean-Claude André, CNRS research director and Azran Mostapha, Malaysian student at ENSIC, who tell us more.

3D printing was invented in 1984 at the DCPR lab (which was integrated into the LRGP lab 10 years ago); this digital technology consists of the automated addition of material that transforms locally (e.g. from liquid to solid), hence its name "additive manufacturing". Since, at the time, France did not find the idea interesting, this first patent was put in the public domain. Industrial development was therefore essentially American during this period. But the Nancy site remained active with work, partly in common with the LRGP, IJL (Institut Jean Lamour) and LORIA (Laboratoire Lorrain de Recherche en Informatique et ses Applications) labs. The pioneering spirit is robustly maintained.

Factuel: *What is 4D printing?*

Jean-Claude André and Azran Mostapha:

> 3D printing makes it possible to create objects of very complex shapes with passive materials (polymers, ceramics, metals, alloys, etc.). The new idea (which came from the University of Bath in the UK in 2012) is to add functionality to materials (called active, stimulable, smart, intelligent), which can thus see their shape or functionality change under external stimulation (see Figure 2.3). It took about 25 years for additive manufacturing to develop (market of 30 billion €/year) and today we are at about 100 million €/year for 4D printing, but with a publication rate increasing by 40%/year. However, we are still far from the Terminator 2 movie, with global (and not localized) stimuli and materials that are too soft and response times that are too low, but it works and it is spectacular. What we are looking for today (with the same accomplices from the University of Lorraine, but also from the Belfort-Montbeliard University of Technology (UTBM), Sorbonne-University, Southern Brittany University, INSA-Lyon, Grenoble, Marseille, etc.) is to overcome the obstacles presented above. But there are (too) many of them.

Factuel: *What is the purpose of this survey?*

Jean-Claude André and Azran Mostapha:

> With such an international competition, even if we work in interdisciplinarity with partners from often disjointed disciplines,

either we do a little better than our international competitors, or we break down the barriers presented in the previous paragraph, and we win. This is what we want to do (and are doing). But, the potential application targets are (too) numerous. If we have some ideas on possible applications (which you will discover by answering the survey), we may miss some happy ideas.

Indeed, in such fields, we will not be able to cover everything from a scientific research point of view with a view to applications. That is why we are asking you to help us and see if we are right. According to our own ideas and the results of the survey, we will choose lines of action oriented toward the satisfaction of applicative needs (desires?). Several patents are already in the process of being filed, but it is still possible to orientate our work.

Factuel: *How will you format the results?*

Jean-Claude André and Azran Mostapha:

Our survey, carried out by Azran, a third-year student at the ENSIC, does not fall within the classical methods used by our friends in the Human and Social Sciences. It is open and can be used as an indicator for future actions for the University of Lorraine research units involved in 4D printing. First of all, the survey will allow us to identify ideas or potential ideas associated with applications. But, at the same time, it is possible to take into account social and/or socio-professional components to estimate whether the proposals depend on the social status, age, etc. of the participants in the survey. This is why you should not hesitate to distribute it widely.

The blank French survey is included in Appendix 1 (section 2.6) of this chapter (see https://qfreeaccountssjc1.az1.qualtrics.com/jfe/form/SV_1LWHO6ZiCrDpcRD or http://factuel.univ-lorraine.fr/node/16013). An English survey was also conducted (see https://qfreeaccountssjc1.az1.qualtrics.com/jfe/form/SV_1Ul2Geqlg06ZiQK).

2.2.2. *Items not transmitted*

The misfortune of being exploited by capitalists is nothing compared to the misfortune of not being exploited at all. (Galbraith 1980)

Considering the spatial deformation of a 4D objects under stimulation in this chapter, we are essentially concerned by the following embodiments:

– Homogeneous 4D printing: it corresponds effectively to 3D printing with an elastic material distributed in a homogeneous way. But during an energy stimulation, only the active voxels which are concerned by the stimuli take part in the transformation, the others behave like passive voxels (but voxels that will however take part in the global transformation).

– Hybrid 4D printing: the same logic applies as long as the materials have elastic properties; on the other hand, if one of the materials is not elastic and acts as a structural element, the effects of a stimulus must take this into account and have no clear relationship with the cases where all the materials have deformation possibilities. We would move from the mechanics of flexible materials to the mechanics used in robotics.

– Heterogeneous 4D printing: with different materials and separate fabrication techniques, one of which contributes to the realization of the 4D object from a 3D printing, it is possible to be in both situations: flexible system and rigid system (with joints).

Criterion	Homogeneous 4D printing	Hybrid 4D printing	Heterogeneous 4D printing
Manufacturing	One material and a 3D printer	Several materials and a 3D printer	Several materials and several machines including at least one 3D printer
Design	Simple for simple objects	Same as 4D Homogeneous except for the selection of materials and their distribution in space	Problem of the assemblies to be done
Stimulus	External	External	External or internal
4D deformation modeling	Difficult	Difficult	Possible (see work carried out in robotics)
Mechanical performance	Elastic material – Young's modulus of less than 10 MPa for organic materials	Same as Homogeneous 4D printing	Possibility of higher mechanical strength
Response time	Long (>1 minute)	Same as Homogeneous 4D printing	It can be short (seconds)
Reversibility	Yes	Yes	Yes
Fatigue	Possible fractures	Same as Homogeneous 4D printing	Best fit
Applications	POC	POC	Transformation of what currently exists by exploiting the performance of 3D printing

Table 2.1. *Quick and reductive comparison of the performance of different 4D printing strategies*

What the published results and modeling show is the existence of unaddressed problems or scientific barriers, which vary according to the form of manufacture. The summary of these difficulties is shown in Table 2.1.

In addition, Figure 2.3, taken in part from Kanu et al. (2019), highlights the set of subtopics to be mastered in order to optimize 4D printing.

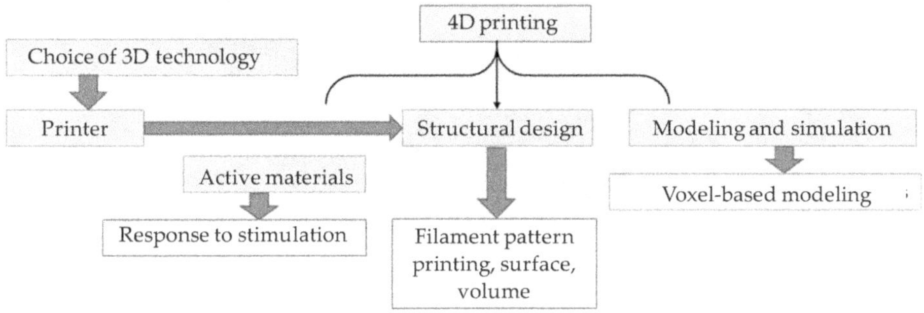

Figure 2.3. *Relevant subtopics in 4D printing to be mastered*

This table recalls the association between a 3D printing method, active materials and a stimulus to act on the form and/or the functionality of the created objects. Each contributor to the field brings their own constraints, leading in the end to the realization of devices whose operationality is questioned. For the authors, the aim is to understand the reality of the questions posed and the answers given by science and technology, so as to enable the fundamental debates to focus on the hierarchy of the arguments involved and, as far as possible, on the most relevant choices for enhancing the value of technology. The scientific or technical aspect of the observations is a matter for the sciences, epistemology and project management, but also for the human and social sciences, particularly economics and sociology.

Moreover, in Demoly and André (2021), from the reading of the titles (and many abstracts) of the various articles associated with the terms "4D printing" in the CNRS database, it was possible to realize Figure 2.4. Several keywords were selected, corresponding to: (1) organic materials and their stimuli; (2) inorganic materials and their stimuli; (3) review articles; (4) modeling; (5) miscellaneous; (6) applications; and (7) processes (including 4D origami).

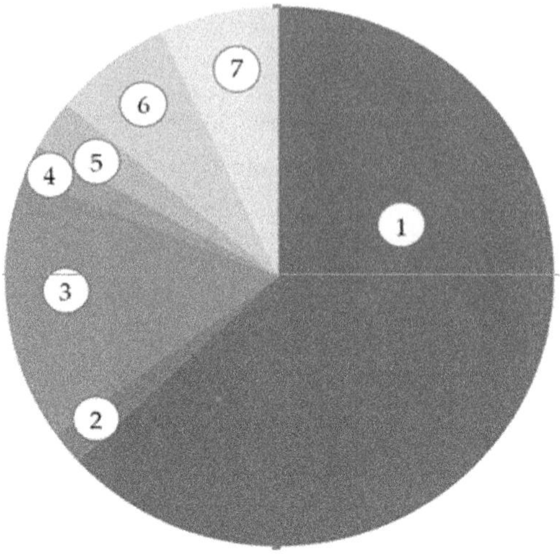

Figure 2.4. *Topics covered in publications on 4D printing (1. Organic materials and their stimuli: 63%; 2. Inorganic materials and their stimuli: 2%; 3. Review articles: 16%; 4. Modeling: 2%; 5. Miscellaneous: 3%; 6. Applications: 7%; 7. Processes (including 4D origami): 7%). For a color version of this figure, see www.iste.co.uk/demoly/4Dprinting1.zip*

COMMENT ON FIGURE 2.4.– *The classical applications mentioned (apart from bio-printing) deal with soft robotics, 4D shoes, sensor and drug delivery applications, mechanical engineering and optics.*

What is very important to underline is the very large unity of the works focused on the stimulation of organic materials that are pure (polymeric) and charged (in particular with nanoparticles and fibers (e.g. Liu and Moran 2021)), which correspond to 63% of the published activities. But if we remove 16% of review articles (for an emerging field where the number of publications is in the hundreds), this percentage becomes 79% of the total. And when we remove the miscellaneous documents and applications, this figure is even higher. We therefore find ourselves inside the cone defined in Figure 2.2, without any divergent proposals being expressed.

This observation leads to the preponderant involvement of a dominant scientific field constituting a body of work whose applicability of results remains modest and which can therefore raise questions (but which respects the definition of 4D printing combining active materials with 3D printers). Is this due to the fact that the coupling of 3D printing with programmable or stimulable material can only be achieved in this type of example? Or are we exploiting the old principle of the reverberator?

There are some works on inorganic materials that come close in terms of mechanical performance and response time to what can be required from systems (actuators) realized by classical ways. Yet, these represent only a few percent of the total. Is this way to be favored, or others, exploiting other paradigms (which will be discussed later), other visions (Heterogeneous 4D Printing, e.g., Jeong et al. (2019), Li et al. (2019), Zolfagharian et al. (2019) and Ding et al. (2018)) or physical and design properties of materials (Coulais et al. 2016; Hu et al. 2020)? Should not we foster the real creativity that seems to be absent in conservatism and perpetuated habits? Basically, is this main direction satisfactory?

However, we have shown in Chapter 1 that the number of potential applications is very large for 4D printing, which requires going beyond this focus. The survey, in the hopes of the promoters, aimed at the possibility for the respondents to use their imagination to propose ideas that may be outside of the box, not related to scientific knowledge. It was, for the authors, a means of creating a form of social demand.

> But as soon as the environment changes, these efforts are jeopardized. On the contrary, competition favors adaptability. Natural selection directly chooses the individuals best adapted to the current environment from the biodiversity of individuals present. As soon as the environment changes, it simply chooses others, kept in reserve. (Roddier 2012)

Frost and Sullivan conducted an analysis of 4D technologies in 2014 by thinking about the possible application areas (see Mangiola (2014), Carmel (2014) and Khoo et al. (2015)). These are presented in Figure 2.5 with a five-year view. This was "at the time" a considerable potential market that could still open up and that would likely have had to support an exploration on the side of academic research (because it would have been necessary to anticipate this research in order to lead to credible technological proposals). Let us note, however, that these promises have not really been fulfilled in 2021 (from 500 to 700 scientific publications for 4D printing against just over 170,000 for additive manufacturing). Though Ray Kurzwell (quoted by Jackson (2016)) wrote: "By 2020, there will be a whole range of products available immediately to buy for a few pennies and print immediately. It will become the norm for people to have printers in their homes," it is even more unlikely that this will be the case for 4D printing! And that is in 2021!

By approaching a vision of this kind, 4D printing would then have an interdisciplinary trajectory for the future, provided that it considers the aforementioned risks, goes beyond the current POCs (with various soft materials) based on the principles of transformation without external srimulation with conventional 3D printers, and meets the requirements of sustainable development policies (particularly in terms of disassembling, recyclability, reversibility, reconfigurability, reprogrammability, etc.).

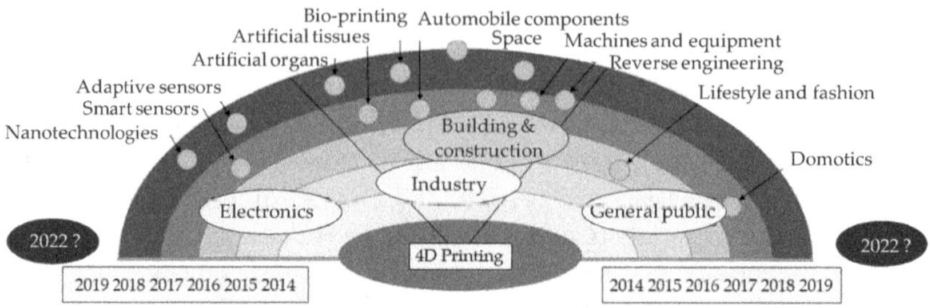

Figure 2.5. *Possible (potentially optimistic) anticipated applications of 4D printing methods. For a color version of this figure, see www.iste.co.uk/demoly/4Dprinting1.zip*

These are some of the questions that led to and justified this survey in an attempt to make current scientific knowledge "profitable" in relation to a potential demand that we could not have imagined. Indeed, on the basis of this demand, in a teleological approach, governed by the ends, it would have been easier for us, given the field of possibilities, to imagine the most robust possible ways of achieving it.

Box 2.1. *Future of 4D printing already considered*

2.2.3. *Some general survey results*

> It is one thing is to recognize that external circumstances are forces with which evolution must reckon, it is another thing to maintain that they are the directing causes of evolution. The latter thesis is that of mechanism. (Bergson 2019)

Several avenues were chosen to launch the survey; first of all, the traditional work links (University of Lorraine, UTBM, etc.) were selected, which represented a minimum of 5,000 people likely to respond. At the same time, social networks via the Internet were exploited, as well as the address lists of letter-writing relations, associations, the Lorraine Academy of Sciences, trade unions, etc. We imagined a relay via these links to broaden the base of the survey. For the scientific part, we called upon the e-mails of authors of scientific articles related to 4D printing (about 500 individuals), the French embassies abroad, etc. About 10,000 people were thus likely to respond with a modest return of about 300 responses (about 3%) for the French respondents, much more modest for our foreign colleagues.

A first remark is presented in Figure 2.6, regarding the distribution of respondents. According to the OECD (2020), about 14% of the French have only primary education, 59% secondary education and 14% higher education (of which less than 1% have a doctorate).

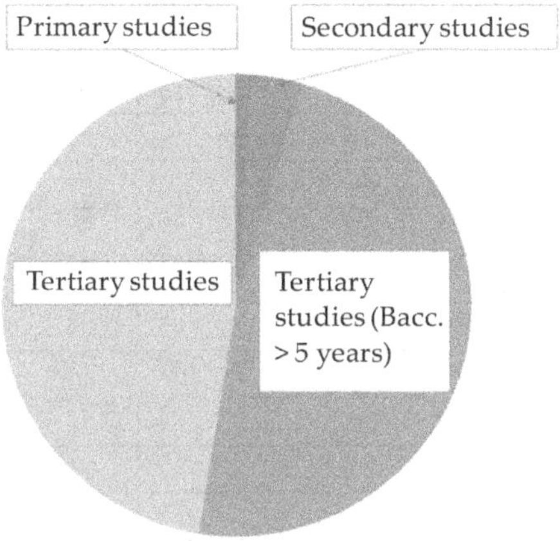

Figure 2.6. *Educational attainment of respondents (to compare with OECD data). For a color version of this figure, see www.iste.co.uk/demoly/4Dprinting1.zip*

In addition, to confirm the correlation in the data between respondents and the national educational structure, Figure 2.7 shows the respondents' fields of activity; they come largely from the educational and research world. Table 2.2, taken from INSEE (see: https://www.insee.fr/fr/statistiques/2381478), clearly indicates the existence of gaps between the occupations in the general population and those of the respondents.

Socio-professional categories in France (%)	
Farmers	0.8
Craftsmen, tradesmen, company managers	3.5
Artisans	1.7
Traders and similar	1.5
Managers of companies with 10 or more employees	0.3
Executives and higher intellectual professions	10.1
Liberal professions	1.1
Public service executives	1.0
Teachers, scientific professions	1.6
Information, arts and entertainment professions	0.7
Administrative and commercial managers of companies	2.9
Engineers and technical managers of companies	2.9

Intermediate occupations	13.6
School teachers, primary school teachers and similar staff	1.9
Intermediate occupations in health and social work	3.1
Clergy, religious	0.0
Intermediate administrative occupations in the public service	0.9
Intermediate administrative and commercial occupations in enterprises	4.1
Technicians	2.5
Foremen, supervisors	1.2
Employees	15.0
Civilian employees and public service officers	4.7
Police and military	1.0
Corporate administrative employees	2.9
Commercial employees	2.6
Direct services to individuals	3.8
Workers	11.4
Skilled industrial workers	2.1
Skilled craft workers	2.7
Drivers	1.4
Skilled workers in handling, storage and transport	0.9
Unskilled industrial workers	2.0
Unskilled workers of the artisanal type	1.6
Agricultural workers	0.6
Not in the labor force but having worked in the past	32.5
Former farmers	1.1
Former craftsmen, shopkeepers, company managers	2.4
Former executives, senior professionals	4.0
Former intermediate occupations	6.7
Former employees	10.2
Former workers	8.1
Persons who have never worked	12.9
Unemployed persons having never worked	0.7
Pupils or students	7.8
Various non-working persons under 60 years of age	2.2
Various non-working persons aged 60 years or over	2.2

Table 2.2. *Distribution of socio-professional categories in France*

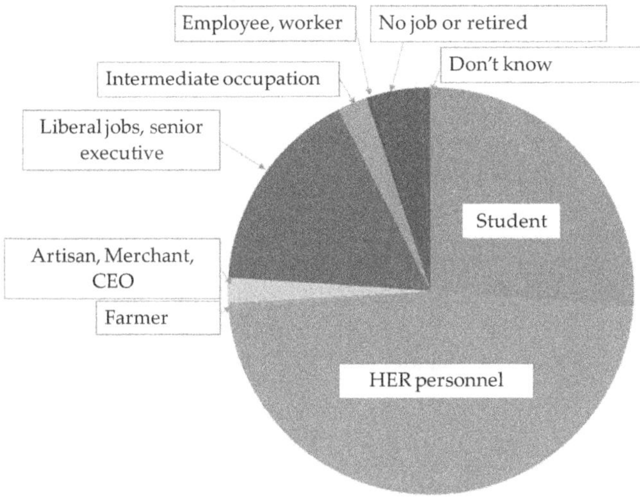

Figure 2.7. *Distribution of respondents by activity (HER for Higher Education and Research). For a color version of this figure, see www.iste.co.uk/demoly/4Dprinting1.zip*

Moreover, Figure 2.8 completes this assessment insofar as a majority of respondents consider that 4D printing could be useful to them.

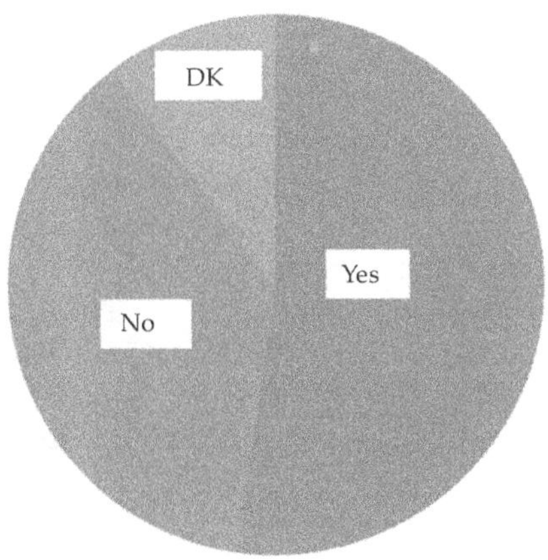

Figure 2.8. *Usefulness of 4D printing for respondents in the future (DK: Don't know). For a color version of this figure, see www.iste.co.uk/demoly/4Dprinting1.zip*

2.2.4. *Note: English language survey*

> A class of experts is inevitably so cut off from common interests that it becomes a class with private interests and private knowledge, which in social matters is no knowledge at all [...]. [It is] an oligarchy managed for the interests of the few. (Dewey 1954)

The e-mail addresses of scientists who have published in the 4D field in large numbers (about 500 individuals) were extracted from the publications made in the last three years. As mentioned above, the French embassies received the same request from Azran. However, the results are meager: 16 returns (including some from direct friends of Jean-Claude André). It may be interesting to reflect for a moment on this result.

First of all, the regimes of knowledge production in the sense of Pickstone (2001) have modified the forms of scientific discussion, the rationalization of administrative procedures, funding, the structuring of research, etc. This standardization, which is constantly progressing, constitutes a process of unification through standards and norms of the ways of producing new knowledge (Busch 2011), of procedures and forms of administration (Gardey 2008). All these tendencies go toward the unifying formatting of intellectual or scientific contents. For Ruano-Borbalan (2017), it is a convergence of individual behaviors and interactions limited by a standardized egoism. This unifying and formatted process translates into standardized training on a global scale, scientific frameworks and procedures for the use of instruments, exploitation of theories and paradigms, as highlighted in the work of Françoise Waquet (2016). On this basis, what will make the difference between two research works, thus two publications (or even more), is the small differentiating spark corresponding to the initial idea (as well as the consecutive research work). In this international competition, responding to the survey and/or providing insights into the field and its future is therefore possibly losing some of its leadership. This standardization of the professional behaviors of researchers in a specific field is the condition for the possibility of counterpoint discourse around the promotion of creativity and individual identity. In fact, the French students and scientists outside the scope of the survey responded to it (in admittedly modest numbers) by "playing the game" because they felt little direct concern for the subject.

> And why, then, should I love my neighbor or your future humanity, which I will never see, which will not know me and which, in turn, will disappear, leaving neither trace nor memory. (Dostoyevsky 1935)

To move forward in understanding non-responses, it may be useful to return to the "theory" of rational choice corresponding to approaches based on the "principle of rationality" that Popper (1957) stated as follows: "Agents always act in a manner appropriate to the situation in which they find themselves" (see also Bouveresse (1978)). But, "To say of individuals who behave in an apparently appropriate manner that they are seeking to maximize something is a normative, not a descriptive, view" (Moessinger 1996). Rational choice theory assumes that:

– choosing is central to social life;

– personal choices do not interact from person to person;

– all choices are based on self-interest. They are guided by a principle of utility maximization.

The maximization calculation takes into account the uncertainty linked to the near future that we anticipate in our own way (competitive spirit). In order for there to be sharing between scientists concerned with the same subject, it is therefore necessary to ask the question of what the contributor gains or fears to lose relative to the receiver who launched the investigation. The Pareto principle does not tell us how to find the right approach to sharing, but its optimum, disconnected from any ethics or morality, is simply based on the very general value that it is better to have more (for ourselves).

The 80-20 Principle or 80-20 Law is an empirical approach that has been observed in some fields: about 80% of the effects are the product of 20% of the causes.

Box 2.2. *Pareto's Principle*

Strictly speaking, such a principle is amoral (Moessinger 1996), but this is what is probably observed in the absence of responses or disruptive ideas produced on the part of our French and foreign colleagues working on the theme of 4D printing. There is no prior contractual commitment between scientists working on the same subject, the competition imposes a constraint that protects the researcher against an anticipated change in their preferences (and their independence). When each party, be it the requester of distant and low-risk information or the respondent to the survey, anticipates the altruism of the other, behaviors and representations may fluctuate, and sharing could become accepted, unpredictable or repulsed.

So, perhaps we need to return to the writings of "*Self-Reliance*", published in 1841 by Emerson (2001) who spoke of intuition, characterized by resistance to a constraint imposed from outside and which highlights creativity, "capable of questioning at any moment the traditional delimitation of knowledge as well as the forms of surveillance and control that derive their legitimacy from it" (Duplay et al. 2020).

Emerson writes: "A man should learn to detect and watch that gleam of light which flashes across his mind from within, more than the luster of the firmament of bards and sages" (Emerson 2001). If the intuition that compels the scientist in their international competition forces them to function by themselves and for themselves, relying on a freedom of thought and action, they must quickly step out of their freedom of choice to comply with the normative rules presented above. Emerson, again wrote: "We lie in the lap of immense intelligence, which makes us receivers of its truth and organs of its activity. When we discern justice, when we discern truth, we do nothing of ourselves, but allow a passage to its beams".

In short, a very long development to try to explain politely that empathy (much like humor) is the most widely distributed thing in the world!

> When we speak of the higher orders, we should always say intellectually superior – But who, before the aspect of the earth, covered with the most beautiful savannahs and forests, would dare to say that intellectuality "is the only goal in this world"? (Darwin 2020)

2.3. Results of the survey

> A continuous process of creation, an incessant invention of forms constantly meeting changing needs. (Lippmann 1913)

The Eiffel Tower, the Palace of Optics, velocipedes, electric lights, moving walkways, photography, luminous fish, scientific toys, vaccines, fairy tales, women with beards or three heads and learned seals.... This is what was on offer at the more-or-less universal exhibitions of an era when everything seemed possible, at the beginning of the previous century. There were also a number of newspapers talking about science, such as "*La Science illustrée*", specialized weeklies from the 1880s to the 1900s that promoted a positive, entertaining and spectacular popularization. It was also the time of Jules Vernes who, in his adventure novels, supported (exploited) the taste for scientific wonder that other novels of anticipation would bring to a climax.

These publications supported an open field where science brought technological well-being to the inhabitants of the planet. The newspapers of the time, by providing links between dreams and reality, created a marvelous imagination where creativity was in full play. This science (and associated technology) is associated with the power of nations. If we can speak for the newspapers of popular science, which had difficulty in freeing themselves from their image as "traffickers of science" (according to Auguste Comte) to produce a popular science capable of covering all of the scientific fields with enthusiasm, and involve the readers in the same attitude where tomorrow would be better than yesterday because imagination was in power. Then science was everywhere, it could do anything, as demonstrated, for example, at the great universal exhibitions of 1899 and 1900.

Today, the situation is different because of omnipresent information that no longer makes people think, to the extent that they confuse reality with "anything goes". Science-fiction movies made us dream, then we reached the age of "Fake-News" where the real is cleverly mixed with the fake and we see that the separating power of people tends to disintegrate. One guide that can give us some control over this again is indeed to play together, as part of a collective work (Stiegler 2019): "Control, whether you like it or not, is an immense collaboration" (Lippmann 2015). Normally, in this space linked to this survey where we can freely say everything and its opposite, from the respondents, we hope for a liberated creativity full of original proposals (apart from the refusals to respond from French and foreign colleagues for the reasons mentioned above). The open survey as constructed must exploit mental or cognitive strategies, very quickly and in theory very effectively, while freely holding as true what is considered credible. The time of the investigation, which was deliberate, was to prevent the inhibition of these elements of belief by the introduction of more logic and rationality, to make room for unconstrained creativity (Houdé 2019). This is what we discuss below with a willingness to share and naturally a desire for cooperation.

2.3.1. *Specific ideas and proposals (open questions)*

> The organism is not a spectator facing the screen of the world and being bombarded with raw data, it is an actor who takes part in the world. (Madelrieux 2016)

Table 2.3 brings together the basis of the proposals made by respondents to the French survey. This table has been constituted as follows:

– definition by the authors of open themes corresponding to possible fields of application in 4D printing;

– launch of the survey;

– receipt of results;

– analysis of 20 scientific articles expressing the keywords "4D printing" and "applications" in their title (they are referenced in section 2.8 at the end of this chapter).

What this table shows is the great general unity between the themes proposed under these three headings. The proposal of breakthrough ideas (even unfeasible ones) did not occur.

Title	Comment	Feasible?	References	Survey?	Answers?
Deployable elements	Space applications involving the use of folds (origami)	Yes	Mitchell et al. (2018)	Yes	Yes
Deployable habitat	Ditto (kirigami)	Yes	Mitchell et al. (2018); Singholi and Sharma (2020)	Yes	Yes
Interconnection	Bringing together of two elements under stress	Yes	Mitchell et al. (2018)		
Flexible robotics; exoskeleton	Large subject with modest mechanical possibilities	Yes/No	Mitchell el al. (2018); Hann et al. (2020); Zolfagharian et al. (2020); Ploszajski et al. (2019); Singholi and Sharma (2020)	Yes	Yes
Actuators	Likewise, sensitivity to pH, temperature, humidity, magnetic field, piezoelectric – μ-fluidic; chemistry	Yes/No	Mitchell et al. (2018); Rayate and Jain (2018); Grinberg et al. (2019); Ryan et al. (2021); Chu et al. (2020)	Yes	Yes
Home automation		Yes/No		Yes	Yes
Means of transport	Self-adaptation (mechanical problems)	No	Netouanoglou et al. (2018); Pei and Loh (2018); Xin et al. (2019)	Yes	Yes
Self-repair	Not a 4D subject but need a material that can be repaired	Irrelevant	Netouanoglou et al. (2018); Ryan et al. (2021); Pei and Loh (2018)	No	Yes

Adaptive seals	See actuators	Yes	Rayate and Jain (2018)	No	No
Implants	Shape change and biocompatibility	Yes	Javaid and Haleem (2019); Melly et al. (2020); Melocchi et al. (2019)	Yes	Yes
Intelligent textiles; packaging	Color change; waterproofing	Yes	Shie et al. (2019); Xin et al. (2019); Headrick (2015)	Yes	Yes
Applications in dentistry	?	?	Khorsandi et al. (2020)		
Dispensing of drugs	Localized delivery by 4D opening of a carrier	Yes	Melly et al. (2020); Melocchi et al. (2019); Singholi and Sharma (2020)	No	Yes
Self-organizing system	Getting closer to what the living do	No	Pei and Loh (2018)		
Adaptive optics	Photochrome type system	Yes		Yes	Yes
Gadgets		Yes/No?		Yes	Yes

Table 2.3. *Comparison between proposed themes and survey results*

Table 2.4 presents the specific quotations of note made by the French respondents.

A large part of the proposals actually concern applications of active materials without strictly answering 4D questions. Some ideas are already being developed (e.g. the 4D tire from Michelin, General Motors (Carnorama 2014; Hankook 2019)). Furthermore, no truly disruptive or even revolutionary ideas emerged from the survey. However, many proposals reinforce the (non-disruptive) ideas that we are supporting, in particular on the medical and industrial side. On this subject, we could reflect on the use of this sentence by Bayard (2007), concerning a culture that conditions minds; he writes: “Paralyzed by the respect due to texts and the prohibition of modifying them, constrained […] to know what they ‘contain’, too many students lose their inner capacity to escape, and forbid themselves to call upon their imagination”. And, by continuity, this same phenomenon occurs when they move into academia!

Theme	Examples of ideas	Current opportunities
Daily life	Encapsulation of food and active principles (health) – encapsulation of harmful products (waste); waterproofing of electronic equipment in contact with a liquid (very fast response time required); protective clothing for "high-risk" sports such as modified airbags: reaction to a shock causing significant deformation (e.g. swelling) to protect the user; music: modification of the timbre or tone of the acoustic spectrum of a given instrument to broaden its range or even modify the instrument; self-adapting watch strap and self-adapting anti-Covid-19 mask; adaptive mattresses and pillows (but how?); clothing that releases drugs; safety for electrical outlets	Yes; No; Yes; ?; ?; No; Yes; ?
Agriculture	Self-adapting greenhouses (temperature and humidity dependent)	?
Instrumentation	Adaptive microscopes (how?)	No
Habitat	Roofs of houses that change color according to the climate; earthquake-proof constructions; transformable objects (e.g. for storage); stairs that turn into a ramp for the disabled	?; ?; ?; ?
Art	Sculptures that change shape	Yes
Medicine	Adaptive medical beds; drug delivery	?; Yes
Industry	Membrane pollution control; servo-controls (compensation of expansion, pressure variations, etc.) and actuators/sensors; frugal production; micro-actuators	Yes; Yes; ?; Yes
Transportation	Adaptive tires according to road conditions	?

Table 2.4. *Specific notable quotes issued*

The following figure illustrates the distribution of proposed responses in terms of feasibility based on current scientific knowledge.

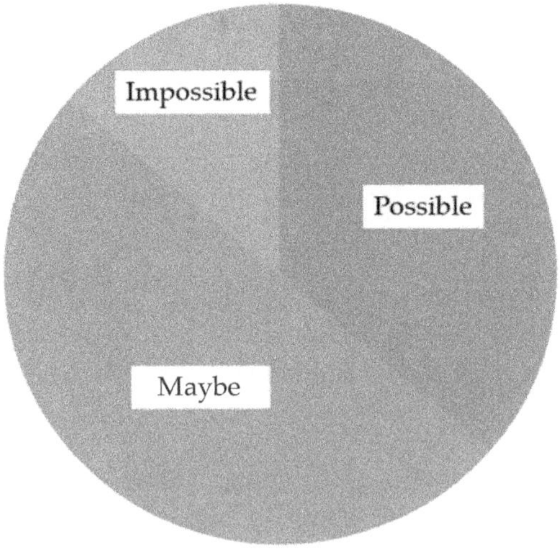

Figure 2.9. *Distribution of spontaneous responses according to their current feasibility. For a color version of this figure, see www.iste.co.uk/demoly/4Dprinting1.zip*

This figure shows, on the one hand, the small proportion of ideas that are strictly unfeasible (which is a pity, since spontaneous forms of creativity were sought), and on the other hand, the possibilities of opening up to relevant and unexpected subjects. For the most part (50%), the proposals fall strictly outside the 4D framework or cannot be applied in the short term. Finally, the respondents, locked in a fear of breaking the mould, are reasonable in their proposals.

Other comments

Other elements were provided by the respondents and are shown in Table 2.5.

Targets	Comments
Stimulation	Use of multiple stimulation modes
Frugality	Developing frugal processes with low energy consumption
Self-repair	Consider self-healing materials
Ordering	Possibility to control deformations by voice

Table 2.5. *Other comments*

2.3.2. *Presentation and analysis of the quantified results of the survey*

> For as soon as we reflect on an act that has been undertaken with the conviction that we are a free agent, it seems to fall under the sway of two kinds of causality, on the one hand the causality of internal motivation, and on the other the causal principle that governs the world. (Arendt 1972)

The questions asked in this survey were mainly addressed to educated laymen, who, although mostly scientists, answered by appealing to common sense, in a language adapted to everyone's understanding. In a very uninformed and simplified view of this type of survey, it is unrealistic to expect answers that could convince a specialist. Arendt (1972), already quoted, writes: "The answer is unlikely to convince the scientist because he has been forced by facts and experience to give up sense perception and hence the common sense by which we coordinate the perception of our five senses into a total awareness of reality".

2.3.2.1. *Priority axes*

Two sets of questions are developed in this paragraph; they correspond to the vision that respondents have of the use of 4D printing today and in the future. The answers to all the questions are presented in Appendix 2 (section 2.7).

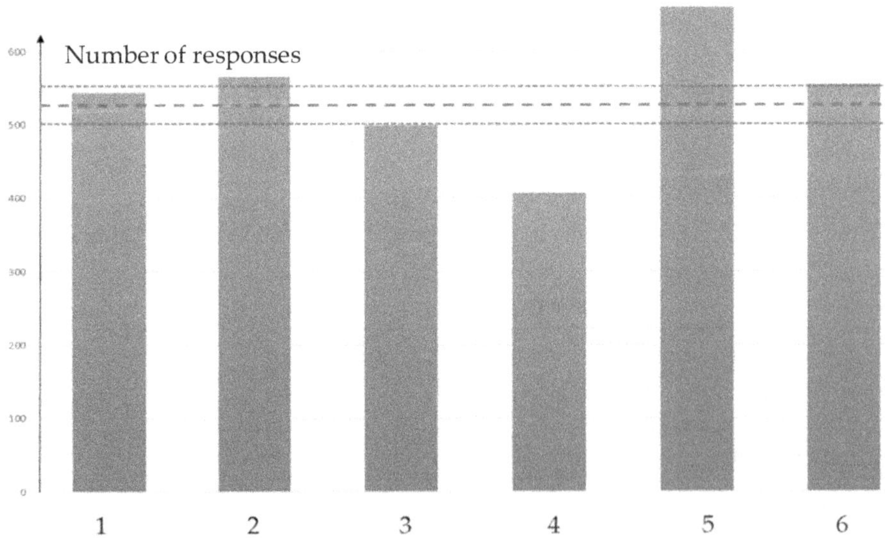

Figure 2.10. *Sum of responses concerning current uses (1: not interested with a gradation up to very interested (5) – Don't know: 6)*

Generally speaking, the questions asked about examples are almost all favored by the respondents, with no clear difference between "not interesting" and "very interesting" opinions. This relatively homogeneous response pattern is also found in the choice of topics. Figure 2.10 shows the sum of the answers given for all proposals combined. If the choices between interested and very interested were combined, all the results would fall between the mean and plus or minus the standard deviation.

On the other hand, the elements of Figure 2.11, which corresponds to the same questions for future applications, show an effective confidence in technological activities, with a continuous growth between the "low interest" and "high interest" domains with variations which go beyond the mean increased or decreased by the standard deviation.

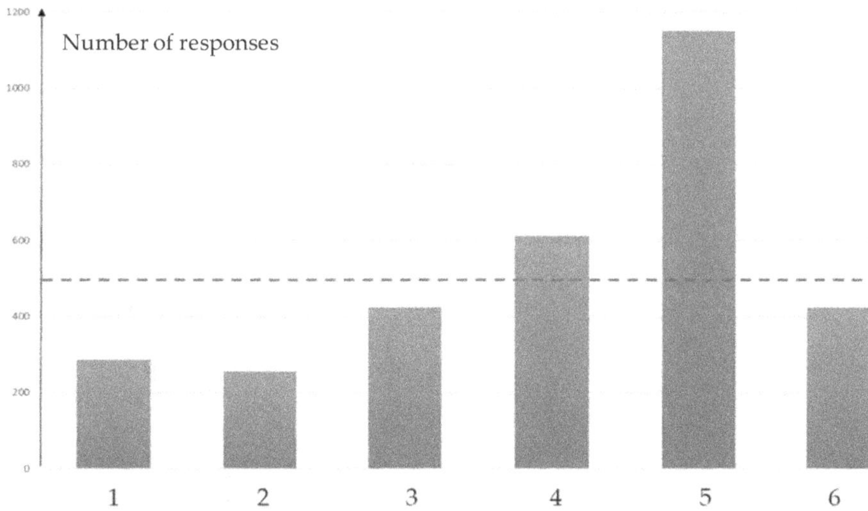

Figure 2.11. *Sum of responses concerning future uses (1: not interested with a gradation up to very interested (5) – Don't know: 6)*

If we try to go a little further, on the areas that seem to be priorities to be explored, Figures 2.12–2.16 present the results in the form of histograms. By category (1 to 5), it is possible to make a comparison between present and future.

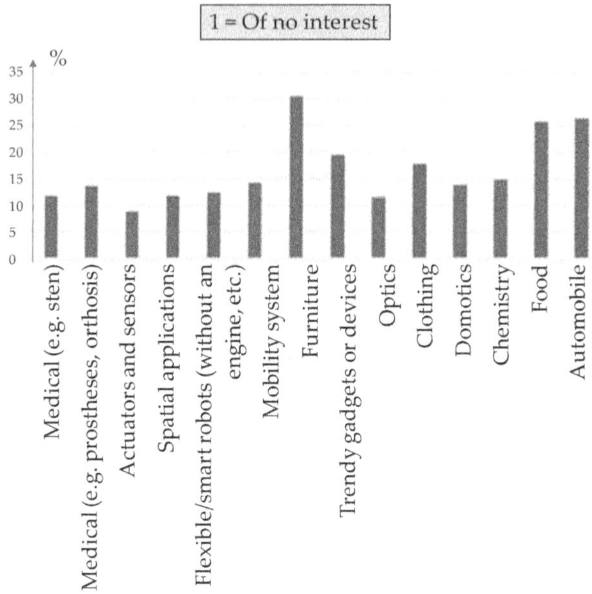

a)

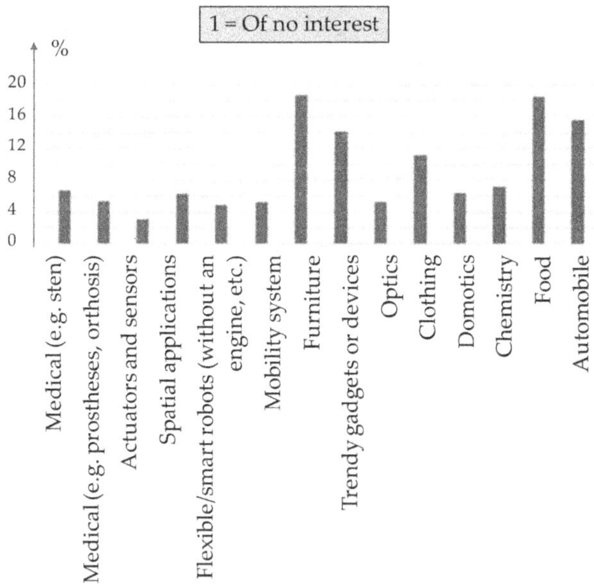

b)

Figure 2.12. *Responses by category (1: not interested – A: now; B: future). For a color version of this figure, see www.iste.co.uk/demoly/4Dprinting1.zip*

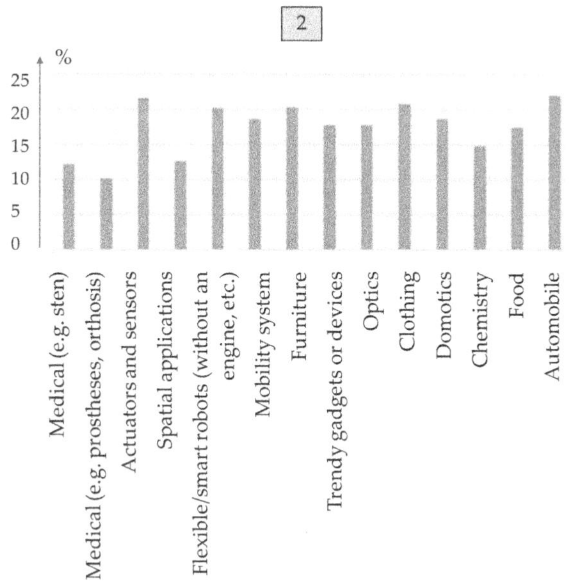

a)

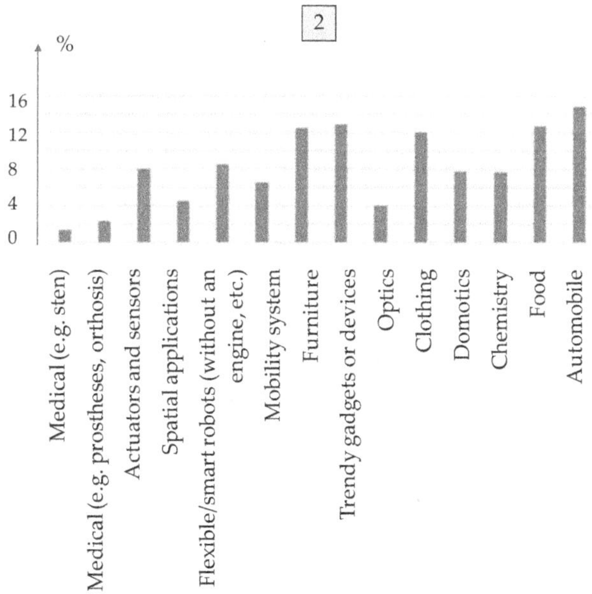

b)

Figure 2.13. *Responses by category (2: not interested – A: now; B: future). For a color version of this figure, see www.iste.co.uk/demoly/4Dprinting1.zip*

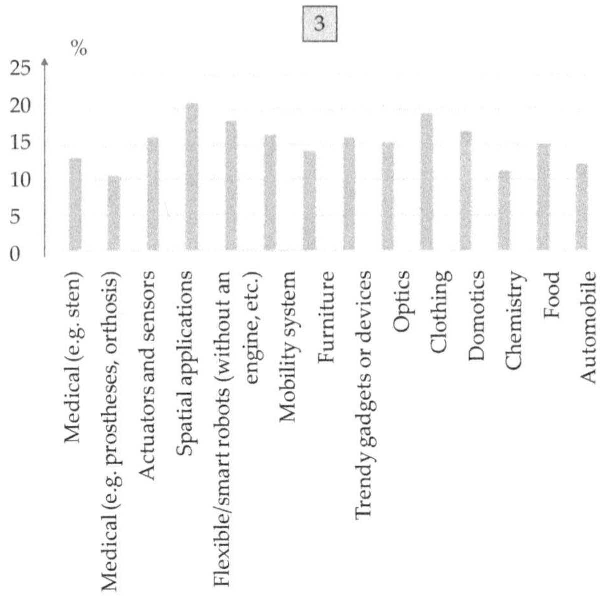

a)

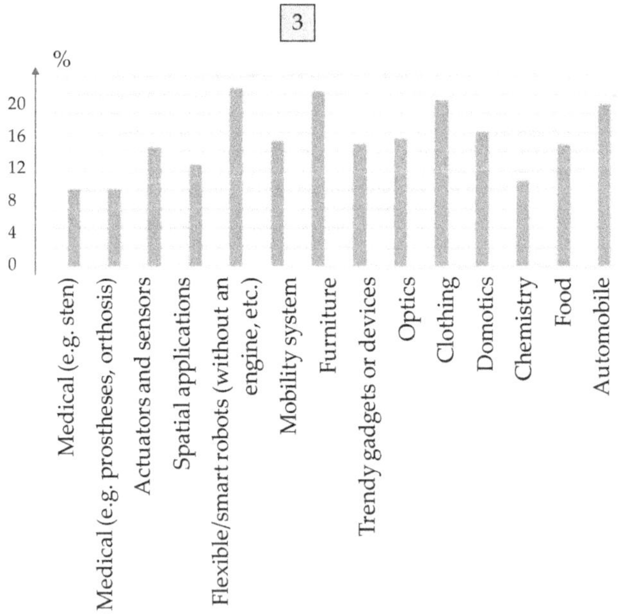

b)

Figure 2.14. *Responses by category (3: moderately interested – A: now; B: future). For a color version of this figure, see www.iste.co.uk/demoly/4Dprinting1.zip*

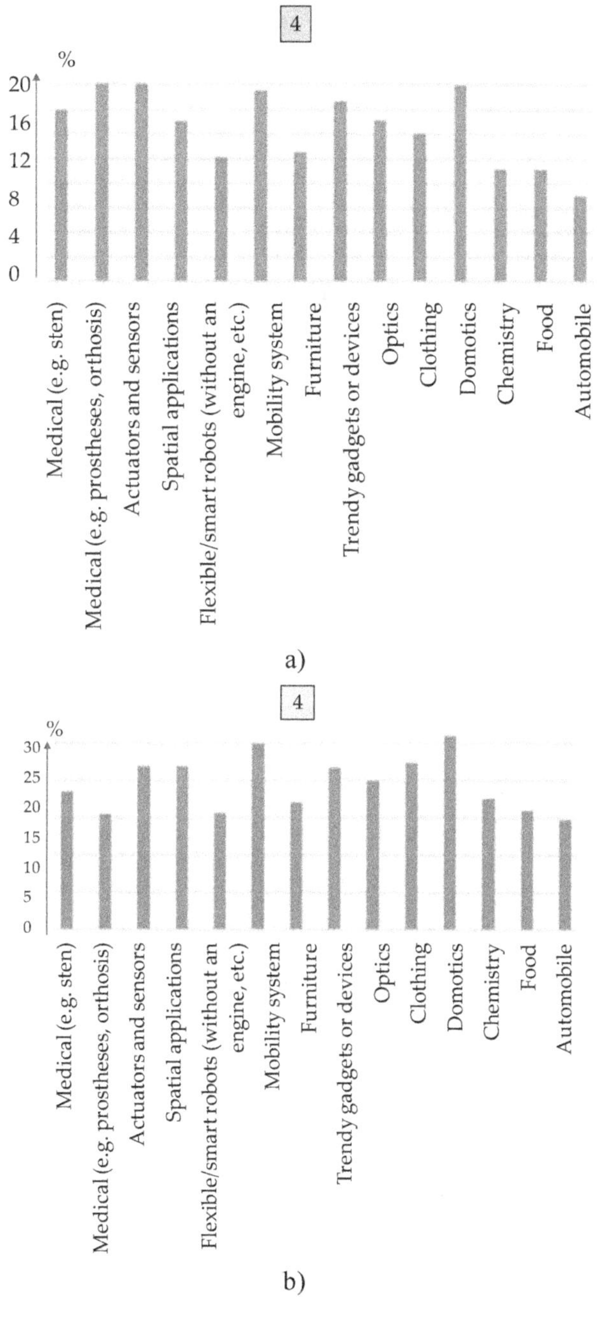

Figure 2.15. *Responses by category (4: interested – A: now; B: future). For a color version of this figure, see www.iste.co.uk/demoly/4Dprinting1.zip*

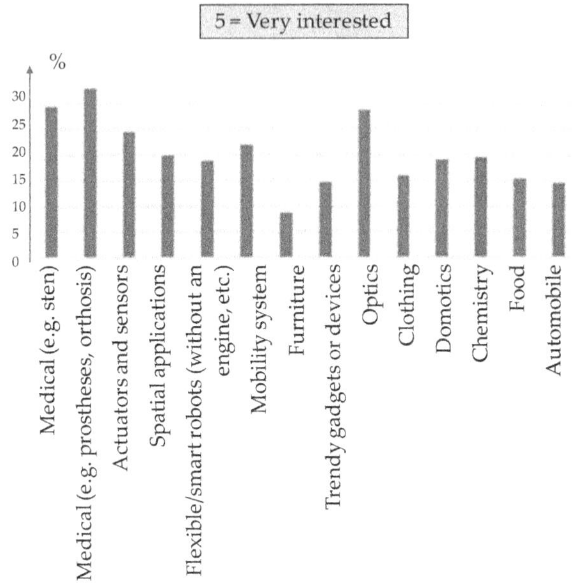

a)

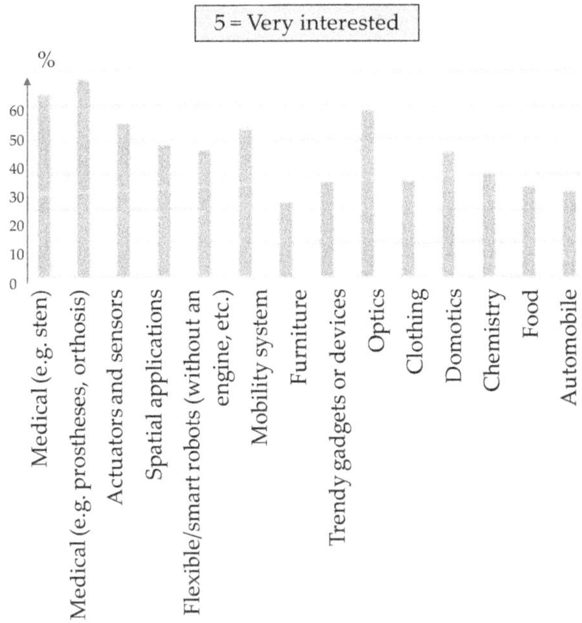

b)

Figure 2.16. *Responses by category (5: very interested – A: now; B: future). For a color version of this figure, see www.iste.co.uk/demoly/4Dprinting1.zip*

Reading these different graphs, we get the feeling that all the domains have roughly the same importance. Some fields that are beginning to be invested scientifically are emerging, corresponding to medical and optical applications, while soft robotics is not in the top three priorities. This situation of non-choice, already noted above, is a reality, but does not lead to a preference for one direction rather than another. Thus, there is no appearance of "revolutionary" ideas or preferred directions; just confidence in the future.

> Any model is […] an approximation whose relevance also depends on the person who develops it. The model is only a large hypothesis […]. It is based on choices, conscious and unconscious, and in each of these choices lie our own perceptions that result from who we are, our knowledge, our scientific profiles and so many other things. (Jean 2019)

2.3.2.2. *Application specificities*

Figure 2.17 (also in Appendix 2, section 2.7) shows some new elements; in particular, high-quality functionality is privileged in relation to other properties such as ergonomics and especially aesthetics.

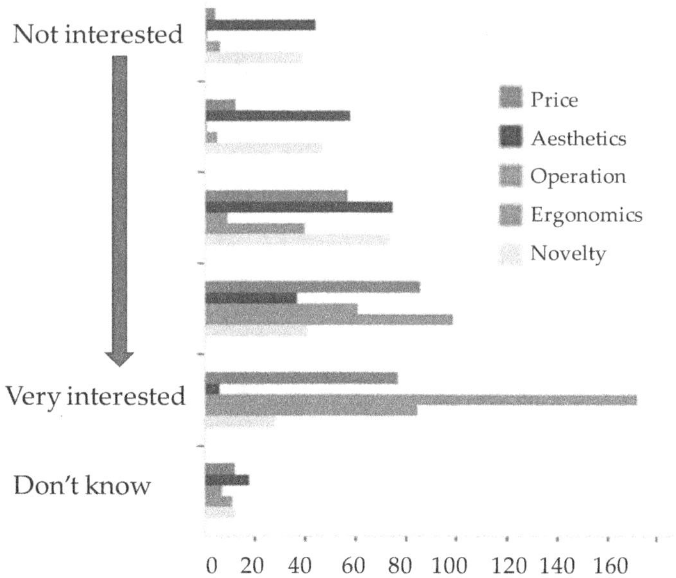

Figure 2.17. *Specificity of choice. For a color version of this figure, see www.iste.co.uk/demoly/4Dprinting1.zip*

2.3 2.3. *Comparison of results between generations*

> The consequence of this is that, at present, the latter [principles] deviate very appreciably from the intuitive ideas one may spontaneously have. Such difficulties undoubtedly explain that a compartmentalization – which is again neither in the mind nor in things – nevertheless conditions our thinking activities. (D'Espagnat 2015)

Since the respondents were from roughly the same intellectual background, it was difficult for us to see if any salient facts could be drawn from cross-comparisons. The only criterion that could be used was age. Figure 2.18 shows the data provided by the respondents.

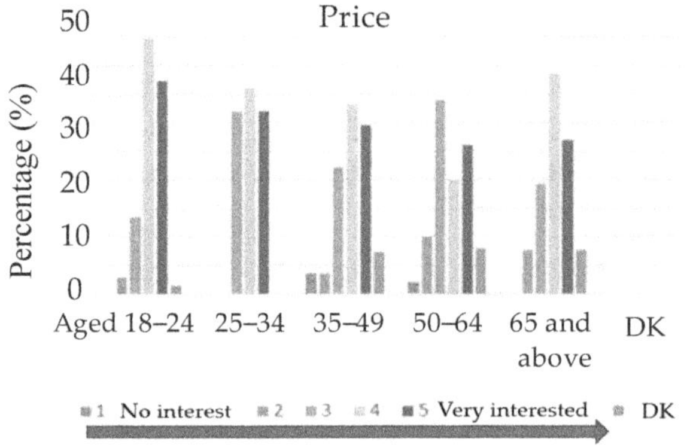

a)

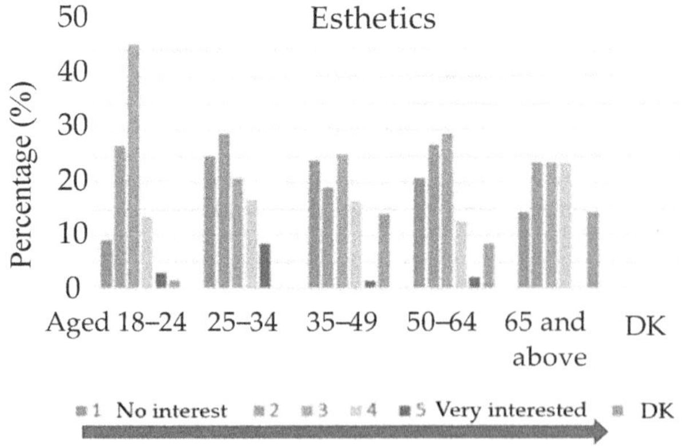

b)

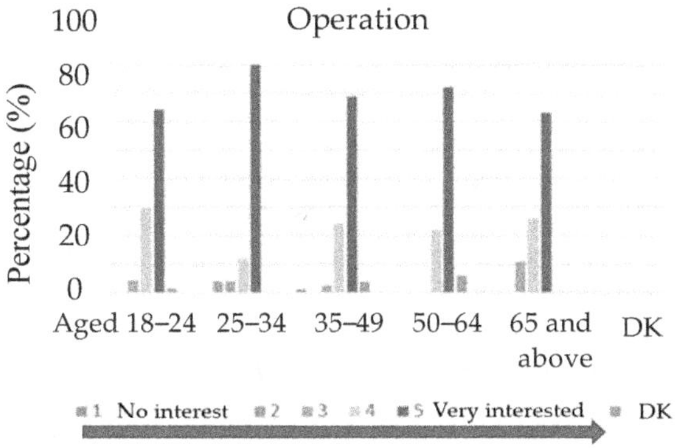

c)

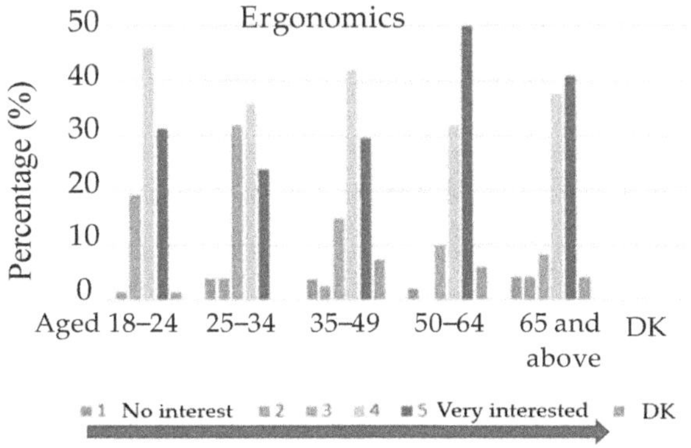

d)

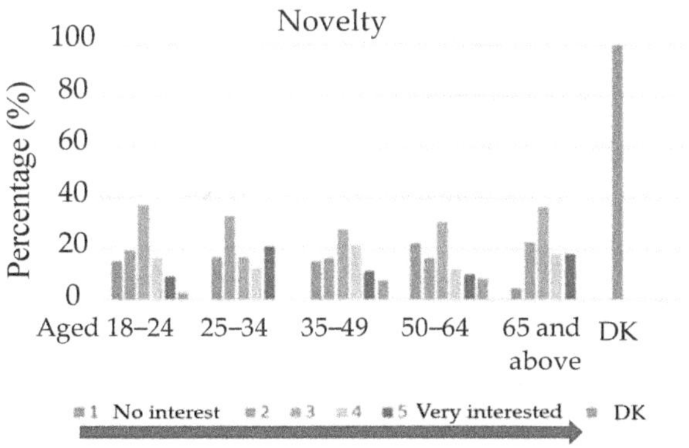

e)

Figure 2.18. *Effect of age on responses (A: Price; B: Esthetics; C: Function; D: Ergonomics; E: Novelty). For a color version of this figure, see www.iste.co.uk/demoly/4Dprinting1.zip*

Regarding price, the answers are reasonable with wishes not to pay too much for the innovation, whatever the age of the respondents. Aesthetics seem to be more important for young people, but functionality is favored by all age groups. On the other hand, ergonomic aspects seem to be more important for seniors, and there is no clear intergenerational difference with regard to novelty. In short, we find ourselves in the position of having to make compromises in order to design a commercial device desirable enough to be the object of a purchase, either reasoned or for pleasure.

2.3.2.4. *What kind of stimuli should be used?*

> Man has in his hands a fragment of reality. More or less consciously, abused or fooled, he assimilates this fragment of the observed object to its totality. He then engages in actions that are destined to fail from the start. He speaks of "Cartesian myopia", an incurable disease of the researcher of the third millennium. (Algoud 2014)

Figure 2.19 shows the "very favorable" responses and does not allow us to conclude that there is a clear preference for a particular direction.

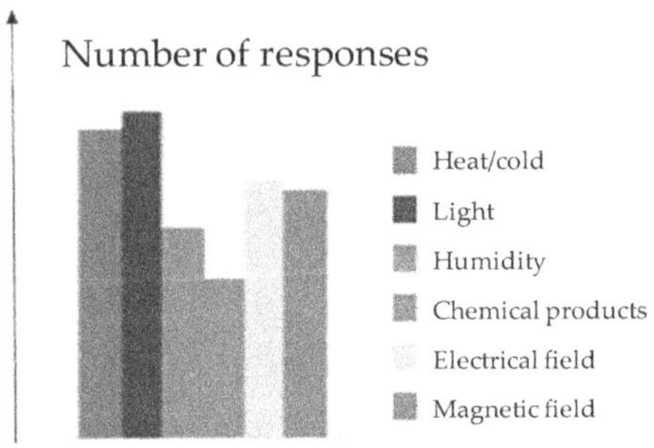

Figure 2.19. *Responses (very favorable) regarding preferred stimulation methods. For a color version of this figure, see www.iste.co.uk/demoly/4Dprinting1.zip*

2.4. Discussion

> Anticipation is not merely forecasting; it is much more than predicting what the next event will be. It is more than dealing with the next event, it is creating it. (Follett 1949)

Our era is that of "the great transformation" (Polanyi 2009) because we have to rethink our relationship with a world experiencing heavy tendencies that are not very optimistic: resource depletion, population growth, migration flows, forced individualization, global warming, etc. (André 2019). This situation is reflected in two antagonistic phenomena, one related to "business as usual" and the other to the reinforcement of a tyranny of urgency (Laïdi 1999). For several decades we have been approaching a fourth era of humanity, "caught between the omnipotence of technologies and the finitude of our natural resources, which make us both powerful and vulnerable" (Guilhot 2018). Current frameworks where several worldviews are mixed, between technology capable of solving the problems it has created and a more integrated ecological retreat, force a change of temporal horizons. In this new unstable and unpredictable world, no one is safe from a change in the rules of the game. "Yesterday's solutions make today's bankruptcies and Business Models have become [...] temporary" (Cheminade 2018). Technological developments, increasingly rapid since the digital revolution, are disrupting the whole of society (Stiegler 2016). The urgency is there, forcing humans to change their conduct, or at least their conduct in innovation processes.

2.4.1. *Non-response (voluntary)*

> I contend that the sole purpose of science is to relieve the pains of human existence. When Men of Science, intimidated by selfish Men of Power, are content to amass knowledge for knowledge's sake, science may be mutilated, and your new machines may only mean new torments. You may discover in time all that can be discovered, but your progress, however, will only be a progression that will take you away from humanity. The gulf between you and humanity may one day become so great that your cry of joy at some new conquest that may be answered by a universal cry of horror. (B. Brecht cited in Gras 2003)

Creativity, which is often put forward with its many clichés or Epinal images, is a mandatory starting point for innovation, in particular, that resulting from research exploring teleological approaches, such as 4D printing, because it forces us to think differently in increasingly complex and uncertain systems. This situation, which has become urgent, without being very new in substance, induces a new phenomenon in research structures, particularly academic ones, that of taking the risk of failure (but in the event of success, the one who arrives first wins everything). However, it is clear that researchers are recruited on the basis of their ability to go deeper (with very low selection rates; their scientific competence is therefore certainly not brought in to question), with financial support defined (to put it simply) by personalities who have shown their relevance in the old world, whether at the national or European level. Incrementality therefore reigns. While there is no question of eliminating these access routes to forms of progress, the authors believe that it is necessary to better support these "obstacles to thinking in circles", that is creative people, who in any case will need, in order to bring their idea to a "saleable" conclusion, affirmed specialists, those who today explore continuity and the serious exploration of scientific paradigms.

> Only that which lasts through the centuries can finally claim to be a cultural object. (Arendt 1972)

Naively, with this survey, a form of remote interaction, we thought we had found a way of exchanging with our colleagues, as well as reaching an audience that was certainly educated but nonetheless more profane. In this open-ended approach, we felt that the survey was not an obstacle to the development of our own skills, nor did it erode our sense of belonging to our research team. Perhaps the health crisis and the gloom of the likely economic outlook had a role to play in the non-responses because, on a global scale, in the spirit of heightened competition, the keystone of the work of free exchanges, even at a distance, does not rest on a global managerial transition, favoring progress for the inhabitants of the planet.

Clearly, "we are in a society that liberates, as no other has done, all the forces, the potentialities that lie dormant in the individual, but also that pushes each individual to disassociate himself from others while using them" (Godelier 1996). This apparent freedom, subject to the possible understanding of forms of domination considered as charismatic and psychological, is dominated, according to Ramonet (2004), by subliminal forms of unconstrained seduction, by a certain thirst for pleasure. He takes up the writings of Beigberger (2000) who reminds us that "we live in the first system of domination of Man against which freedom itself is powerless. On the contrary, it bets everything on freedom, that is its great discovery [...]. The system has achieved its goal: even disobedience has become a form of obedience".... We leave Orwell's 1984 to enter Burgess' 1985... Thus, as Neyrinck (1990) reminds us, in the context of growing complexity, control or manipulation and certain associated forms of distrust of technical progress, "the elaboration of new norms appears to be a precarious undertaking, afflicted by considerable uncertainties". The same author mentions the risk of mistakes; he also writes that "any blunder has such far-reaching consequences that it is always cheaper, in financial and political terms, to prevent than to explain and apologize after the fact". For the researcher engaged in the same activity, the evaluation of choices depends on the relationship between means and ends or self-interest, which must be "efficient" and experienced positively (but rather with an a posteriori judgment). Under these conditions, "the moralization of everything translates [...] into the assumption of responsibility for the moral dimension of damage: not only must compensation and material reparation be made, but moral reparation must be made" (Michaud 2002). Bourrelly (2007) proposes several methods of manipulating opinion that rely on the passive attitude of the world concerned; thus, among his proposals we should mention:

- diversion strategy (diversion of attention by lateralization to irrelevant information);
- artificial creation of problems to impose pre-prepared solutions;
- application of an incremental strategy (slowly, if possible insensitively, engaging in irreversible processes with application of the "frog" paradigm);
- temporal lateralization of problems (agreement in the present for application in the future);

– exploitation of infantilizing speeches, of a dialectic on the impossibility of explaining and encouragement of the social body concerned to engage in "mediocre" activities;

– replacing the will to revolt with guilt;

– and so on.

Box 2.3. *False freedom?*

> With the advent of modernity, the status of the individual changes radically. He is recognized for himself, independently of his inclusion in the collectives. (Castel 2003)

> Confidence is inseparable from the perception of a potential future and a perspective of expansion. (Damien 2006)

In this complex space, we must now try to compare positions on the same phenomenon and work out, from its description, what is relative to the point of view (and therefore dependent on social understanding, which is by nature evolving) and what is absolute (and therefore invariant over time). "In fact, what makes it difficult [...] is actually the awareness of the very existence of different points of view, and the acceptance of their equal dignity" (Levy-Leblond 1996). But how can we avoid rigid positions of the mind by relying on previous modes of reasoning that have worked, modes that have become unsuited to new situations defined in other fields of technology? "The danger of belief in its dogmatic drift is to think once in order to not have to think again" (Perrin and Weiss 2003).

In the struggle for an international position, science plays an increasingly important role vis-à-vis other countries, but also internally (this can be observed singularly today with the crisis associated with Covid-19, where the French media and citizens consider it unacceptable that France has not produced its own vaccine). It is thus a means of educating and accessing a social role, with scientists (still) belonging to a kind of intellectual elite. The increasingly unfounded image of elevated spirits is still imprinted in the mental representations of citizens, giving the knowledgeable a role as great sages that they probably do not deserve. It is thought that the scientific social body is subject to different conditions from the rest of society. So the image of the scientist is still subjectively associated with the isolated scientist (when the reality is closer to small and medium-sized companies operating more and more in a liberal regime), a scientist who produces material or immaterial products in isolation, possible forms of financed social "kitsch", the quintessence and testimony of what participates in making society selfish. It is therefore a somewhat dubious honor to be recognized in this way, especially since hard science and technoscience on the one hand, and philosophy on the other, have separated.

Moreover, in this troubled context, the social body expects immediate answers to questions in a logic of equally immediate utility.

Livet (2006) studied the relationship that can exist between an approach to the future and the emotional impact that can result from the different proposals considered. For this author, forms of anxiety emerge whenever:

– the emotional gap between the present and the projected future is large;

– negative emotions are predominantly present;

– the way to reach the proposed goal is not "well" mapped out, nor is it clear how traditional routines can be "revisited";

– the stakes are potentially high;

– the objective is not the norm or the usual;

– the fear is great for the proposal to prove highly likely.

When the questions are put to researchers, there can be a certain temporal relaxation of opinions. On the other hand, the presence of third parties (such as a hierarchy) is likely to generate movements of opinion, mediated by the relational system and the organization. On this basis, it is even possible to show ways of anticipating possible crises through the adapted use of messages... (Heiderich 2007). Thus, they consider the existence of three times associated with three requirements: the time of the emotion linked to the impulse, that of the questions and finally that of the controversies leading to decisions. We are, in fact, only in the time of emotion.

Box 2.4. *Emotions*

"Panis et circenses", the "industry" of science is confronted with national egos, and scientists who have become professionals must constantly provide new publications. In these conditions, where predation is normally forbidden but can be envisaged without too much risk, it is possible to plunder ideas; this is through the use of the bibliography, even recommended. This knowledge from international colleagues cannot be exploited as it is: it is possible to make recensions, in which case they must be prepared in order to respect the standard rules of publications. This partial transcription work is time-consuming, is part of the imposed competitive dynamic (if one wants to receive subsidies), participates in the creation of a competence that we do not want to share, but avoids launching into risky operations. This is indeed what is observed in the non-responses from our foreign colleagues, but also from the members of the national networks contacted. For once, the French exception is not visible!

Indeed, in a fundamental difference with the distant past (until the end of the 19th century), we have moved to mass scientific research, which has become a mass "commodity" satisfying quantitative criteria just as much as a submission to the extraordinary (like 4D videos), which leads to the realization of POCs, but without going too far for reasons of time and the classic difficulties of setting up interdisciplinary projects. The scientist, in this context, provides information respecting an agreed code, not intended for the world. The creative act, when it exists, therefore remains hidden, except for the experts who help finance its realization. It is an element of a certain "virtuosity", but not of freedom, where it is the proposed execution that is often praised. So, the unexpected is not considered, we are in fact oriented according to a principle of causality, of following the external demand toward the classical. Science has thus entered the consumer society and an absence of empathy and a spirit of non-cooperation are part of these new rules. In fact, the products of science are indeed consumed like other consumer objects, giving the impression that they participate in the vital need of society, even if their use in terms of usage is perhaps not at the level of the propagated beliefs. So, for the scientists interviewed, the question of applications may seem incongruous because of a disconnect between what they produce and the reality in the field, which they may (still) be unaware of. In the non-answers we would see a double reality: "I have some ideas, but I don't want to share them" and/or "I don't understand the question, it is outside my field of interest"!

> Our natural tendency is to imagine and judge at the same time when we think, yet we must learn to separate these two ways of creative thinking. (De Brabandère 2017)

2.4.2. *Survey responses*

> Faced with the unknown, Man's natural tendency to idealize and his natural fear unite to achieve the same goal: to intensify the unknown through the imagination and to consider it with an intensity not usually reserved for obvious realities. (Simmel 1991)

As Guilhou and Lagadec (2002) point out,

> when it comes to signals, we are faced with a basic dilemma: either we give priority to sensitivity, and we have alarm trigger thresholds that are drawn down (with the disadvantage of having false alarms [...]); or we give priority to specificity [...] but with the risk of not detecting certain serious phenomena.

In the elements presented in section 2.4, the possible social demand remains unclear and if choices must be made, it will ultimately be up to the researchers themselves to determine them according to their own parameters. An important point, however, is that for those who agreed to respond to the survey, there is a tacit and benevolent agreement to the development of 4D printing technology in the future. All the answers given in this survey, whether they come from real laymen or from scientists largely outside their scientific field of excellence, are rather non-scientific, without it being possible to define orientations to be privileged in a robust way. Their proposals, a form of exchange of opinions, are in the end rather conventional and fit themselves poorly to scientific assertions. Should we conclude that the respondents, in their answers, trusting the applicants and their common sense, are not, on the specific theme of 4D printing, far from reality?

This analysis is, of course, biased by the non-answers of specialists who may have kept justified ideas to themselves that could have guided choices, allowing a faster promotion of very young 4D printing technologies. But we have already expressed our regrets on this point. Gass (1994) deplored a narcissistic turn in the vision of the future, accusing his colleagues of vanity and egoism, with a return on ourselves (Chapuis 2000); this is undoubtedly possible for the voluntary non-respondents, but it is not conceivable for the others. There is no doubt that in this type of survey, which must be completed quickly, we have not been able to bring out reflections free of all conformism; social correctness may have reigned.

However, not all the manifestations of indiscipline or simple freedom expressed are justifiable by an aversion to arbitrariness. We did not deal with subjects who under a rational appearance let irrepressible or impulsive moods shine through. But these spontaneous impulses could have manifested an instinctive sense of kairos and appeared as a motor of self-promotion, or even a certain transgressive source of enjoyment. We are obviously not there yet ....

DEFINITION.– *Kairos*: a concept that allows us, if not to define time, at least to situate events according to that dimension. Doing the right thing at the right time is part of kairos.

> The traditional approach in science is based on a mechanical order in which the basic elements […] are external to each other and exist separately. (Böhm 1979)

2.5. Conclusion

> One does not know, gentlemen, one never knows what one is doing by touching the great inventions that have honored the human mind… No, no one can say; but everyone will say that inventions of the material order have been the most powerful vehicle of ideas, of the efforts of intelligence in the world. (A. de Lamartine cited in Papon 2020)

The introduction or reinforcement of competition is supported by many powers as a means to solve the problems of universities and research bodies. For this reason, considerable energies (with associated resources, whether individual, collective or national) are now largely invested to promote scientific competitiveness (aiming for excellence). For researchers, and this has been the case for several decades, this effort to maintain or increase competition extends to the way in which researchers behave and organize their existence. They are obliged to be concerned about their chances of success in areas of intellect that are translated into scientific productions, real professional work. What is referred to as "improving the economic environment" of the research world is a form of struggle to maintain our personal value in relation to so-called criteria of excellence.

> The problem for individual and collective actors is that their rivals (whether a competitor for a job or another nation-state) are also investing ever more resources and energy to increase their own competitiveness. Competition thus becomes a zero-sum game, despite the increasing stakes of the players. Do we really need more competition? (Rosa 2019)

"The transformation of the way resources, goods, positions, privileges and social esteem are allocated is a fundamental aspect of the transition from traditional to modern society" (Rosa 2019). What makes modern society a competitive society is the way it organizes allocations of positions and funding, thus an element of power and influence, which must be maintained by any means possible (including voluntary non-responses). The definition of scientific goals is one of the decisive elements in the competitive struggle, especially in a rapidly emerging field with more than a 40% increase in the number of publications per year. We can therefore observe the importance of a principle of competition that goes beyond the economic sphere alone. This is the rule of the New Public Management… "Publish or Perish"…

But in this reference to our own responsibility in committing to risky strategic choices (loss of time, risk of failure, etc.), there is a separation between a few scientists who are obliged to be responsible and laymen, both groups of whom may be far from the truth and the right strategy. But for the first group, in its privileged field of activity, simple questions such as those asked in this survey, betray forms of anxiety. Arendt (1972) explains that: "The man of science has not only left behind the layman and his limited understanding, he has abandoned a part of himself and his own power of understanding which remains human understanding when he goes to work in his laboratory".

Figure 2.20 reminds us of some elements concerning the quality of interactions between possible partners, willing to exchange (or not), but with awareness. What we wished for, but no doubt expressed badly, was a willingness to open up to others in order to gain access to and deepen knowledge together, with a view to operational purposes of societal utility. The answers expressed leave us free to engage in any possible way, as often, we will find ourselves in front of our blank sheet of paper to imagine (in small number) a possible future in which 4D printing, we are convinced, will have its place.

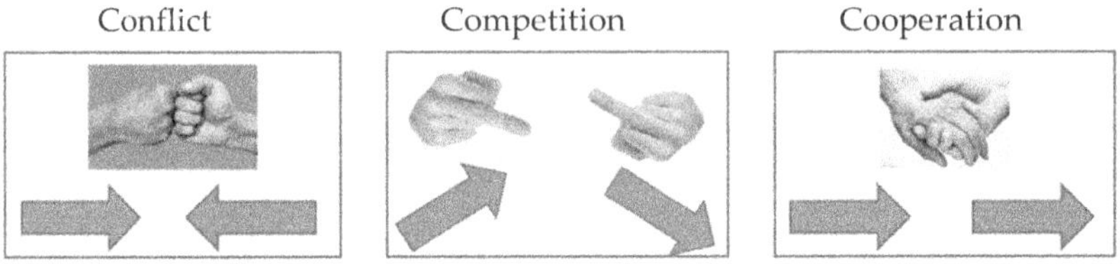

Figure 2.20. *Possible modes of interaction between colleagues*

However, in European University Association (2021), it is explicitly stressed that universities will need to cooperate with a wide variety of partners. It is proposed that:

> While basic, curiosity-driven research will be a prerequisite for knowledge-based solutions, it will also be fully recognized as an end in itself. Universities will provide a space for lateral thinkers, who test and develop new ideas that are not yet recognized by their research colleagues or by society at large,

all the while facilitating dialogue between disciplines and encouraging interdisciplinary research. For innovation, they will engage in co-creating solutions with a wide range of partners and with the aim of addressing common challenges and making a demonstrable difference to society through technological and social innovation.

> They will bring stakeholders together around a common vision, building bridges between different cultures, ranging from academia, business and start-ups to civil society and the social and cultural scene. They will also strengthen their contribution to the development of knowledge and skills with ecosystem partners.

So the road is still to be built… But, we are not yet in 2030 (vision proposed in this document).

> I've met many of these inventors who invent something that cannot work, that is theoretically impossible. But they spend three years developing it, running a motor without electricity, with magnets. You

> explain to them it won't work. It violates the second law of thermodynamics. And they say: "Don't give me your goddamn Washington law!" (Robinson 2013)

> Self-doubt can be an ally. This is because it serves as an indicator of aspiration. It reflects love, love of something we dream of doing. (Pressfield 2012)

2.6. Appendix 1: The blank survey

> If the identity of the object is no longer discernible, no community of nature, still less the unnatural conformity of a mass society, will prevent the destruction of the common world, usually preceded by the destruction of many aspects in which it presents itself to human plurality. (Arendt 1994)

2.6.1 *What is 4D printing?*

3D printing (1984) uses digitized coordinates of an object to manufacture it in different materials (from plastic to metal). 4D printing (2013) uses 3D printing with materials whose shape or functionality can change in the presence of a stimulus (thermal, chemical, light, electromagnetic field, etc.).

By adding a complementary parameter (time), 4D printing opens up the possibilities of creating the objects of the future. For example, a seat that adapts to your shape and evolves according to your behavior, a self-adapting plaster cast that evolves with the evolution of a fracture, clothes that change their properties when it rains, etc.

An example of a 4D part is shown below; it is a 4D printed clamp, which changes its shape depending on the temperature.

Figure 2.21. *3D clamp (according to Ge et al. (2016))*

However, for industrial application, this science-based technology still suffers from some flaws:

a) *Moderate Young's modulus* (mechanical strength)

Figure 2.22. *Situation of 4D "polymer" printing in the spectrum of Young's moduli. For a color version of this figure, see www.iste.co.uk/demoly/4Dprinting1.zip*

b) *Response time to a stimulus*: about 5–10 minutes for a size of 1 cm, one hour for 10 cm

1) TODAY, do you think that this technology, **under the current conditions,** can be useful to you?

- ☑ Yes
- ☐ No
- ☐ DK (Don't know)

TOMORROW, if now, we wish to get closer to industrial conditions with new stimulable materials and new processes (e.g. greatly improved mechanical performance and greatly reduced response time):

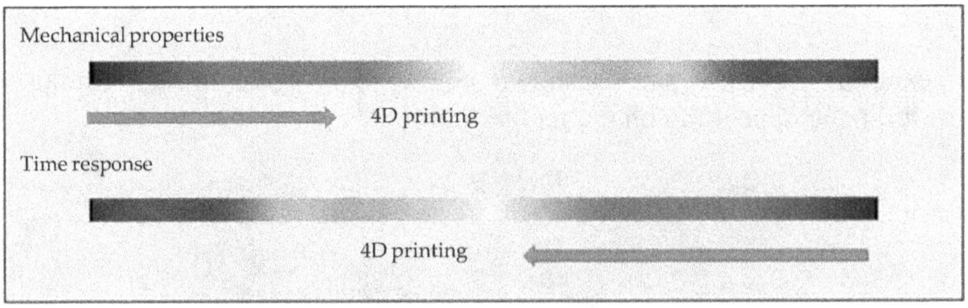

Figure 2.23. *Position of 4D printing in terms of mechanical properties and time response to stimulation. For a color version of this figure, see www.iste.co.uk/demoly/4Dprinting1.zip*

2) So, what areas of application are you interested in?

	TODAY						TOMORROW					
	1= No interest	2	3	4	5 = Very interested	N/A	1= No interest	2	3	4	5 = Very interested	N/A
Medical (e.g. Sten)	☐	☐	☐	☐	☐	☐	☐	☐	◉	☐	☐	☐
Medical (prostheses, orthoses)	☐	☐	☐	☐	☐	☐	☐	☐	☐	☐	☐	☐
Actuators and sensors (see clamp above)	☐	☐	☐	☐	☐	☐	☐	☐	☐	☐	☐	☐
Space applications (e.g. antenna opening)	☐	☐	☐	☐	☐	☐	☐	☐	☐	☐	☐	☐
Flexible robots (without conventional motor)	☐	☐	☐	☐	☐	☐	☐	☐	☐	☐	☐	☐
Mobility system (adaptive seats according to its shape and constraints, e.g. acceleration, self-deploying wheelchair)	☐	☐	☐	☐	☐	☐	☐	☐	☐	☐	☐	☐
Furniture (furniture that is self-assembling or self-fabricating right out of the box)	☐	☐	☐	☐	☐	☐	☐	☐	☐	☐	☐	☐

	TODAY						TOMORROW					
	1= No interest	2	3	4	5 = Very interested	N/A	1= No interest	2	3	4	5 = Very interested	N/A
Gadgets or fashionable devices (e.g. a smartphone that self-repairs when scratched or that compacts for carrying purposes)	☐	☐	☐	☐	☐	☐	☐	☐	☐	☐	☐	☐
Optics (self-focusing and self-filtering glasses; augmented reality)	☐	☐	☐	☐	☐	☐	☐	☐	☐	☐	☐	☐
Clothes (change of color with temperature, waterproofing in the presence of rain, etc.)	☐	☐	☐	☐	☐	☐	☐	☐	☐	☐	☐	☐
Home automation (openings, ventilation)	☐	☐	☐	☐	☐	☐	☐	☐	☐	☐	☐	☐
Chemistry (self-adapting reactors)	☐	☐	☐	☐	☐	☐	☐	☐	☐	☐	☐	☐
Food (smart packaging)	☐	☐	☐	☐	☐	☐	☐	☐	☐	☐	☐	☐
Automobile (car that changes its shape according to need)	☐	☐	☐	☐	☐	☐	☐	☐	☐	☐	☐	☐

3) Do you have any other ideas for applications?

4) There are now several types of stimulus that allow changes in the form and/or functionality of 4D systems; which ones do you think are most relevant:

	1 = No interest	2	3	4	5 = Very interested	N/A
Heat/cold	☐	☐	☐	☐	☐	☐
Light	☐	☐	☐	☐	☐	☐
Moisture	☐	☐	☐	☐	☐	☐
Chemical products	☐	☐	☐	☐	☐	☐
Electric field	☐	☐	☐	☐	☐	☐
Magnetic field	☐	☐	☐	☐	☐	☐

5) Do you have any comments?

6) Does 4D printing have a possible link with **your** present or future **activities**?

☐ Yes

☐ No

☐ N/A

7) If YES, how does this relate to **your interests** or professional **activity?**

☐ Student

☐ Fab-Lab

☐ Search

☐ Industry potential user

☐ Industry potential manufacturer

☐ Other

8) When purchasing 4D products or devices, on what criteria do you base your choice?

	1 = No interest	2	3	4	5 = Very interested	N/A
The price	☐	☐	☐	☐	☐	☐
Aesthetics	☐	☐	☐	☐	☐	☐
How it works	☐	☐	☐	☐	☐	☐
Ergonomics	☐	☐	☐	☐	☐	☐
What's new	☐	☐	☐	☐	☐	☐

9) For example, the company Adidas has just launched a new model of T-shirt that becomes waterproof when it rains. This T-shirt is said to cost 300 euros. Would you be likely to buy one at that price?

☐ Yes, it is possible

☐ No, not at all

☐ Don't know

10) We will cross-reference your opinions with personal (anonymized) information as follows:

What age group do you fall into?

☐ Under 18 years old

☐ 18–24 years old

☐ 25–34 years old

☐ 35–49 years old

☐ 50–64 years old

☐ Over 65 years old

☐ Don't know

11) Socio-professional category that best corresponds to you

- ☐ Student
- ☐ Higher education and research staff
- ☐ Farmer
- ☐ Craftsman, trader, company director
- ☐ Liberal profession, senior executive
- ☐ Intermediate profession (technician, supervisor…)
- ☐ Employee, worker
- ☐ Not working and retired
- ☐ Don't know

12) What is your level of education?

- ☐ Self-taught, Certificate of Primary Studies, junior school certificate, NVQ/vocational diploma
- ☐ Baccalaureate
- ☐ Technology degree, advanced technician's certificate and equivalent
- ☐ Higher education (up to Bac +5)
- ☐ Higher education (more than five years of higher education)
- ☐ Don't know

13) Which annual income bracket do you fall into?

- Less than €30,000
- From €30,000 to €50,000
- From €50,000 to €100,000
- More than €100,000
- Don’t know

14) If you are interested in the results of this survey, we will be able to send it to you at the end of February or beginning of March 2021. If so, please provide us with an **email address**:

2.7. Appendix 2: Answers as of February 16, 2021

TODAY, do you think that this technology, under the current conditions, could be useful to you?

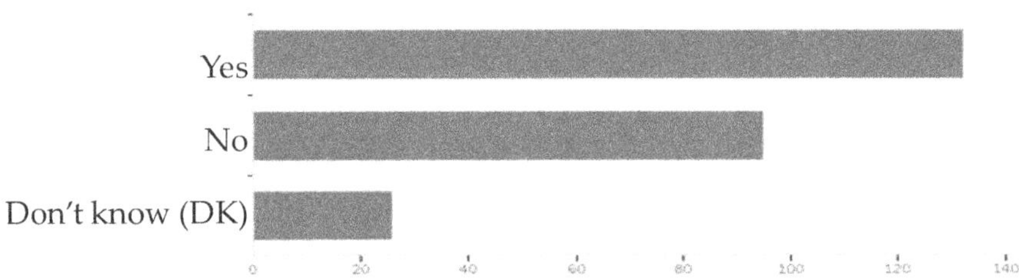

Figure 2.24.

TOMORROW, if now, we wish to get closer to the industrial conditions with again…

- TODAY (see figures)

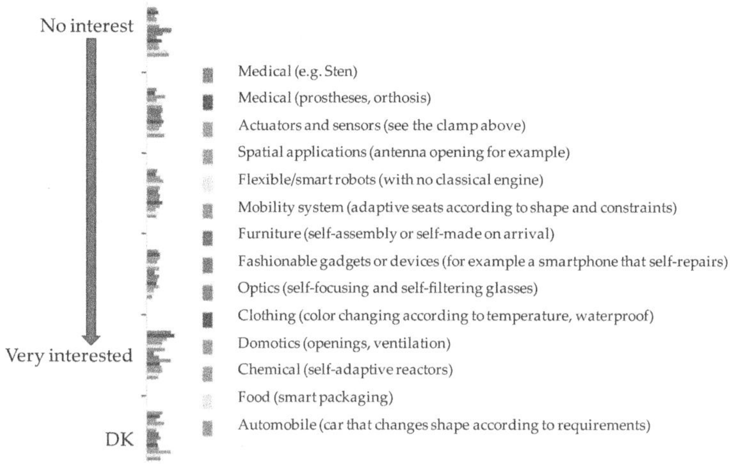

Figure 2.25. *For a color version of this figure, see www.iste.co.uk/demoly/4Dprinting1.zip*

TOMORROW, if now, we wish to get closer to the industrial conditions with again...

- TOMORROW

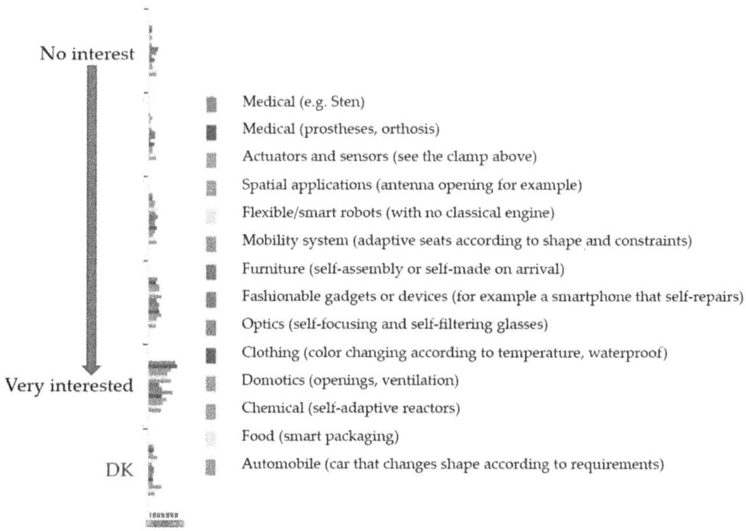

Figure 2.26. *For a color version of this figure, see www.iste.co.uk/demoly/4Dprinting1.zip*

There are now several types of stimulus that allow changes in the form and/or functionality of 4D systems; which ones do you think are the most relevant?

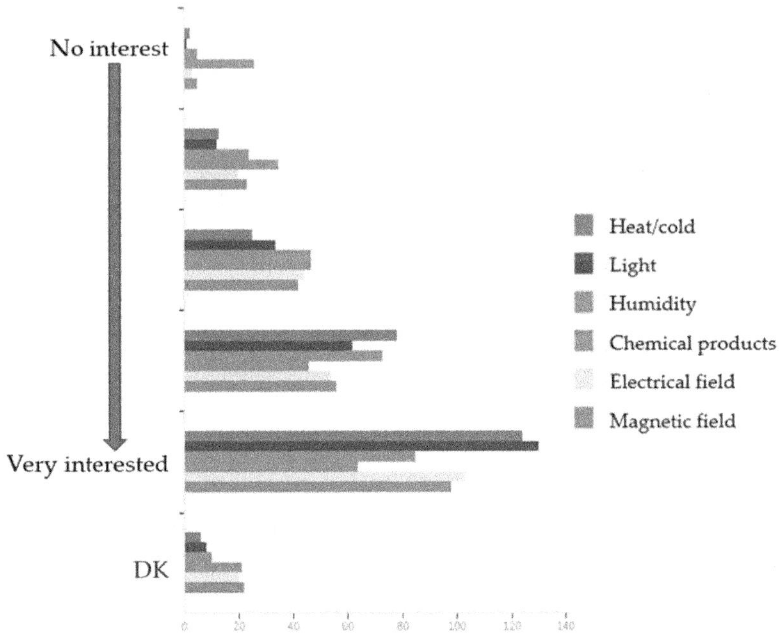

Figure 2.27. *For a color version of this figure, see www.iste.co.uk/demoly/4Dprinting1.zip*

Does 4D printing have a possible link to your current or future activities?

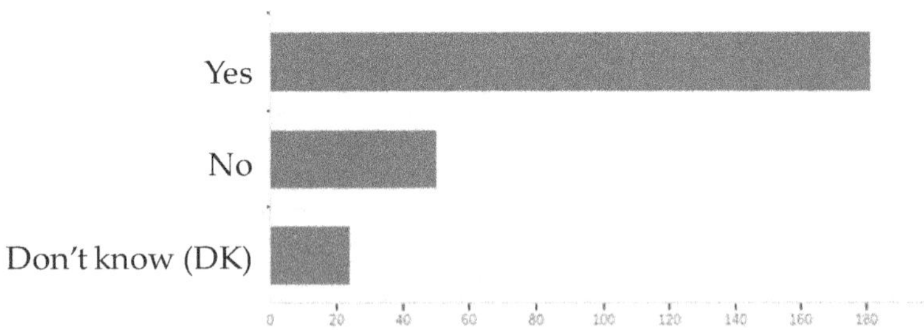

Figure 2.28.

If YES, how does this relate to your interests or professional activity?

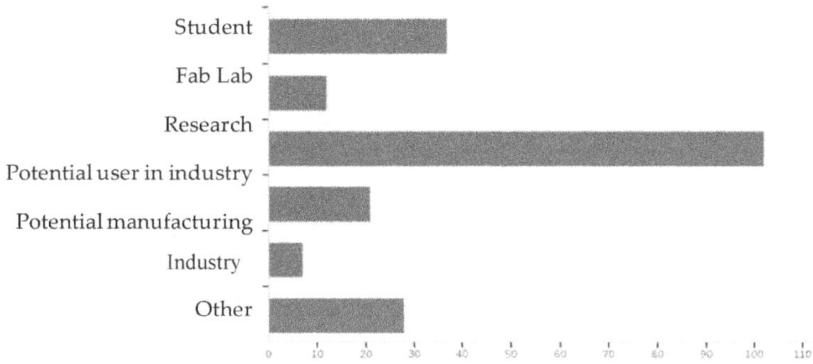

Figure 2.29.

When purchasing 4D products or devices, on what criteria do you base your choice?

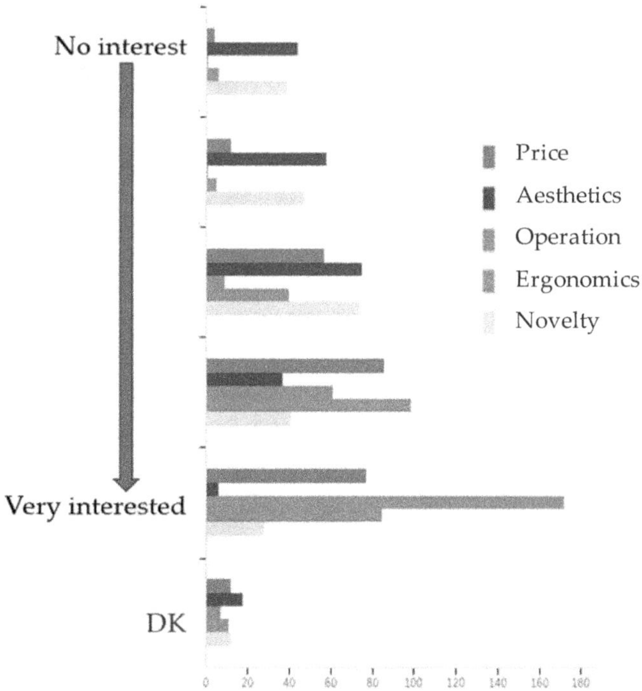

Figure 2.30. *For a color version of this figure, see www.iste.co.uk/demoly/4Dprinting1.zip*

For example, the company Adidas® has just launched a new model of T-shirt that becomes waterproof when it rains. This T-shirt is said to cost 300 euros. Would you be likely to buy one at that price?

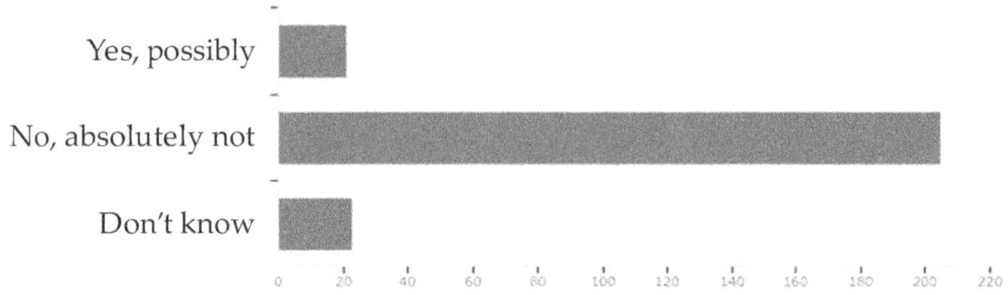

Figure 2.31.

We will cross-reference your opinions with personal (anonymized) information as follows: what age range do you fall into?

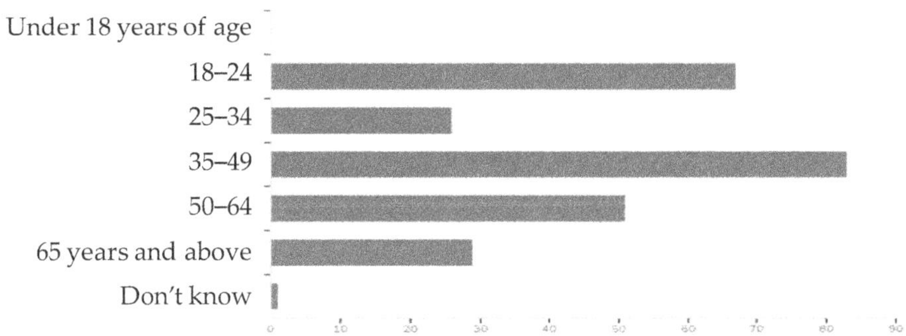

Figure 2.32.

The socio-professional category that best suits you:

Figure 2.33.

What is your level of education?

Figure 2.34.

2.8. References of scientific articles with "4D printing" or "applications" in their titles

Chu, H., Yang, W., Sun, L., Cai, S., Yang, R., Liang, W., Yu, H., Liu, L. (2020). 4D printing: A review on recent progresses. *Micro-machines*, 11, 796.

Grinberg, D., Siddique, S., Le, M.Q., Liang, R., Capsal, J.F., Cottinet, P.J. (2019). 4D printing based piezoelectric composite for medical applications. *Journal of Polymer Science, Part B: Polymer Physics*, 57, 109–115.

Haleem, A., Javaid, M., Vaishya, R. (2018). 4D printing and its applications in orthopedics. *Journal of Clinical Orthopedics and Trauma*, 9, 275e276.

Han, M., Yang, Y., Li, L. (2020). Energy consumption modeling of 4D printing thermal-responsive polymers with integrated compositional design for material. *Additive Manufacturing*, 34, 101223.

Hann, S.Y., Cui, H., Nowicki, M., Zhang, L.G. (2020). 4D printing soft robotics for biomedical applications. *Additive Manufacturing*, 36, 101567.

Headrick, D. (2015). 4D printing transforms product design. *Research Technology Management*, 58, 7–8.

Javaid, M. and Haleem, A. (2019). 4D printing applications in medical field: A brief review. *Clinical Epidemiology and Global Health*, 7, 317–321.

Khorsandi, D., Fahimipour, A., Abasian, P., Saber, S.S., Seyedi, M., Ghanavati, S., Ahmad, A., De Stephanis, A.A., Taghavinezhaddilami, F., Leonova, A. et al. (2020). 3D and 4D printing in dentistry and maxillofacial surgery: Printing techniques, materials, and applications. *Acta Biomaterialia*, 26, S1742-7061(20)30767-4.

Melly, S.K., Liu, L., Liu, Y., Leng, J. (2020). On 4D printing as a revolutionary fabrication technique for smart structures. *Smart Materials and Structures*, 29, 083001.

Melocchi, A., Uboldi, M., Inverardi, N., Briatico-Vangosa, F., Baldi, F., Pandini, S., Scalet, G., Auricchio, F., Cerea, M., Foppoli, A. et al. (2019). Expandable drug delivery system for gastric retention based on shape memory polymers: Development via 4D printing and extrusion. *International Journal of Pharmaceutics*, 571, 118700.

Mitchell, A., Lafont, U., Hołyńska, M., Semprimoschnig, C. (2018). Additive manufacturing – A review of 4D printing and future applications. *Additive Manufacturing*, 24, 606–626.

Netouanoglou, K., Stavropoulos, P., Mourtzis, D. (2018). 4D printing prospects for the aerospace industry: A critical review. *Procedia Manufacturing*, 18, 120–129.

Pei, E. and Loh, G.H. (2018). Technological considerations for 4D printing: An overview. *Progress in Additive Manufacturing*, 3, 95–107.

Ploszajski, A.R., Jackson, R., Ransley, M., Miodownik, M. (2019). 4D printing of magnetically functionalized chainmail for exoskeletal biomedical applications. *MRS Advances*, 4, 1361–1366.

Rayate, A. and Jain, P.K. (2018). A review on 4D printing material composites and their applications. *Materials Today: Proceedings*, 5, 20474–20484.

Ryan, K.R., Down, M.P., Banks, C.E. (2021). Future of additive manufacturing: Overview of 4D and 3D printed smart and advanced materials and their applications. *Chemical Engineering Journal*, 403, 126162.

Shie, M.Y., Shen, Y.F., Astuti, S.D., Lee, A.K.X., Lin, S.H., Dwijaksara, N.L.B., Chen, Y.W. (2019). Review of polymeric materials in 4D printing biomedical applications. *Polymers*, 11, 1864.

Singholi, A.K.S. and Sharma, A. (2020). Review: Recent advancement and research possibilities in 4D printing technology. *Materialwissenschaft und Werkstofftechnik*, 51, 1332–1340.

Xin, X., Liu, L., Liu, Y., Leng, J. (2019). Mechanical models, structures, and applications of shape-memory polymers and their composites. *Acta Mechanica Solida Sinica*, 32, 535–565.

Zolfagharian, A., Kaynak, A., Kouzani, A. (2020). Closed-loop 4D-printed soft robots. *Materials and Design*, 188, 108411.

2.9. References

Algoud, J.P. (2014). *Systémique – vie et mort de la civilisation occidentale 1150–2050*. L'Interdisciplinaire, Lyon.

André, J.C. (2017a). *From Additive Manufacturing to 3D/4D Printing 1: From Concepts to Achievements*. ISTE Ltd, London, and John Wiley & Sons, New York.

André, J.C. (2017b). *From Additive Manufacturing to 3D/4D Printing 2: Current Techniques, Improvements and their Limitations*. ISTE Ltd, London, and John Wiley & Sons, New York.

André, J.C. (2017c). *From Additive Manufacturing to 3D/4D Printing 3: Breakthrough Innovations: Programmable Material, 4D Printing and Bio-printing*. ISTE Ltd, London, and John Wiley & Sons, New York.

André, J.C. (2019). *Industry 4.0 – Paradoxes and Conflicts*. ISTE Ltd, London, and John Wiley & Sons, New York.

Arendt, H. (1972). *La crise de la culture*. Folio Essais, Paris.

Arendt, H. (1994). *La condition humaine*. Pocket, Paris.

Bayard, P. (2007). *Comment parler des livres qu'on n'a pas lu ?* Editions de minuit, Paris.

Beigberger, F. (2000). *99 F*. Grasset, Paris.

Bergson, H. (2019). *L'évolution créatrice*. PUF, Paris.

Bezes, P. (2014). Administration. In *Dictionnaire des politiques publiques*, Boussaguet, L., Jacquot, S., Ravinet, P. (eds). Presses de Sciences Po, Paris.

Böhm, D. (1979). Unicité de l'ordre enveloppé. Colloque science et conscience – Cordoue. *Le Monde*, 24 October.

Bourrelly, L. (2007). Techniques de manipulation de l'opinion publique et de la Société [Online]. Available at: http://www.adicie.com/archives/57.

Bouveresse, R. (1978). *Karl Popper ou le rationalisme critique*. Vrin, Paris.

Busch, L. (2011). *Standards, Recipes for Reality*. MIT Press, Cambridge.

Carmel, M. (2014). Soft robotics: A perspective – Current trends and prospects for the future. *Soft Robotics*, 1, 5–11.

Carnorama (2014). Automotive smart memory materials [Online]. Available at: https://www.carnorama.com/1225/automotive-smart-memory-materials/ [Accessed 25 March 2022].

Castel, R. (2003). *L'insécurité sociale*. Le Seuil, Paris.

Chapuis, S. (2020). Little failure, great success: The writer's memoir as a fail-safe genre. *Revue française d'études américaines*, 163, 59–71.

Cheminade, B. (2018). Histoire vécue de disruptions [Online]. Available at: https://www.benjaminchaminade.com/histoire-vecue-disruptions.

Coulais, C., Teomy, E., de Reus, K., Shokef, Y., van Hecke, M. (2016). Combinatorial design of textured mechanical metamaterials. *Nature*, S35, 529–535.

Csikszentmihalyi, M. (2014). Implications of a systems perspective for the study of creativity. In *Handbook of Creativity*, Sternberg, R.J. (ed.). Cambridge University Press Ed., Boston, USA.

Damien, R. (2006). La confiance rationnelle. In *Les moments de la confiance – connaissance, affects et engagements*, Ogien, R. and Quéré, L. (eds). Economica, Paris.

Darwin, C. (2020). *Le corail de la vie – Carnet B 1837-1838*. Rivages Poche, Paris.

D'Espagnat, B. (2015). *A la recherche du réel – le regard d'un physicien*. Dunod, Paris.

De Brabandère, L. (2017). Gaston Trouvetou, recrue de choix pour DRH. In *Philosophie Magazine – Special Edition "Gaston – un philosophe au travail"*, Paris.

Debray, R. (2020). *Eloge des frontières*. Gallimard, Paris.

Dewey, J. (1954). *The Public and its Problems*. Swallow Press, Athens.

Ding, Z., Weeger, O., Qi, H.J., Dunn, M.L. (2018). 4D rods: 3D structures via programmable 1D composite rods. *Materials & Design*, 137, 256–265.

Dostoïevski, F. (1935). *L'adolescent*. Gallimard, Paris.

Duplay, M., Quanquin, H., Rouquet, C. (2020). Disciplines/indisciplines. *Revue française d'études américaines*, 165, 3–9.

Emerson, R.W. (2001). Self-reliance. In *Emerson's Prose and Poetry*, Porte, J. and Morris, S. (eds). W.W. Norton, New York.

European University Association (2021). Universities without walls – A vision for 2030. [Online]. Available at: https://eua.eu/downloads/publications/universities%20without%20walls%20%20a%20vision%20for%202030.pdf [Accessed 25 March 2022].

Follett, M.P. (1949). *Freedom and Coordination – Lectures in Business Organization Management*. Publication Trust Limited, London.

Galbraith, J.K. (1980). *Théorie de la pauvreté de masse*. Gallimard, Paris.

Gardey, D. (2008). *Ecrire, calculer, classer – comment une révolution de papier a transformé les sociétés contemporaines (1800–1940)*. La Découverte, Paris.

Gass, W. (1994). The art of self: Autobiography in an age of narcissism. *Harper's Magazine*, May.

Ge, Q., Sakhaei, A.H., Lee, H., Dunn, C.K., Fang, N.X., Dunn, M.L. (2016). Multi-material 4D printing with tailorable shape memory polymers. *Scientific Reports*, 6, 31110.

Gévart, P. (2018). Prospective et science-fiction – le dialogue nécessaire. *Voir loin – prospective et stratégie*, 9, 89–97.

Godelier, M. (1996). *L'énigme du don*. Fayard, Paris

Gras, A. (2003). *Fragilité de la puissance : se libérer de l'emprise technologique*. Fayard, Paris.

Guéniot, A. (2020). Les projets face à la catastrophe environnementale : saturation du présent ou ouverture de l'avenir ? *Revue française d'éthique appliquée*, 10, 57–66.

Guilhot, N. (2018). Transitions : évolution, révolution, disruption ? Les transformations du monde, entre crises et opportunités / 14 June 2019, Journée de recherche interdisciplinaire de l'IUT de Sceaux [Online]. Available at: https://ahmo.hypotheses.org/2609.

Guilhou, X. and Lagadec, P. (2002). *La fin du risque zéro*. Eyrolles, Paris.

Hankook (2019). Hankook's futuristic hexonic smart tire [Online]. Available at: https://glitchmind.com/hexonic-smart-tire/ [Accessed 25 March 2022].

Heiderich, D. (2007). Persistance rétinienne et rémanence sociale. *Magazine de la communication et de crise sensible*, 15, 15–19.

Houdé, O. (2019). Les promesses des nouvelles méthodes d'apprentissage scolaire. In *Les promesses de la science*, Chauvières, F. (ed.). Flammarion, Paris.

Hu, F., Wang, W., Cheng, J., Bao, Y. (2020). Origami spring-inspired metamaterials and robots: An attempt at fully programmable robotics. *Science Progress*, 103, 1–19.

Jackson, B. (2016). Futurist Ray Kurzweil makes 3D printing predictions [Online]. Available at: https://3dprintingindustry.com/news/futurist-ray-kurzweil-reiterates-3d-printing-predictions-100504/ [Accessed 25 March 2022].

Jardim-Goncalves, R., Romero, D., Grilo, A. (2017). Factories of the future: Challenges and leading innovations in intelligent manufacturing. *International Journal of Computer Integrated Manufacturing*, 30, 4–14.

Jean, A. (2019). *De l'autre côté de la machine*. Editions de l'Observatoire, Paris.

Jeong, H.Y., An, S.C., Seo, I.C., Lee, E., Ha, S., Kim, N., Jun, Y.C. (2019). 3D printing of twisting and rotational bi-stable structures with tuning elements. *Scientific Reports*, 9, 324.

Kanu, N.J., Gupta, E., Vates, U.K., Singh, G.K. (2019). An insight into biomimetic 4D printing. *Royal Society of Chemistry Advances*, 9, 38209.

Kaynak, A. and Zolfagharian, A. (2019). Stimuli-responsive polymer systems – Recent manufacturing techniques and applications. *Materials*, 12, 2380.

Khoo, Z.X., Mei Teoh, J.E., Liu, Y., Chua, C.K., Yang, S., An, J., Leonga, K.F., Yeong, W.Y. (2015). 3D printing of smart materials: A review on recent progresses in 4D printing. *Virtual and Physical Prototyping*, 10, 103–122.

Laïdi, Z. (1999). Pourquoi vivons-nous dans l'urgence ? *Etudes*, 390, 775–788.

Lévy-Leblond, J.M. (1996). *Aux contraires*. NRF Essais, Gallimard, Paris.

Li, S., Stampfl, J., Xu, H., Xu, E., Diaz, E., Malkin, V., Rus, D., Wood, R. (2019). A vacuum-driven origami "Magic-ball" soft gripper. *IEEE* [Online]. Available at: https://dspace.mit.edu/handle/1721.1/120930 [Accessed 25 March 2022].

Lippmann, W. (1913). *Politics*. Mitchell Kennerley, New York.

Lippmann, W. (2015). *Drift and Mastery – An Attempt to Diagnose the Current Unrest*. The University of Wisconsin, Madison.

Liu, Y. and Moran, B. (2021). Crack tip fields in a neo-Hookean sheet reinforced by non-linear fibers. *Journal of Mechanics and Physics of Solids*, 152, 104406.

Livet, P. (2006). Confiance, émotion et manifestation de valeurs. In *Les moments de la confiance ; connaissance, affects et engagements*, Ogien, R. and Quéré, L. (eds). Economica, Paris.

Mangiola, S. (2014). 4D printing on the rise [Online]. Available at: http://machinedesign.com/3d-printing/4d-printing-rise [Accessed 25 March 2022].

Michaud, Y. (2002). *Changements dans la violence et dans la peur*. Odile Jacob, Paris.

Moessinger, P. (ed.) (1996). Choix rationnel et rationalité. In *Irrationalité individuelle et ordre social*. Droz, Paris.

Neyrinck, J. (1990). *Le huitième jour de la création : introduction à l'anthropologie*. Presses Polytechniques et Universitaires Romandes, Lausanne.

OECD (2020). Regards sur l'éducation 2020 – les indicateurs de l'OCDE [Online]. Available at: https://www.oecd-ilibrary.org/docserver/7adde83a-fr.pdf?expires=1613237180&id=id&accname=guest&checksum=ACBF74284F4FC9A2DE48C3B311E48CBA.

Papon, P. (2020). *La démocratie a-t-elle besoin de la science ?* CNRS, Paris.

Perrin, D. and Weiss, J. (2003). *La croyance*. Ellipses, Paris.

Pickstone, J.W. (2001). *Ways of Knowing, a New History of Science, Technology and Medicine*. The University of Chicago Press, Chicago.

Polanyi, K. (2009). *La Grande Transformation – aux origines politiques et économiques de notre temps*. Gallimard, Paris.

Popper, K. (1957). *The Poverty of Historicism*. Routledge, London.

Pressfield, S. (2012). *The War of Art – Break through the Blocks and Win Your Inner Creative Battles*. Black Irish Entertainment LLC, New York.

Ramonet, I. (2004). *Propagandes silencieuses*. Folio-Actuel, Paris.

Robinson, J. (2013). *Creativity – The Psychology of Discovery and Invention*. Harper Perennial, New York.

Roddier, F. (2012). *Thermodynamique de l'évolution*. Parole, Paris.

Rosa, H. (2019). La compétition comme mode d'interaction – la société concurrentielle et ses conséquences sur la culture et sur les structures sociales. *Sociologie*, 10 [Online]. Available at: http://journals.openedition.org.bases-doc.univ-lorraine.fr/sociologie/5933.

Ruano-Borbalan, J.C. (2017). La "grande standardisation" et les transformations du savoir dans les sociétés technoscientifiques. *Innovations*, 52, 17–41.

Simmel, G. (1991). *Secret et sociétés secrètes*. Circé, Paris.

Stiegler, B. (2016). *Dans la disruption. Comment ne pas devenir fou ?* Les liens qui libèrent, Paris.

Stiegler, B. (2019). *Il faut s'adapter – sur un nouvel impératif politique*. NRF Essais, Paris.

Tibbits, S. (2013). The emergence of 4D printing [Online]. Available at: https://www.ted.com/talks/skylar_tibbits_the_emergence_of_4d_printing [Accessed 25 March 2022].

Tibbits, S. (2016). *Self-Assembly Lab: Experiments in Programming Matter*. Routledge, New York.

Waquet, F. (2016). *L'ordre matériel du savoir, comment les savants travaillent (XVIème–XXIème siècle)*. CNRS Editions, Paris.

Weil-Dubuc, P.-L. (2014). Dépasser l'incertitude – le pari hasardeux de la médecine prédictive. *Esprit*, 406, 20–29.

Zolfagharian, A., Kaynak, A., Kouzani, A. (2020). Closed-loop 4D-printed soft robots. *Materials and Design*, 188, 108411.

3

Who Would Prevail Today from Lamarck or Darwin to Help the Controlled Evolution of 4D Printing?

> What is striking is that many discoveries can often appear to be the result of "happy coincidences". But history teaches us that only minds prepared by their intellectual and scientific past can benefit from these chances, allowing them to be surprised by an unusual fact that escapes the majority. (Denis 2007)

Preamble

With the concept of informed matter, there should be the possibility of changing the form and functionality of objects over time (4D printing), to print living matter (bio-printing), etc. This last area will not be covered in this book, but the interested reader can find information in André (2017c, 2021). By adding time to the mastery of the spatial approach already allowed by 3D printing, research and breaking with traditions of incremental innovation can lead (thanks to a little creativity and epistemic deepening) to huge markets (relative to the current "modest" market of the order of a few tens of billions of euros/year). However, if it is possible to build objects containing different materials that can be activated by additive manufacturing technologies (homogeneous, hybrid or heterogeneous), the question of mastering a 4D transformation that meets the designer's specifications is raised here. At the same time, as this chapter deals with the control of temporal evolution, a parallel with natural processes and associated theories is made, allowing a comparison and the emergence of some action points.

For a piece of information to enable the transformation of a system, it is necessary to have specific information, related to the objective to be achieved (Keller 2011). The following question then arises:

How do we define it and how do we know how much of all the available structural information is effective for this purpose?

The question is how and where information is "encoded" in the material. Harel and Pnueli (1985) introduced the term "reactive system" to designate a class of complex systems to distinguish them from the sequential (or transformational) systems familiar to traditional computing; they write: "A transformational system accepts inputs, performs transformations on them, and produces outputs...". A system of this type does not compute or execute a function, but is expected to maintain some continuous relationship with its environment. Moreover, the very form in which an evolving structure is found gives it the capacity to actively inform other structures, thus triggering the chain of interactivity on which Lehn's (2004) vision of a smooth progression from the Big Bang to thinking beings is based.

The evocation of the different attractors of disruptive innovation in 4D printing is therefore part of the spirit of this chapter, beyond the scientific and technical aspects. Transposing the reality as experienced by the authors with its possible biases of understanding is to remake a little the history of hopes associated with these new processes of additive manufacturing which could, in case of success, populate or even invade our daily life in a few years. It is in the spirit, to find limits before making a story, to internalize it by trying with the means and the time available to restore them with a dual vision, with the risk of making mistakes, of having missed the fertile idea... But it is the price to pay. One must try and dare to advance.

3.1. Introduction

> What is worthwhile is a certain arrangement of things. (Saint-Exupéry 1953)
>
> To search is to invent the world, to set new rules for an ephemeral world. Not like tyrants who also invent another world, but impose it on others. The researcher does not remake the world, he undoes it to make it. He imagines one, then confronts it with the real world to enlighten it, not to exhaust it. Research is a never-ending quest. (Rose 2001)

Fabrications, whatever their nature, are generally made up of elements (voxels in 3D printing) assembled in a precise and often complicated way, from a scan, respecting an algorithmic approach to construction. For 3D components alone, they are based on deterministic principles by respecting the production program desired by the designers. On the other hand, according to Doursat et al. (2012), self-organization in natural systems (physical, biological, ecological or social) relies on identical agents and on essentially stochastic dynamics that are likely to create order. Collective behaviors and non-trivial patterns can occur because we explore behaviors of complex systems. For Doursat et al. (2012),

> yet the vast majority of these naturally emergent patterns (spots, stripes, waves, clusters, etc.) are random and are only modified by boundary conditions. They can be described with a few statistical variables, such as order parameters, but do not have an intrinsic architecture as industrial machines and systems do.

It is therefore a question of programmable self-organization. A central question, considered in this chapter in an open and general way, is to see if we can find links between natural self-training capacities and 4D systems that are designed by engineers. Are there, from this attempted approach, proposals for the creation of such morphogenetic systems that are applicable to 4D transformation, the latter understood as complex, with active materials associated with a matrix that is normally passive, but made complicit by actuation? Therefore, to do this, we have planned, with some effort to analyze the bibliography, to wander between determinism and randomness, in short, a little in play, between Lamarck and Darwin.

3.2. General considerations

3.2.1 *The 4D fabrications concerned by this chapter*

> The notion of theory changed its meaning. It no longer designated a system of reasonably assembled truths, which, as such, had not been made but given to reason and sense. Rather, it became the modern scientific theory, which is a working hypothesis that changes according to the results it produces and depends for its validity not on what it "reveals" but on whether it works. (Arendt 1972)

In the general introduction, we defined domains corresponding to the different forms of 4D printing. They correspond to the realization of homogeneous and/or heterogeneous objects, realized with one or several materials and with several types of energy stimulation, whether global or local. In this chapter and Chapter 4, we will show that the effects of a stimulation are not instantaneous, with variable modifications according to their amplitudes, and spatially distributed. Considering the spatial deformation of a 4D object under stress in this chapter, the following modes of realization will be essentially concerned:

– *Homogeneous 4D printing*: this corresponds effectively to 3D printing with an elastic material distributed in a homogeneous way. However, during an energetic solicitation, only the active voxels which are concerned by the stimulus take part in the transformation, the others behave like passive voxels (which will, however, take part in the overall transformation).

– *Hybrid 4D printing*: the same logic applies as long as the materials have elastic properties; on the other hand, if one of the materials is not elastic and acts as a structural element, the effects of a stimulation must take this into account and will have no clear relationship with the cases where all the materials have deformation possibilities. We would pass from the mechanics of flexible materials to a mechanics as used in robotics.

– *Heterogeneous 4D printing*: with different materials and separate fabrication techniques, one of which contributes to the realization of the 4D object from a 3D fabrication, it is possible to be in both situations: flexible system and rigid system (with joints).

If we want to look for relationships between 4D printing and biological processes, it is clear that in the first situation we would be approaching the functioning of "soft" organs, whereas in the second situation we would be approaching areas where, for example, the bones of a vertebrate are involved. We also have two different forms of mechanics in these two situations. However, as shown in Figure 3.1 from Sommer (2016), presented at a CNRS conference, all the important characteristics that a biological tissue must have must be taken into consideration: definition of the dominant elements by tissue (adipose (presence of proteoglycans, glycoproteins, etc.), muscle (elastic fibers), cartilage (collagen fibers), bone or tooth (calcified tissues)), by function (information, support, movement, etc.) with specificities (elastic and mechanical properties, porosity, granulometry, vascularization, anisotropy, cell adhesion, fiber orientation, etc.). To these elements, Sommer (2016) adds scaling elements that should also be able to be taken into account, as shown in Figure 3.1 in the example of tendons (see also Soman and Vijayavenkataraman (2020) for nerves with the same scaling issue approach).

Obviously, with typical resolutions of the order of 0.1 mm, this biological complexity is far from being achieved. At best, there will be approximations between models of transformation under stress.

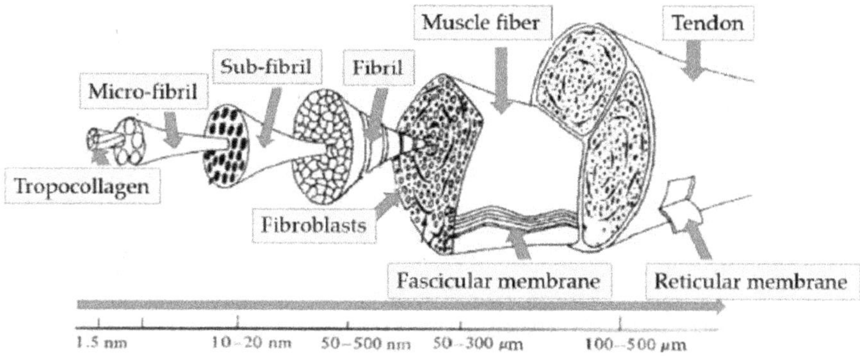

Figure 3.1. *Scale elements to be considered (case of the tendon)*

3.2.2. *Toward a transposition between theories of nature and 4D printing*

> In a civilization saturated by technology, there is no more adventure, only the risk of being late. (Sloterdijk 2006)

Emmanuel Kant (2000) defines the living as "an organized and self-organizing being. An organized being is not merely a machine, for the machine possesses only a driving force; but the organized being possesses a formative force which it communicates to materials that do not possess it. It organizes them". Auguste Comte, for his part, wrote in 1842 (Comte 2012): "It is certain that all the phenomena, whether mechanical or chemical, which take place in gross bodies, are observed in living bodies, in a very special order of phenomena, the vital phenomena properly so called, those which have to do with organization." Bichat, quoted by Le Biez (2021), who fits into these two definitions, highlights an important element, that of the arrangement that exists between the components of a living structure. It is this arrangement that provides all properties. Now, as we pointed out in the introduction to this book, the distribution of active and passive parts in a 4D structure represents an arrangement, which is certainly created by a designer (it has no DNA) and which needs to be stimulated by specific energy inputs (like any living being). This is where the comparison should end, because in the current scientific situation, 4D printing is still in a state of strong, even major, rusticity in relation to the complexity of living things. Nevertheless, some parallels and comparative comments can be made, allowing some questions and some scientific obstacles that are applicable to 4D printing to emerge.

The transposition of this field can be developed, or simply thought out, from the theories of Lamarck and/or Darwin for aspects of evolution, of Galton for relationships with eugenic selections; it can also be based on general concepts of thermodynamics and reductionism of scientific and technological difficulties. However, to approach the general functioning of a 4D structure, it is necessary to observe it from its different levels (as one can do for an animal system). Indeed, the only knowledge of the effects of a particular stimulation on a homogeneous object, already complex, is probably not enough to master the spatial or functional modifications of an arrangement including it. The multi-component 3D object is a matter of holistic thinking, for which the behavior of the whole is more than the sum of the behaviors of all its constituent parts (except of course in "simplistic" cases with a single material). Indeed, with several types of materials (at least one of which is active), distributed in a 3D object, with at least one energy stimulation process positioned in certain regions of space, the 4D parts, in the absence of good modeling, will probably have to be made by successive improvements in order to achieve a useful objective as intended by the designer. Should we then imply a Lamarck-like finalistic vision, which therefore be teleological, or a Darwin-like teleonomic vision, with a Galton-like selection? However, if we have to go back to the 4D arrangement, on what pretext can we rely to say that step n is better than step n – 1 in an incremental improvement process? Indeed, many questions arise regarding the path between an initial arrangement and one that is expected, based on an undefined or ill-defined objective function?

DEFINITION.– *Teleology*: study from the final causes, from the finality; to go up the scientific and technological path from the final objective.

DEFINITION.– *Teleonomy*: concept introduced by Jacques Monod concerning a mechanical finality of evolution implemented by living beings according to a program they would have.

3.3. General considerations

> Fashion always combines a taste for imitation and a taste for change, conformism and individualism, the desire to blend in with the social group and the desire to differentiate oneself from it, even if it is in small details. (Simmel 2003)

The object must, so to speak, bear the consequences of the inconsequentialities of its creators and those of the effects of stimuli, involved in transformations in time and space, which can modify the conditions that affect its own existence: the deformations can be uncontrolled by an approximate design (with an illusory or even impossible fixation of the ends) affecting local movements that have an effect on the object in its entirety.

At the scale of each voxel, each isolated element is subjected to stimulation according to deterministic laws. However, for an object composed of a large number of voxels, how can the same overall systems be described by equally deterministic laws? This is due to the immense number of voxels involved in these 4D systems. Because of the law of large numbers, does the variability become partially negligible (sensitivity problem) or should we take into account all the voxels involved in the transformation? To try to go further, should we not think of a cumulative process reconciling the stability of the old (3D printing) and the emergence of the new?

There are already "advanced" computational techniques that have been developed for this purpose: multi-scale approaches (Ben Dhia 1998; Ladevèze and Dureisseix 2000), parallel approaches (Farhat and Roux 1991; Mandel and Brezina 1993; Cresta et al. 2007); enrichment methods (Moës and Belytschko 2002), etc. These methods allow the introduction of more physico-chemical knowledge in the exploitation of the algorithms and, consequently, the treatment of problems with more degrees of freedom. Based on the observation that, in many situations, the critical phenomena requiring detailed analyses are localized, a common practice which allows the enrichment of a global model is "*sub-modeling*". This top-down approach consists of applying boundary conditions deduced from a relatively rough model to a detailed model of the area of interest. According to Guguin (2014), this method would ignore the overall redistributions associated with a fine description of local effects, which could lead to prohibitive errors, as shown in the case of buckling by Cresta et al. (2007). We are thus confronted, in order to model displacements in a 4D system, by the need for an overall algorithm associated with local knowledge of physical, mechanical and/or physico-chemical natures. Moreover, with voxels made of different materials, the deformations generated by the stimulations must not generate cohesive losses between voxels.

3.3.1. *The question of arrangements and the control of the arrow of time*

> The mistake of Social Darwinism was to focus on the selection of individuals rather than institutions and practices, and on the selection of individuals' innate abilities rather than culturally transmitted ones. (Hayek 2013)

One can obviously, when the field of possibilities is modest, realize all possible configurations and make comparisons. However, as shown in Figure 3.2, with an initial structure composed of n = 10 square boxes, containing four smart boxes and six inert ones, the number of configurations to be considered is 10 × 9 × 8 × 7, corresponding to 5,040 possible trials (some of which are duplicates). Of course, common sense allows us to eliminate situations that we know will not lead to the right result, but if it takes 1 hour per trial, this leads to a time of about 7 months for a 24/7 activity. It may therefore be useful to examine whether biological evolution can help us to go faster in reaching the goal, a new way of "feeling" time.

In this example, we start from an easily quantifiable objective to find an operational optimum, to consider a less random decision tree, by considering two parameters (for the readability of Figure 3.2) which inform on the value of the objective function. This assumes the existence of a single optimum and, in this case, whatever the path or its length, it can be reached, as signal processing methods have shown for several decades (see, for example, André (2017a, b and c)). However, this ignores local optima, good accuracy and the existence of many nonlinearities and recursivities, possibly induced by the effect of energy stimuli and associated diffusive processes (heat, matter, etc.). For some, this vision would have the advantage of playing with interdisciplinarity by juxtaposition, without the need to seek a major integration of knowledge (insofar as it exists). Figure 3.3 shows an illustrative example of this in a 2D space. The idea is to successively vary a variable to reach the local minimum, then change it, etc., hence the zigzag path shown on the right-hand side of the figure. In a noiseless system, with all the variables identified, if there was only one optimum, it is conceivable that this could be reached… But is this sure with nonlinear effects, time lags, etc.?

> Let us think of the factitious courtesy that makes the sterile tedium of so-called "interdisciplinary" meetings. (Stengers 2020)

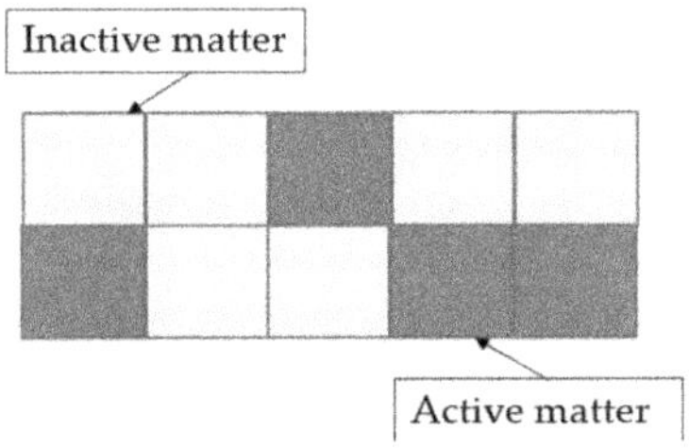

Figure 3.2. *Random distribution of 4 active elements in a set of 10*

However, with the possible existence of nonlinearities, and with a large number of arrangements with voxels with different properties in terms of actuation, preferentially oriented in certain favored directions of space, is there not a risk of finding situations where cooperation with the global movement sought can exist at the same time as certain forms of exaltations in the deformations? We must consider that a typical object is made up of voxels of 0.1 mm on average and of a size of the order of magnitude of a decimeter. Under these conditions, the number of voxels is of the order of 10^8–10^9 voxels, which indicates a fundamental impossibility of processing each active element and its interactions with its neighborhood! In this shifting unknown, it is possible to find, in certain situations of arrangements and stimulation on the same 4D object, an ambiguous "game" of altruism and egoism (to make an analogy with human behavior) with voxels which behave, for example, as independents that are harmful to the desired change or as "stowaways"?

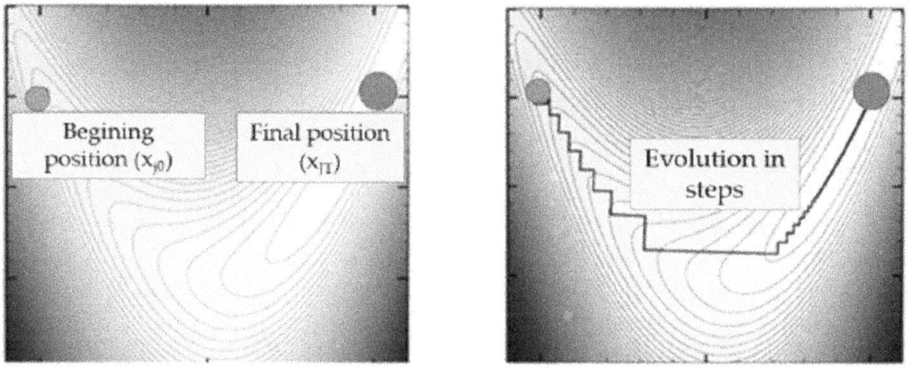

Figure 3.3. *Search for an optimum (supposedly unique) by successive changes of variables that are considered as independent (in the case of Figure 3.1, positioning of the active elements in the support matrix): Alternating path with unidirectional search for the local optimum. For a color version of this figure, see www.iste.co.uk/demoly/4Dprinting1.zip*

> Finally, it is possible to "play" on "serendipitous" luck by a method of trial and error by limiting the influencing variables and choosing them randomly in order to collect the results and then control the robustness; after as many trials as possible, one can carry out a mapping of the system like the one proposed in Figure 3.3 (as long as one has a measurable definition of the objective function). Experimental designs can be exploited to try to reduce the number of trials (assuming that one has a small idea of the approximate position of the optimum that is sought and provided that it exists and is unique)...

Box 3.1. *A trial and error method*

In a way, we are close to what Salviati, Galileo's spokesman, wrote (quoted by Julien (2020)):

> The difficulties and objections that you have just put forward are so well founded that it seems to me impossible to dismiss them. [...] I grant that the conclusions established in the abstract are modified in reality, and are shown to be so inaccurate that neither the transverse motion can be uniform, nor the natural acceleration can be carried out

> according to my admitted proportion, nor the trajectory of a projectile have the shape of a parabola.

How then to go a little further on this subject?

3.3.2. *Complexity induced by the stimulation*

> It is not the ability to change form in accordance with external pressure [...]. It is essentially the ability to learn from experience [...]; it means the power to modify actions on the basis of the results of previous experiences. (Dewey 1997)

The concept of "phronesis" was coined by Aristotle to designate a mode of knowledge and action required when the mechanical application of scientific knowledge, routines or standardized protocols would be defeated by the complexity and singularity of the situation or problem to be addressed (it is just a little more thorough than empiricism informed by common sense). The consequences of this complexity and singularity are the irreducible uncertainty of the results of human scientific action and, consequently, the particular difficulties encountered in formalizing these actions. It cannot therefore claim to be universal. Recognizing that a situation is a matter of practical wisdom means accepting a certain amount of subjectivity. However, this does not authorize doing just anything. On the contrary, showing "phronesis" has several implications:

Analyzing a case in its complexity before deciding on action, trying to keep an overview.

– Allowing collective deliberations on this case.

– Being alert to signals that could be clues to hidden problems or difficulties.

– Finding a justifiable balance between requirements that the complexity of the cases handled may bring into tension ("efficiency" of a complex 4D manufacturing experiment), etc.

However, if we can agree on an understanding of an approximate future, will we have reached the goal of the project, which is to manufacture a material with a precisely known evolving form, in the real operational time (which is still very far from the resolution of the inverse problem; see, e.g. Sherman et al. (2020))? The experimental verification of the hypotheses made to build the model would then have to be based on a cycle of back-and-forth experiment-calculation verification, a validation that calls into question the validity of the experiments, their own precision, the validity of the calculations with their programming errors, the existence of bifurcations and the discrete approximations of mathematical expressions that are generally continuous. The restitution of a set of singular experiments is necessary to access the average that the model aims to define. Thus, indeterminacy and the application of laws of large numbers can contribute to excluding singular events or events involving non-locality and dependence on verification processes from the field of models. The epistemology of the model approach may thus prove to be a singular limitation on the explanatory ambitions of the sciences called for in 4D printing; but it does allow scientific knowledge to be considered in a more balanced and rigorous framework (Perrier 2003).

Several types of complexity have been revealed by the practice of modeling: the complexity of the arrangements of voxels in space and the complexity of the dynamics of growth, transport of matter and energy, etc. The number of possible cellular and inert edifices produced from simple elements, their random or deterministic arrangement with structural scales, clearly introduces a diversity that is difficult to control. Could the modeling use fibrous, laminated aggregates alternatively with structures made of non-living resorbable material? The mathematical tool of choice to go from small to large scales could be the homogenization and identification of "fractal" shapes, that is, with repeatability or multi-scale shape differentiation (Perrier 2003; Pavé 2006; Heard 2013).

3.3.2.1. *An example*

> A mathematician is a machine that turns coffee into a theorem. (Erdös 2000)
>
> The pessimist sees the difficulty in every opportunity, the optimist sees the opportunity in every difficulty. (Churchill 2017).

Let us consider a simple planar structure such as the one shown in Figure 3.4. The expansion induced by the stimulation can lead a priori to a deformation either upwards or downwards of the initially flat shape, without it being possible, in this example, to know in which direction the curvature will occur. As it happens in nature, complexity is an element to be taken into account when it comes to the control of deformations in view of an operative end.

> The set of genes draws the landscape and the probability of following a given path is generally very high, the other paths or creodes being unlikely, which does not eliminate their existence. A single genome therefore allows for several such creodes, which means that there are reserve phenotypes, distinct from the majority genotype, and novel environmental conditions may favor one of these reserve phenotypes… (Prochiantz 2012)

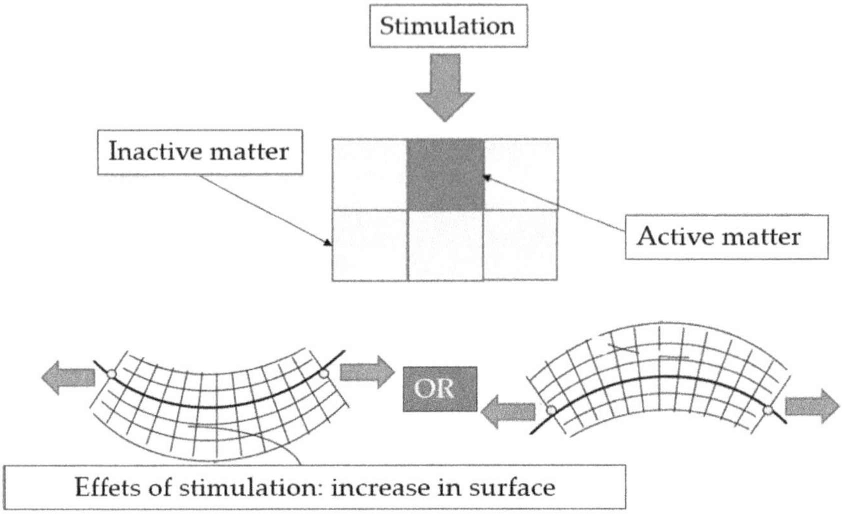

Figure 3.4. *Demonstration of stimulation-induced bifurcations. For a color version of this figure, see www.iste.co.uk/demoly/4Dprinting1.zip*

Would not there be some connection between 4D printing and biology?

> As a reminder and according to Sheldrake (2003), Blanchoin et al. (2014) and Heard (2013), the following words (found in the literature) are defined as follows:
>
> – Entelechy: phenomenon comprising in itself its end and its objective.
>
> – Creode: morphogenetic field (*cre*: it is necessary; *odos*: path).
>
> – Epigenesis: theory of development by progressive elaboration of forms.
>
> – Motility: refers to the ability to move spontaneously or in response to stimuli and actively, consuming energy in the process.

Box 3.2. *Some definitions used in biology*

3.3.2.2. *A return to complexity*

> The current metamorphosis marks the beginning of a new era, in all fields, and not even the law is exempt from needing to be reformulated. For beyond a production technique, 3D Printing is a paradigm shift, impacting the design of products, their creation, their consumption, and the business models that follow. (Charles 2016)

In operations of interest to the public, there is an increasing emphasis on creating immersive experiences. Examples of research use digital technologies, such as virtual and augmented realities (Wojciechowski et al. 2004; Carrozzino and Bergamasco 2010; Jung et al. 2016). To date, there is little evidence of how physical objects can contribute to these experiences (Hone 2018). Yet, there are ways to create robotic movements via electronics, pneumatics or hydraulics (Kalnad 2016) that can make objects feel alive (at least with movement). However, realistic natural motions are difficult to achieve due to the complexity of coding and the reduced motions of mechanical parts; the movements can sometimes seem jerky. Soft robotics, such as the Tetrapod (Shepherd et al. 2011), uses pneumatic chambers to create smoother movements. This technology is intended for practical applications (Rus and Tolley 2015). 4D printing on its own could give shape and animate objects. This raises the question of where to place active elements in a passive matrix, as conventional actuators do in robots and automata. This aspect is addressed here.

Bergson (2019) wrote:

> But this is what we have great difficulty in understanding because we cannot help representing organization as fabrication. Yet it is one thing to manufacture, and another to organize. The first operation is proper to man. It consists in assembling parts of matter that have been cut in such a way that they can be inserted into each other and obtain from them a common action. They are arranged, so to speak, around the action which is already the ideal center. Thus, the production goes from the periphery to the center or, as the philosophers would say, from the multiple to the singular. The work of organization, on the contrary, goes from the center to the periphery. It begins at a point that is almost a mathematical point, and spreads around this point in concentric waves that are always expanding. The more material is available to this fabrication process, the more efficient it becomes. It proceeds by concentration and compression. On the contrary, the act of organization has something explosive about it: it needs, at the beginning, the least possible space, a minimum of matter, as if the organizing forces were only reluctantly entering the space. If we replace centration by layers, we can follow what is written [...]. But, fabrication and/or organization pose the problem of the contribution of active voxels placed "in the right place" in the elastic support matrix [...].

Therefore, it is natural to look for a mathematical rigor, algorithmic procedures that can ensure, at the interface between technology, engineering sciences and 4D printing, a conceptual body of doctrine that could ensure a new knowledge for the production of deformable objects or whose functionality evolves according to a given specification, knowing that no specialist can, in principle, exceed a culture much wider than that of their discipline. In a desire to make things as simple as possible, the mobilizing idea is to found a robust theoretical/empirical model whose surplus of descriptive content is minimal compared to what is essential to account for the phenomena. But is it that easy to implement?

In the case of 4D printing, several disciplines are at work (3D/4D printing, various materials, energy stimuli), which leads to "modeling" mixtures/ mutualizations of domains with behavioral equations that are likely different, multiple and singular interfaces, nonlinearities, scale changes associated with possible fractal or metamaterial forms, etc. (Nottale 2008; Chaline 2012). Complexity invades the field by several routes: that of geometric arrangements, of complexity of dynamics (initial arrangement, spatially and temporally resolved stimuli, tensions in arrangements for example) and that of functionality… and all probably still greatly incomplete.

Abdelmalek (2004) refers to the "systems movement" as a set of activities concerning the dynamics of natural and even cultural systems. They are based on a number of presuppositions, the most important of which are:

– the existence of common, transdisciplinary general laws governing complex and highly interactive systems;

– the relational (or cybernetic: internal or external interactions) nature of laws;

– the "holistic" character of certain properties, in the sense that they concern the whole system. Some emergent properties only exist and make sense at the level of the system as an indivisible whole; the degree of autonomy depends on the structure in space and time and on the logical organization of the whole system involved.

Finally, the existence of general laws and invariants would not imply that systems are deterministic and predictable. On the contrary, systems composed of various pairings of passive and active voxels would be very sensitive to the interplay between local contingency and relational necessity. The systemic approach is thus a grid that prepares for a paradigm shift. But what do we actually mean by complexity? Ricard (2003) considers a set of two subsystems X and Y for which he introduces functions H(X,Y), H(X) and H(Y) expressing the number of degrees of freedom.

If H(X,Y) = H(X) + H(Y), the properties of the set can be deduced from "only" the components X and Y (it behaves as a result of the union of the two subsystems). If H(X,Y) is less than the sum H(X) + H(Y), the potential richness of the set results from a reduction in terms of the degree of freedom following an integration phenomenon; if H(X,Y) is greater than this sum, new properties emerge from the association. These two situations are considered as complex (and not complicated in the sense of Ferrari (2002)).

In the case of 4D printing, this complexity appears in situations such as:

– The coupling (space, "inert" materials, active materials, energy flow) results in various pieces of information that have a role in the desired dynamics; the active part is able to modulate this information according to signals emanating from outside its environment (and perhaps also inside it); the complexification increases if the number of active voxels or the number of stimuli increases.

– The principle of superposition (one voxel activated at a time) is not valid because of interdependencies.

– Is an experiment with a population of N active voxels then the same as two experiments conducted with N/2 active voxels?

– The dynamic behavior of the system cannot be described by its local dynamic properties that play on the other subsets and by its collective properties that are characteristic of the overall system.

– The system is neither loosely nor fully organized at different spatial scales.

– The system is open, exchanges matter and/or energy (as well as possible information) with its environment; it is therefore not in equilibrium.

– The system has its history and its state depends on it (chaos?).

– As a result, such a system exhibits nonlinear effects.

This problem becomes even more complicated according to Ricard (2003), because in these systems where there can be successions of bifurcations, threshold events, etc., the number of interdependence parameters to be taken into consideration increases exponentially (combinatorial explosion). In an attempt to go beyond this pessimistic observation, Nicolescu (2002) reintroduces the concepts of levels of reality previously developed by Heisenberg in 1942; they are three in number:

– The state of the quantifiable elements, independently of the knowledge process (a priori).

– The state of the elements that are inseparable from the knowledge process.

– The state of the elements created in connection with the knowledge process (applications, effects, ethics, etc.).

However, this work assumes that the variables influencing the overall system are under control; in any case, this is a necessary exercise to be carried out, even if it means including it in a reflection with feedback(s). But even if the system has a nonlinear dynamic behavior, can we find conditions so that the spatial and/or functional trajectories are not divergent or strictly out of control? Various situations should be considered:

– There is a "basin of attraction" such that all trajectories remain within a domain that is "acceptable" to the end user.

– There are initial conditions that lead to diverse trajectories, but which end in the aforementioned acceptable domain.

– It is possible to introduce a corrective control structure affecting the system dynamics, such as growth agents (see Box 3.3).

– The system admits a stable trajectory (apparent causal model), as is the case in simple situations, such as the one shown in Figure 3.5.

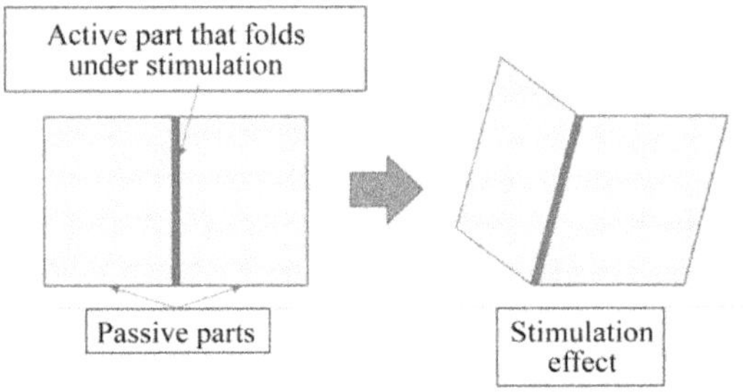

Figure 3.5. *Causal deformation in a simple case of origami deformation*

Let us remember that to qualitatively explain the origin of a shape, independently of the determinisms of 3D printing, we can resort to the presence of internal or external forces applied to a set of elements. However, the explanation of a shape by forces often proves impossible, because it frequently happens that the acting forces are either unknown or are of such complexity that they escape the possibility of predictive calculation. For Lesne (2002):

> In open systems, traversed by flows of matter, energy or charges, the observed forms are explained by the underlying dynamics, of which they are stationary states. This dynamic point of view naturally links the notion of form to that of self-organization. This term summarizes the fact that a dynamic equilibrium is established spontaneously and, implicitly, that the assembly of the various elements through exchanges, interactions, and various couplings leads to a new result, qualitatively different from each of its parts.

Moreover, in nature, it is possible to recall the existence of a paradox between continuous and discrete approaches whose resolution by infinitesimal calculus and differential equations can be approximated. The notion of chaos comes from this deterministic approach and is associated with the large variations of solutions of equations by perturbation of some parameters. "This de facto uncertainty is opposed to the uncertainty inherent in the quantum model, which is effective for the description of the small scales of Nature, where the uncertainty relations induce permanent fluctuations that make evolution ever-possible" (Dubois 2002).

Chaos can arise from differential equations as shown by E. Lorenz (1960) and D. Ruelle (1988). For "well-chosen" systems of differential equations, a small perturbation of the initial conditions can lead, after a finite time, to a deviation on the solution of the system which is of the order of magnitude of the size of the explorable configuration space. Motion prediction is impossible from a practical and algorithmic point of view; a sufficiently accurate discretization of the system is too complex and does not allow a satisfactory prediction. According to Dubois (2002), phenomenology is analogous to discrete systems and the Hénon (1976) attractor would be an example.

Box 3.3. *Chaos*

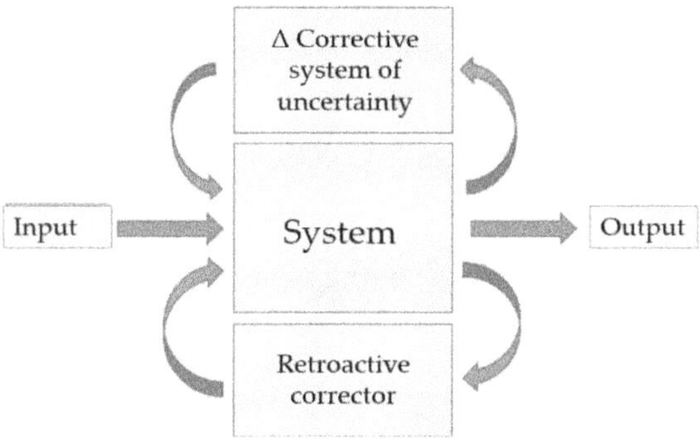

Figure 3.6. *Control model. For a color version of this figure, see www.iste.co.uk/demoly/4Dprinting1.zip*

COMMENT ON FIGURE 3.6.– *Figure 3.6 shows a system with feedback control. However, since experimental measurements are fraught with random errors and even biases, this imprecision adds a degree of uncertainty to the model. It is possible to consider the contribution of a model of uncertainty Δ; then, with the model and its structure defined, the objective of the control will be to ensure the stability of the system to achieve the goal of the printing operation. Would it be possible under these conditions to reduce the sensitivity of the control structure in the presence of unknowns, parametric variations and possible disturbances affecting the system in its spatio-temporal and functional evolutions? (Cinquin and Demongeot 2002, 2005).*

As long as one masters the knowledge of the mobile system under stimulation, it is possible to find analogies between systems coming from automatic control and 4D printing (except that it is more complicated than making two electric trains cross each other… See also Ramasubramanian and Taber (2008)).

3.3.2.3. *Bifurcations*

> No single experiment can play a decisive, let alone "crucial", role in tipping the balance between two rival research programs… however, I do not deny that a few experiments have a decisive psychological effect in the war of attrition between two programs. (Lakatos 1994)

It should be remembered that we are working on a body of incomplete knowledge whose degree of generalization to larger or smaller scales may be questioned. The idea, however, is to study the control parameters (temperature, various flows, various heterogeneities, spatial distributions, etc.) to try to examine the existence of attractors and their deformation(s) induced by the environment. Indeed, most of the processes responsible for 4D functioning result from interactions between passive and active voxels in the presence of a spatio-temporal stimulation. "Deepening their understanding therefore implies mastering tools that can represent and analyze these complex interactions" (Vandeginste 2012).

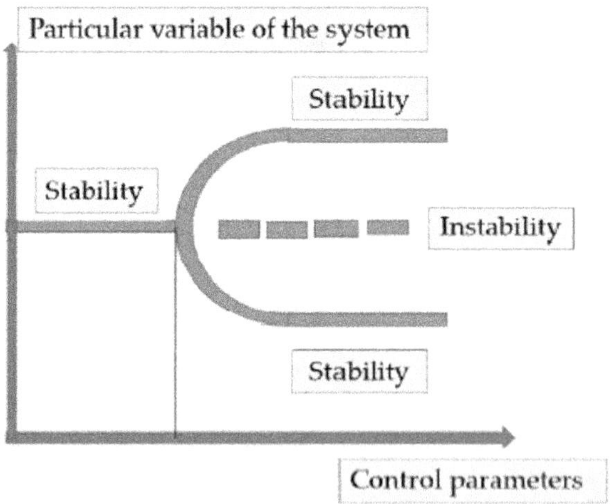

Figure 3.7. *Fork bifurcation. For a color version of this figure, see www.iste.co.uk/demoly/4Dprinting1.zip*

It should also be remembered that in linear systems, changes in parameters lead to quantitative changes, but do not modify the behavior (steady state) of the system. In nonlinear dynamical systems (NLDS), a small variation of some parameters, called control parameters, can, under well-defined conditions, in the vicinity of a critical value, cause a complete change in the equilibrium behavior of the system. This is called a bifurcation. Figure 3.7 shows such a "fork bifurcation".

Thus what a priori appears "simple" is in fact the result of a concerted evolution of various characters, moving toward a real optimization of means. It is therefore, for a 4D object, something very complex and "the primary vision of ancestral simplicity disappears in this reversal due to evolutionary reasoning" (Le Guyader 2008). Figure 3.8 from Delignières (2015) recalls, in a 2D representation, situations that are associated with possible bifurcations (reaching different states from the same law of behavior with multi-stable states).

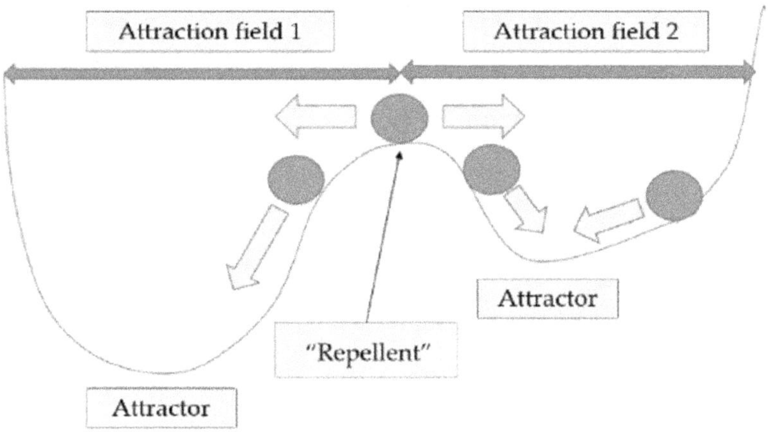

Figure 3.8. *Schematic representation of a landscape of attractors. For a color version of this figure, see www.iste.co.uk/demoly/4Dprinting1.zip*

In this figure, the stability levels are in the attraction basins while the "repellers" correspond to the maximum instability zones; in the figure, the ball can go either to the right or to the left depending on the very low amplitude impulse it receives. The essential characteristic of the attractor is the stability of the order parameter with a strong tendency to reproducibility. According to Delignières (2015), bifurcations are announced by critical fluctuations, that is, a significant evolution of the order parameter in the vicinity of the transition (as in a modification of the orientation of molecules in an organized system as soon as one exceeds a transition temperature). Ultimately, it would be possible to find an applicable approach if, during these evolutions, the number of basins of attraction and their qualities are mostly unperturbed. Or, to put it differently, is it possible to show the possibility of having, as proposed by Thomas et al. (2004), "chaotic labyrinths" that constrain complexity? For bio-printing, starting from a given position of cells in a given environment, does the evolution of the system always converge toward the same solution? In any case, the information resulting from the experimental approach to be undertaken should provide data (if possible, robust data) on the living systems studied and on their evolutionary dynamics, with the possible existence of basins of attraction. If only one exists, it is possible to think that the evolving system can be approached using a deterministic model. But will this be a general case? If we refer to the paper by Olle-Villa et al. (2016) which does not illustrate this potential of having only one deformation path in a 4D system, there is room for doubt (see Figure 3.9).

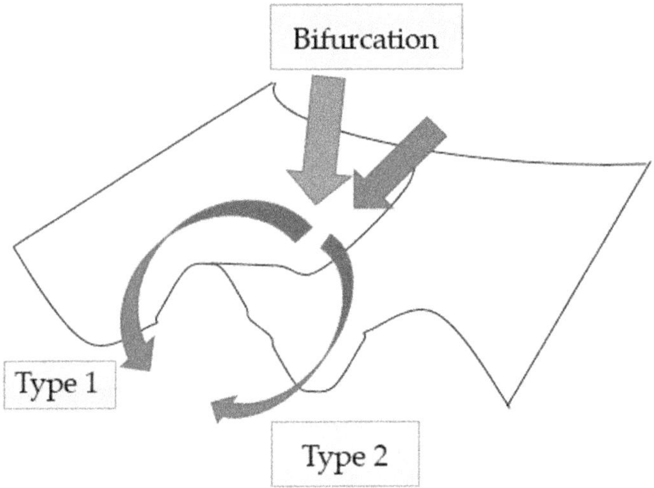

Figure 3.9. *Bifurcations in a 4D system: in the displacement (red arrow from the top) the system arriving at the bottom of the blue arrow can go either left or right. For a color version of this figure, see www.iste.co.uk/demoly/4Dprinting1.zip*

The realization of "custom" and mastered heterogeneous 4D systems may indeed be a way to approach reality (a little) in the sense that it allows a certain degree of (in principle) high complexity (as long as one has, at least in part, a vision of reality and, consequently, a sufficiently robust modeling base) to perform experiments that will indeed reinforce the concept (or disprove it) (Sève 2005; Chang et al. 2011; Laurent 2013; Abbott and Kaplan 2015). Such a scientific approach should make it possible to know whether a certain degree of predictability can be achieved in the making of a 4D organization. It is then indeed a matter of having experimental study tools to validate a conceptual framework for considering a satisfactory solution to a 4D question.

The challenge would therefore be to be able to carry out large-scale scientific research in order to have a knowledge base that would allow us to qualify the causal credibility of the 4D project, regarding the ability to respect a functional specification. Thus, to take an example, if a system that is made up of one voxel (or several) can undergo a bifurcation that is not controlled by the control of external parameters, how can one envisage the relevance of a more complex transformation with credible probabilities?

3.3.3. *Toward a principle of parsimony?*

> As I see it, parsimony is simply the operation of optimizing on a criterion. (Kluge 1994)

"*Entia non sunt multiplicanda praeter necessitatem*"[1] (William of Occam, quoted in (Anderson 2002)). The principle of parsimony was introduced by Occam in the 14th century (Sober 1981; Ohlsson 2012); it states that a hypothesis should not be asserted, or an entity postulated, if it is not necessary in order to explain a phenomenon (Boehner 1957), so the famous razor slices away what is not necessary. Modern justifications of parsimony have presupposed this formulation of the principle, and have sought to justify it on the grounds of probability. Russell (1951) and Quine (1966) consider that removing a claim from a system has the effect of increasing the probability of what is retained. Put simply, this principle requires that we assume that an entity does not exist if its postulation has no explanatory value. In some situations, two seemingly independent components reduce each other. What changes is our understanding of what must exist if the theory or model is true. In fact, this principle applies to theories and probably less to the desired reduction in terms of the number and position of active voxels in a given object for a selected action.

The maximum entropy principle entered physics as a conclusion drawn by Gibbs (1902) from his description of classical statistical mechanics. In its most appropriate form, it was established by the theorem that "if a set of systems is canonically distributed in phase, the average probability index is lower than that of any other distribution in the set with the same average energy" (Rudnicki 2014). Knowing the behavioral laws of the elements under stimulation, it is, in principle, possible to define an "objective" function relying on this principle. For the designer, the "game" then consists of trying to define sets that are as concentrated as possible in active modules to get as close as possible to the theoretical optimum. It is therefore not a reduction in the sense of Occam, but an engineering task relying on parsimony principles (Assis 2015) aiming for the satisfaction of a "least cost" transformation.

1 Entities must not be multiplied beyond necessity.

However, at this stage, the interaction patterns between active elements and with passive elements are not known; moreover, they strongly depend on the stimulation modes and the spatial and temporal localizations applied. In short, there is still a long way to go before mastering a complex transformation.

Pidd (1996) proposes several methods for dealing with this type of problem, one of which consists of dealing with the optimal deformation problem by degree of simplicity, moving toward a complication adapted to the objective (see Figure 3.10). It is a question of learning from simplified (but reasonable) realizations and either rejecting the test for non-conformity with the specifications or improving it (see also Powell (1995)). From one step to the next (as long as the improvement follows a path without local blocking minima), one should obtain a potentially satisfactory result.

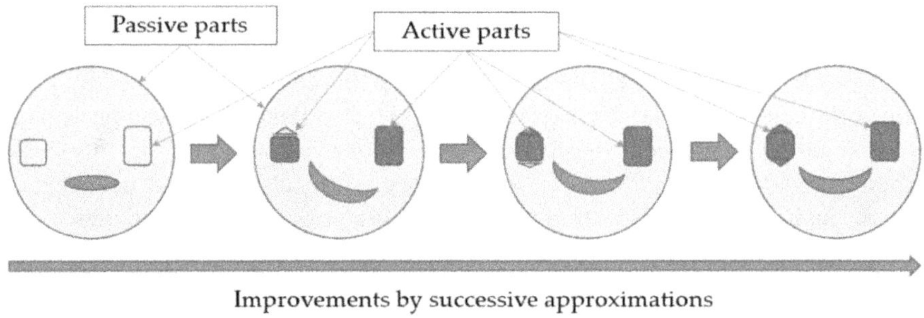

Figure 3.10. *Improvement by incremental approach to the objective (by acting successively on each active element). For a color version of this figure, see www.iste.co.uk/demoly/4Dprinting1.zip*

The irreducible solidarity of the three problems of "catastrophist" structuralism emerges from Thom's work (1968): modeling of the processes of morphogenesis, schematization of structural categories and phenomenological question. From a structuralist perspective, the aim is to improve the formal description of morphology, by highlighting its regularities and hidden symmetries, by showing "its global character through the description of a formal process that generates it axiomatically" (Thom 1968). In order to be applicable, it is presupposed that "phenomena" and "morphologies" are identified and that, in order to be operative, we can access a mathematical idealization of the initial notion of morphology (Petitot 1983). By taking into account possible symmetries, there is already a way to reduce the complication of the system to be optimized.

DEFINITION.– *Morphogenesis*: is normally the set of laws that determine the shape and structure of tissues, organs and organisms; here it is applied to the creation of an object from voxels.

> Richard Feynman said, "You can recognize truth by its beauty and simplicity.... When you get it right, it is obvious that it is right ... because usually what happens is that more comes out than goes in [...] truth always turns out to be simpler than you thought". (Anderson 2002)

3.3.4. *To go a little further*

> The whole theoretical arsenal of mechanics, chemistry, etc. is in principle useful for transforming the rough facts resulting from observations of the effects of a stimulus into scientific facts and, much more, for bringing to light and inventing situations that our senses do not perceive or that nature does not offer us directly. "To explain is to turn a physical problem into a mathematical problem. [...] The process known as mathematization presupposes exactly that all differences between physical intelligibility and mathematical intelligibility have been abolished". (Clavelin 1996)

Starting with a set of selected and spatially distributed materials constituting the object, a 3D printing approach and a deposition programming method, the final step to establish 4D printing is, in principle, to model the entire structure to predict its shape change or functionality behavior in the presence of time and spatially distributed energy stimuli. Shen et al. (2020) believe that a modeling algorithmic shape optimization capability could be performed, allowing users to optimally design specific structures to achieve desired shape changes and associated functions. However, they admit that there are currently few tools to achieve these goals (Chung et al. 2017; Sossou et al. 2018).

According to Champion et al. (2020), machine learning (ML) can provide solutions to the question presented in the previous section in data-intensive scientific and technical domains (Bongard and Lipson 2007; Schmidt and Lipson 2009), provided the following challenges are overcome:

– Physically interpretable models should predict previously unobserved behavior and provide effective predictions (by extrapolation).

– Autonomous systems will necessarily interact with changing and uncertain environments, hence the need for models that can accurately extrapolate based on physical principles (Lusch et al. 2018).

Conventional ML approaches have shown impressive performances in predicting dynamics in an interpolation-based regime, but the resulting models often lack interpretive power and fail to be generalizable. The authors use a regression framework aimed at the use of regularization functions that addresses three challenges:

– Robust handling of outliers and corrupted data in noisy measurements (Pathak et al. 2018).

– Parametric dependencies on the candidate functions used.

– The imposition of physical constraints.

By explicitly addressing these open challenges, the developed integrated and unified algorithm represents a significant advance over current methods. This flexible approach can be adapted to a wide range of applications and datasets, providing a powerful framework for the learning processes of governance models of physical systems based on data. However, in the approach presented briefly in the previous section, what is sought is to be able to work on a minimal number of evolving test forms; this method can be kept in mind, but does not provide useful paths that can be quickly exploited in this phase.

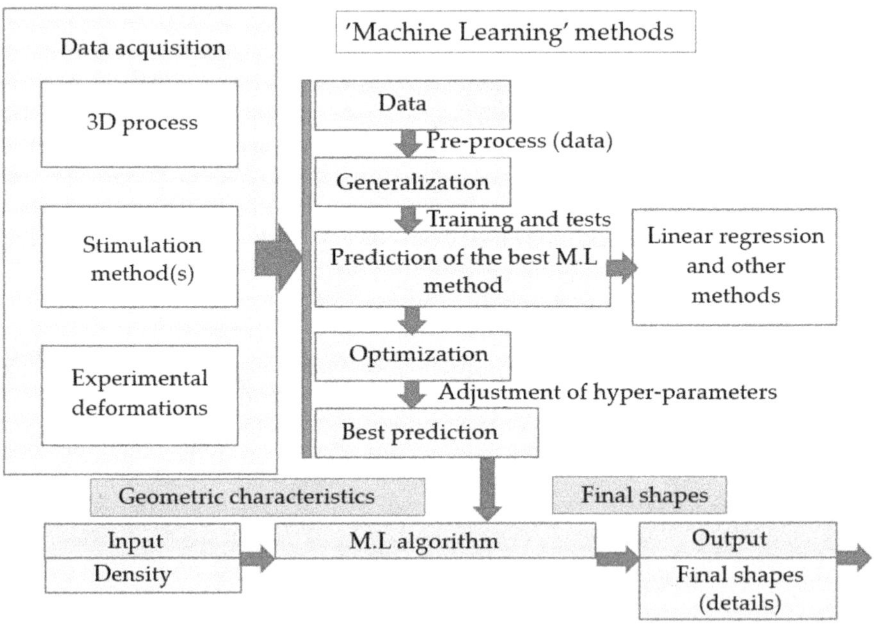

Figure 3.11. *Data-driven modeling flowchart for predicting morphing behaviors of active 3D printed structures (it includes data collection, preprocessing and hyperparameter setting for additive manufacturing model development). For a color version of this figure, see www.iste.co.uk/demoly/4Dprinting1.zip*

Obviously, understanding and predicting the *morphing* response of active 4D structures remains a real challenge. Su et al. (2020) propose an approach based on data and a set of ML algorithms. Used to quantitatively correlate a geometric feature (thickness), they estimate the final "*morphing*" shapes defined by radii and angles of curvature. However, it is necessary to collect a lot of data (150 experiments were involved in the publication). Furthermore, to predict the dynamic response of the printed structures, three time series prediction algorithms have been implemented based on image data during the deformation processes of the 4D structures. This work is still only a proof-of-concept (POC) of using artificial intelligence to delineate and predict the deformations of active printed structures. This path is therefore to be followed in its future extensions and validation on complex 4D objects, as it is only the very beginning, and, as in the previous case, requires a relatively high number of experiments, without knowing where the active volumes and stimuli are located. (see Figure 3.11).

Hydrogels can exhibit satisfactory mechanical performances due to the nature of the long polymer chains that constitute them. In Naficy et al. (2017), hydrogel hinges were fabricated and a simple model was developed to predict their bending characteristics, including curvature and bending angle. This model can be applied to define the optimal parameters for many hydrogel materials (see, in another setting, Deng et al. (2017)).

Box 3.4. *Case of rotation around a fold*

More recently, Li et al. (2021) have designed composite, possibly multi-material, meta-structures with tunable elastic responses under finite deformations. They exploit an inverse problem where the errors between the actual (numerical) and prescribed force-displacement curves are minimized. From numerical design scenarios, it is thus possible to design multi-material structures that realize a variety of programmed load-deformation curves, some of which are physically unfeasible with simple materials.

Box 3.5. *Metamaterials*

In these two more specific but relatively simple situations, it is therefore possible to achieve controlled relationships between materials and their deformations.

3.3.5. *A partial fallback situation*

> Nothing makes the mathematician more uncomfortable than a discontinuity. (Thom 1980)

For Ding et al. (2018), thin 1D structures are strongly present in nature and in engineering. They can serve as building blocks for 3D structures at scales ranging from molecular to architectural. 1D rods with composite cross-sections, programmed to deform into a prescribed 3D shape simply by heating, composed of a glassy polymer and an elastomer, are made by 3D printing. The elastomer undergoes compressive stress during printing and when heated, the rigid glassy polymer softens, releasing the stress in the elastomer and causing the 1D structure to deform into a new permanent 3D configuration. Through the use of a nonlinear thermomechanical computational model, several 3D bar structures are designed and demonstrated, highlighting the potential for increased functionality with material and time savings (see also Weeger et al. (2018, 2019)).

Thus, if we are beginning to be able to move forward on the deformation of 4D wire, the reverse problem is still far from being solved; the same is obviously true for more complex 2D or 3D structures.

3.3.6. *The reverse problem*

The solution of the reverse problem may not exist, partly because the tools available on this subject are limited (Shen et al. 2020). At the moment, analytical calculations are only relevant for certain simple geometries. This is the case for 4D systems involving, for example, hinges. For these systems, an analytical model that explicitly calculates the bending performance of these hinges (e.g. bilayer actuators) could be sensibly used (Gladman et al. 2016) to achieve the targeted actuation performance. By combining simple bending patterns, a series of self-folding structures could be realized, forming somewhat more complex predefined textures and morphologies (Kwok and Chen 2017). With more complex geometry, the analytical approach is no longer possible and numerical methods must be incorporated.

Using finite element methods, a possible design approach is to perform a parametric study and find the parameters that give the closest response (Bodaghi et al. 2016; Su et al. 2018). However, for all that, publications regarding a teleological approach for complex 4D systems, such as volumetric objects, are not abundant. Shen et al. (2020) believe that a design approach exploiting ML could be applied to composite structures with active and passive materials (Hamel et al. 2019). However, for the authors, this field still remains largely to be cleared. Indeed, a more complete approach would have to take into consideration aspects of anisotropy of deposits and the effects of stimuli, functions of time and space, and of nonlinearities present in 4D systems.

3.4. A view from thermodynamics

> The desire to reduce or control risks to an excessive degree is at best an illusory quest, at worst a brake on innovation and performance. (Durand 2020)

It seems difficult to think about the notions of order, progress, frontier, equilibrium or temporal transformation without invoking the concepts of entropy, evolution or scale effect. In the realization of 3D/4D objects, the construct subjected to (at least) one energy stimulation can, in a way, self-organize (this is what happens in the change of form) and see its entropy decrease (while that of the whole increases). The emergence of order makes it possible to evoke the existence of a transformation indicator which is the entropy associated with the object undergoing transformation.

Prigogine and Stengers (2012) remind us that: "Far from equilibrium, correlations arise that require us to modify our conceptions of what it means to be together, even for molecules, that is, to transform our modes of representation of what is the macroscopic level and our criteria of selection about what is or is not meaningful". However, we have seen that being together leads us to work in a more or less incremental way. Therefore if the entropy of the 4D system (without the stimulation) can be defined, if we want to go from initial situation A to final situation B, whatever the path, if in the process of incrementing the entropy is a decreasing value, we effectively reach B. As we will show in the next section, we follow a Lamarck-type process. However, if several minima exist, it is no longer possible to follow an incremental method. Random firing methods can be used, possibly associated with incremental evolutions. We would associate Darwinian selections with Lamarckian sequences. The Darwinian mechanism corresponds to an evolution "where nothing is planned" (Young 2005), whose mechanisms proceed blindly, removing the idea of intentionality, just as his German predecessor was already astonished at the lack of necessity of the forces that govern the universe, at the fundamental absence of ends within the very purpose (Rosset 1994; Schiano 2012).

The Darwinian epistemological approach implies breaking definitively with a conformist, even static and routine conception of action, by opposing, in a realistic way, the trial and error method to the more methodical, even abstract approach of the classical physical sciences, which are still used too often in science as a basic epistemological model. Darwinian evolution translates, as far as 4D printing processes are concerned, into continuous mechanical processes of adaptation. However, as we can see, we have to move forward carefully in the midst of complexity, without taking too much account of methods that have already proved their worth, with a scientific and technological field that has yet to be shaped. It is likely to take some time before we have a robust methodology.

3.5. Darwin, Lamarck and others…

> Evolutionary theory is the most historical subject in the biological sciences, in the sense that its problems extend over the longest time scale. (Sober 1993)

> We must avoid believing that all sciences prove in the same way. Judging another science by projecting one's own system of proof without having verified what was at work there is an unfortunately common error. (Lecointre 2004)

With 4D printing, one of the central problems facing scientists is the problem of the succession of shapes and their control in the presence of a single stimulus (regardless of features). There is probably no single solution to this issue that could be addressed through artificial intelligence, knowing that learning depends on the quantity (and quality) of sampling (Jean 2019). The presence of a good model would permit the numerical performance of experiments in minimal numbers and perhaps a consideration of the treatment of the inverse problem. On this basis, we found it interesting to consider whether knowledge from other scientific fields could be exploited. In this respect, we examine below whether it is possible to consider a model from evolutionary theories that could be used for this purpose.

3.5.1. *Between Lamarck and Darwin*

As long as we do not have a true model of the object being transformed (except in simple cases such as origami or thread transformation (Weeger et al. 2019)), we must proceed by selection and eliminate the "bad subjects" that do not come close to the objective envisaged by the designer. We find ourselves in a principle of improvement of the "4D" species as Galton or Charles Binet-Sanglé could have proposed for the improvement of the animal race or the human species. For the latter scientist, it is a project of scientific improvement of humanity. However, the only thing that makes it possible to distinguish individuals from the point of view of their value is their objective individual potential, and not their belonging to a particular ethnic group, social class or family (see Pichot (1995), Binet-Sanglé (1918), Galton (1869) and Hello (2020)). Fortunately, we are only dealing with non-living matter in this book!

In the absence of a 4D model, we can also start with Plato and the "allegory of the cave":

> Men live in a half-dark cave and believe that what surrounds them is reality, because they have never left their little world. They are

> chained to their ignorance. A pale glow filters through from outside, casting shadows of another life on the walls of their existence. They know that there is something beyond their tiny universe, but unable to give up the certainties that make them feel secure, they prefer to turn their backs on this strange world that scares them. In spite of everything, one of them manages to leave the cave to go, as an adventurer, to meet the unknown. During his journey, he discovers things he does not understand, things that destabilize him but that give him the necessary spirit to understand a life whose complexity he could not imagine. He wants to return to his companions to share with them what he has just experienced. But will they listen to him or will they consider him no longer part of their community now that he has become so different? (Macquart 2020)

This is indeed the impression that can be felt, beyond spectacular but simple situations related to 4D printing that only show us appearances. "The essential is invisible to the eyes" (Saint-Exupéry 2007).

This is the reason for this peregrination – a kind of intellectual game – in other worlds where the theories of evolution take their full dimension with two reasons in mind: the first is related to the conception of the 4D object by successive improvements, the second is associated with transformation under constraints which remind us of these theories (but with temporal bases which have nothing in common). It is therefore an invitation to a journey.

Nature is capable of self-organization, so too can 4D printing be: to simplify: molecules or voxels in a form or container can behave holistically, that is, they follow an overall plan that goes beyond each of them (Demoly and André 2021). As shown earlier, complexity is a property of non-equilibrium and nonlinearity. "Far from equilibrium, nature is more inventive where chaos and order are intimately linked. Far from equilibrium, we discover situations that are sometimes more organized than in equilibrium. There is therefore a self-organization within nature" (Maquart 2020).

According to Lamarck's theory, the organism reacts to environmental influences and adapts. The environment is an active agent of evolution, as in 4D printing, whereas in Darwin's conception, the organism undergoes evolution and has no influence on its future (or on its descendants). However, according to Darwin, the survival of the fittest is assured and is enough to let natural selection take place, as is done in incrementalism, as shown in Figure 3.11 (see Bachelard-Jobard (2001)). But, as has already been mentioned, can we, in a pre-Darwinian logic, with mechanistic presuppositions, by relying on the laws of physics, mechanics and chemistry as much as possible, achieve a form of teleological illusion by imposing the ultimate goal of the whole transformation process? "Are there unpredictabilities in this operation in which we 'simply' introduce time as a parameter?" (Stiegler 2019). On the other hand, Darwinian evolution translates, for what would concern 4D printing processes, into discontinuous mechanical processes of adaptation, corresponding to a very slow evolutionary history, whereas here we want the opposite situation, one that is extremely fast. As already stated, there is a risk of maladjustment. Once again, the complexity imposes upon us the fact that we must consider the modes of control and mastering of the phenomena involved in 4D transformations in a different way.

All new ideas associated with this emerging field are only provisional and necessary fictions with the risk of being drawn into culturally accepted stereotypes (paradigmatic stabilizations that are related to scientific training and disciplinary work). Would we not risk being unable to encompass the full range of problems associated with the movements anticipated by 4D printing and just be able to project a fictitious stability through visual POCs or simple systems that cannot be easily generalized? On these grounds, where there seems to be evidence of an exciting dynamic, one might think that the resulting scientific flow is based on new stereotypes, constituting a form of "rehabilitation propaganda" that should probably be followed. Furthermore, the nature of 4D experiments must/may involve passive components and others that are active, combined with each other in particular ways. Can we know where in space to distribute them in order to achieve a useful objective? Can we consider passive and active elements as separable in action? How can we then understand their interactions? In the complexity associated with the determinism associated with the 4D transformation, should we work in terms of trial and error, and make a selection in the hope that an unforeseen and unpublished piece of evidence will emerge, as long as it can be generalized, creating a field of extension?

The evolution of a living system, self-organizing, is fundamentally heterogeneous with an irreducible heterochrony of evolutionary rhythms, including all shifts in rhythms (ontogenetic, phylogenetic, social, even reproductive, etc.). These elements form, according to Stiegler (2019), "the uncertain and conflicting fabric of any evolutionary history". Moreover, functions and functioning are means and ends of using and being part of the environment. From these interactions with the stimulating exterior therefore emerge singular individualities, combining, in nature, innate hereditary, DNA disruption and those acquired through existence, but singularly not reproducible from one individual or species to another.

DEFINITION.– *Ontogeny*: describes the progressive development of an organism from conception to maturity or death.

DEFINITION.– *Phylogenesis*: the study of the relationships between living beings: between individuals; between populations; between species.

> A well-founded theory can do nothing more than determine in predicative form, in medial thought, what the original giving intuition first posited in a simple way as a being and, in so doing, as a being of a determined "content" or meaning. (Husserl 1982)

3.5.2. *Evolutions*

> Life exhibits a structure that obeys the principles of physics. [...] There are laws for the general framework, forms related to the constraints of construction and function of organisms, and contingency for the detail. (Gould 1998)

> Representative species are arguably found primarily where there are barriers (and what are these barriers if not communication breakdowns). (Darwin 2020)

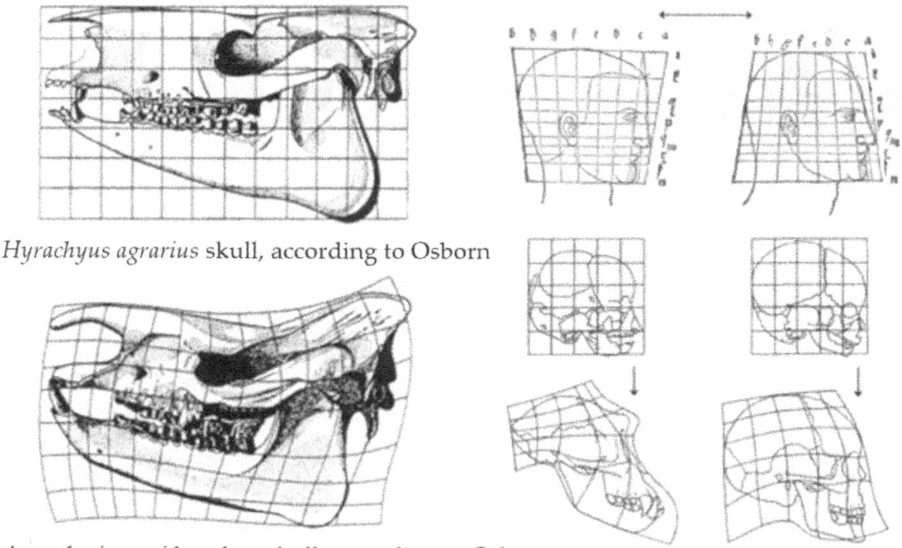

Figure 3.12. *Voxel deformations applied to evolution*

Osborn (1888) (left part of Figure 3.12) and D'Arcy Thomson (1917) (right part of Figure 3.12) are among many scientists (like Darwin and Lamarck) who have given extensive thought to evolution. Darwin asserted that living forms and their distribution obeyed sometimes very simple mathematical laws. If these occur over extended timelines, 4D printing, by contrast, which originates from additive manufacturing, is only seven to eight years old (Tibbits 2013); it involves the very short term. It involves evolutions that are induced by energetic stimuli from complex, possibly multi-material shapes made in 3D printing. An important question is therefore how it is possible to define, from a shape and a spatial distribution of stimulated materials, an overall deformation of an activated object in view of an operative end. According to Thom (1984), "the characteristic of any form is to be expressed by a discontinuity in the properties of the medium". Now, this medium can have a great influence on a transformation. A model is, as has already been specified, essential, in order to envisage linking the expected effects to these different elements, while knowing that it is only a representation of reality that it is as unbiased as possible.

According to Assis (2015), one of D'Arcy Thomson's convictions was that:

> Evolution could not elaborate all the forms that genetic variation alone would have made possible; physical laws (pressure, gravity, energy, etc.) made many of them impossible, and these physical laws subjected evolution to geometrical constraints. The consequence of this conception was, in the author's mind, to bring evolutionary biology closer to physics, and he saw this as another stone in the thesis of the desirable or necessary reduction of biology to physics.

> The waves of the sea, the little ripples on the shore, the sweeping curve of the sandy bay between the headlands, the outline of the hills, the shape of the clouds, all these are so many riddles of form, so many problems of morphology... (D'Arcy Thompson 1917)

> Can we say now in the sense of the widespread notion of "realism", that what is actually perceived (and what, in the first sense of the word, appears) must on its part be regarded as the appearance, as the instinctively sensed basis of something else which in its intimacy is foreign to it and is separate from it? (Husserl 1950)

3.5.3. *Notion of morphogenetic field*

> A "good" model should be able to best represent the data (e.g. in the sense of smallest square to be weighted or unweighted) on the mathematically defined curves it represents. But if interpolation is possible, the prediction of the evolution of the system may be questionable. (Wadington 2017)

3.5.3.1. *General considerations*

Sheldrake, in 2009, developed the concept of the morphogenic field associated with the general concept of "formative causality". Morphic or morphogenetic fields are regions of non-material influences extending in space and time (see Christianity). They are potential organizing patterns of influence, capable of manifesting wherever and whenever the physical conditions are appropriate. When this is the case, they contain a memory of their previous physical existence... (Sheldrake 2009). These morphogenic fields can only be detected by the actions they influence. They are inducers of organization.

Hunding and Engelhardt (1994) on this topic consider that such nonlinear dynamical systems may be able to "read" subtle differences in controller concentration, but with threshold triggers. Some such nonlinear biological systems are subject to multi-stability (a necessary prerequisite for cell differentiation), temporal chemical oscillations, and spontaneous spatial patterning by the Turing (1952) mechanism. The increase in cooperativity in the definition of rate laws greatly facilitates the tendency for such phenomena to occur. However, these important considerations for biology may take us away from the central issue of modeling the desired deformations in 4D printing. This observation means that if connections can be made between two evolutionary systems, this does not mean that knowledge of one, in terms of a model, can directly feed the other. That is a pity, but it does not prevent us from continuing a little further by trying to deepen the theme.

Indeed, it is appropriate, in agreement with Griesemer (2019), to test the effects of various parameters of object construction scale, material plasticity, constraint to change, evolvability, etc., and to see how they fit into the different theories. "Not that theories, fields and practices are not intertwined. But the claims (and counter-claims) must always be clear and clearly distinguished, so as not to miss the others (unless the goal is to miss the others for rhetorical reasons)" (Griesemer 2019).

Waddington (2017) proposed the notion of a "creode" or "developmental channel" on what he would have described as "purely intuitive" grounds. It represents, in biology, a developmental pathway followed by a cell as it develops into part of a specialized organ.

Box 3.6. *Creode*

The morphogenetic field approach aims to understand the organizational principles that define the spatio-temporal order of the generative processes underlying ontogeny (Thom 1976; Oster and Alberch 1982; Goodwin 1999). In biology, they involve relations of physico-chemical processes. What is shown is that the overall order arises from local interactions. Morphogenetic fields are organized hierarchically during ontogeny and interact with each other in various ways (Jaeger 2019). They have the property of being robust in the face of many perturbations (Goodwin et al. 1993; Webster and Goodwin 1996). In other words, a field represents a class or population of individual ontogenetic mechanisms that all share an equivalent structure and parameters defining their model activity (Goodwin 1982a; Oster and Alberch 1982; Alberch 1991).

What Jaeger (2019) shows is that ontogeny can be subdivided into individual morphogenetic fields, each responsible for a distinct local pattern, implying an underlying dynamic modularity (Jaeger 2018; Verd et al. 2018a). It thus allows for the identification of components and interactions within a developmental process, those that contribute to a particular model activity (Verd et al. 2018b). Different local evolutionary modules may have different sensitivities to structural changes: some are very robust, while others are in a critical state. Under these conditions, small changes in some aspects of their structure can lead to large and sudden changes in their dynamic behavior. This is what Kauffman (1993) called "the edge of chaos", which can change the behavior of the overall system in different ways, while maintaining its overall structural integrity. However, for Jaeger (2019), in biological environments, interdependent constraints change by interacting with each other with the risk of bifurcations (Montévil et al. 2016; Mossio et al. 2016; Soto et al. 2016a, 2016b). While this classical theory of dynamical systems is apparently ill-suited to representing such open structural dynamics, in evolutionary biology it may retain interest for 4D transformations of elastic systems.

The dynamics of living organisms are altered during the course of the system (Varela et al. 1974, 1979, 1991; Letelier et al. 2011; Moreno and Mossio 2015; Walsh 2015). Subsystems are created or eliminated, state variables and parameters are reconfigured, the very existence of the system and its configuration space become interdependent. This type of co-dependency goes far beyond a simple mutual causality between two separately existing processes (Walsh 2015, 2018); it must also take into account different flows from its environment.

> Both ontogeny and evolution are governed by a fundamentally dialectical dynamic that depends on constant changes in its underlying structure. This leads to a truly open-ended and unpredictable evolutionary dynamic: new dialectical interactions emerge at different points in time – within the organism and between the organism and its experienced environment – leading to a ceaseless exploration of what Stuart Kauffman calls the adjacent possible, a region of configuration space not only unexplored, but often also non-existent just a moment before (Kauffman 1996; Longo et al. 2012). (Jaeger 2019)

Thus, the co-evolution of transformations is, in evolutionary biology, cumbersome and error-prone.

Indeed, according to Raimbault (2018), complexity should be addressed through the use of models that describe evolving complex systems at multiple levels of abstraction and from different perspectives, combined with automated support for the transformation and analysis of these models. The emerging field of morphogenetic engineering should allow for the design of a complex heterogeneous system focused on the emergence paradigm. This author introduces the possibility of introducing complex adaptive systems forming aggregates at various hierarchical levels into the theory. They correspond to different levels of self-organization and the boundaries are vertically and horizontally entangled in a complex way. In other words, the properties of a morphogenetic system are essentially determined by the co-evolutionary processes between its elements.

3.5.3.2. *From a more practical point of view*

As an example, Cranston et al. (2010) introduced an evolutionary principle from cells that evolve over time by changing their spatial structure. The forces in the cell are considered by these authors as separable into unknown active driving forces and known passive forces. A finite element procedure is used to estimate the field of forces that are applied to passive components to produce their observed incremental deformations. This field is translated into the existence of active forces resolved along user-defined line segments, calculated numerically by an inversion method, whose location, often along the edges of cells, is informed by biology. The effectiveness of the approach is demonstrated using noisy synthetic data from two cell types, time-varying edge forces and a change in neighbor.

These authors hypothesize that, at the deformation rates characteristic of morphogenetic movements, the passive cellular components can be considered to collectively generate an effective viscosity (or viscoelasticity) that can vary with time. The forces required to deform these components contributing to this viscosity can be calculated mathematically. The forces required from the active components are multiple sources resolved along lines of action. They cause the deformation of the fabric (here in two dimensions). Figure 3.13 shows the transition from the action principle to the numerical results.

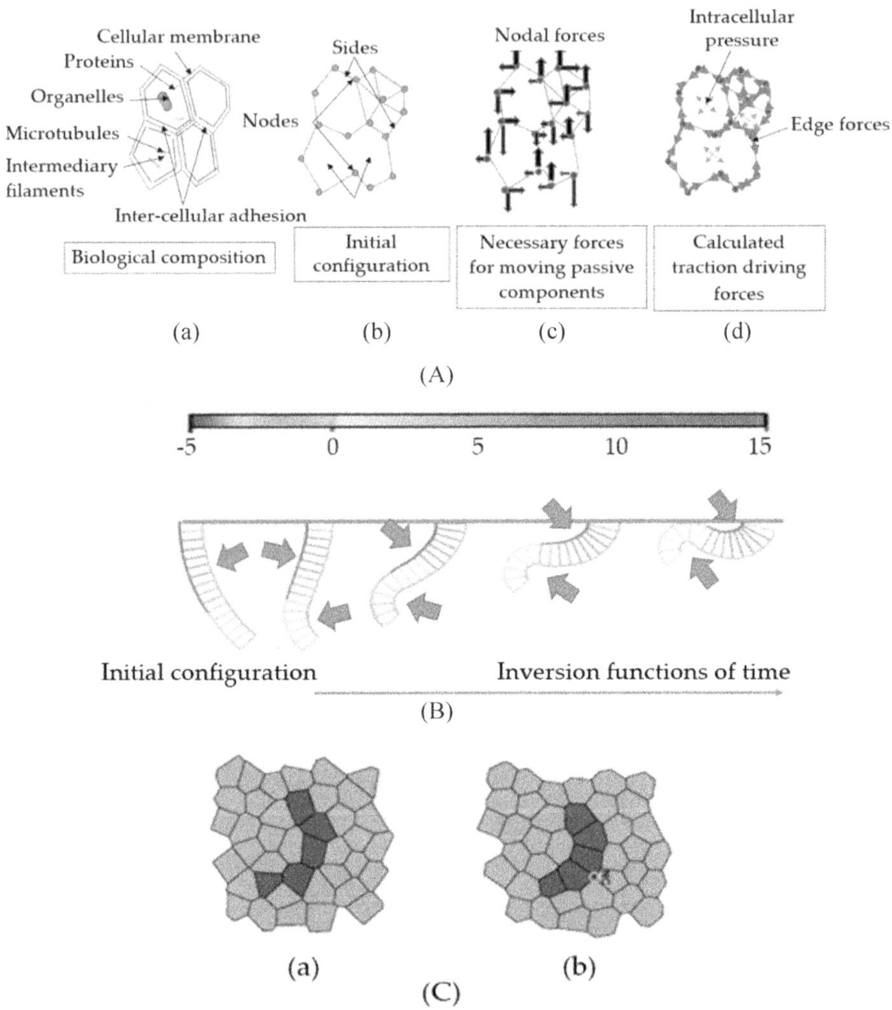

Figure 3.13. *A) Steps of the Cranston et al. (2010) method: (a) Components of direct or indirect structural importance in a typical epithelium; (b) Initial configuration subdivided into discrete regions; (c) Overall forces required to deform the passive components, calculated using the finite element method; (d) Processing of the inverse problem to reveal the forces and pressures to generate the overall forces shown in (c). B) Application to a cross-section of a generic intussusception – Initial configuration. The indicated stresses are those entering the finite element model before deformation – Noise corresponding to 5% of the step displacements was added to the nodal positions before they were passed to the algorithm. C) Aggregate of cells undergoing deformation: (a) Initial configuration of an isolated planar patch, composed of 40 cells of two types; (b) Evolution over time. For a color version of this figure, see www.iste.co.uk/demoly/4Dprinting1.zip*

3.5.3.3. *4D printing?*

The models that we have considered have some analogies with those of statistical mechanics: in both cases, a population of a large number of interacting elements (cells or voxels) is considered, and variables are constructed that describe sets of states of larger elements, subsets of the overall structure. What is interesting in the article by Cranston et al. (2010) is the idea of partitioning the space which can be translated in the domain of homogeneous or hybrid 4D printing into subsets of the object, with the existence of interdependencies between passive and active voxels. However, with viscoelastic materials such as those generally found in the literature, time can intervene both on the deformation and on the self-organization of the macromolecular structures, which can undergo other constraints under the deformation constraint, leading to their creep. As we can see for 4D systems using only materials with elastic behavior, if there is a possible field of reflexive action drawn from biology, a lot of research will still be necessary before finding a programmatic way of controlling the desired deformations in 4D printing. This field of research, which is currently practically untouched, must therefore be approached with polymer specialists and mechanics. In any case, it is largely beyond the competence of the two authors of this book.

3.6. Conclusion

> Only the future will be able to tell us whether Thom's work is merely the late offspring of a bygone era when science and philosophy went hand in hand, or whether, on the contrary, as we would like to believe, it opens up a new era in which science, once again aware of its fundamental theoretical mission, will no longer be satisfied with predicting and calculating reality, but will also seek to understand it. (Boutot 1986)

All the ideas corresponding to the domain of 4D printing are probably only spectacular displays which, like flies attracted to honey, are able to engage scientists in works whose applicative utility can, at least for now, only be modest! From an epistemological point of view, however, if one is able to free oneself from visual and somewhat "advertising" proofs-of-concept to try to explore by the most robust scientific the complexity and interdependencies of the physico-chemical phenomena highlighted in the 4D printing processes by the most robust scientific methods, a great step for future technological projections will have been made. It is, as we have already been pointed out, a way, though a risky one, of avoiding forms of "rehabilitation propaganda". However, everyone knows that the treatment of the "inverse problem", which is considered to be virtuous, is based on other models of complexity, which makes answering the following question even more problematic:

Where should I place such active materials, where should I energetically stimulate the object in order to have the desired effect in a given time? As we can see, the work to be done is particularly vast, is strewn with pitfalls and it is not certain that, given the difficulties, researchers want to confront these problems... Remember, Sisyphus was not a scientist!

> As successive generations of scholars have noted, a suitable mathematics for the dynamics of forms was "unobtainable" and, as Buffon lamented, "absolutely lacking". In the absence of appropriate mathematics, these areas of the natural sciences remained on the fringes of science, and the massive experimental data about them could not be modelled. These disciplines clearly faced a major "epistemological obstacle". (Petitot 2015)

So much for bridging evolution in nature and 4D printing.

> It is in terms of obstacles that we must pose the problem of scientific knowledge. And it is not a question of considering external obstacles, such as the complexity and transience of phenomena, nor of incriminating the weakness of the senses and of the human mind: it is in the very act of knowing, intimately, that slowness and disturbances appear by a sort of functional necessity. It is there that we will show the causes of stagnation and even regression, it is there that we will detect the causes of inertia that we will call epistemological obstacles. (Bachelard 1970)

> Men are conscious of their desires and ignorant of the causes that determine them. (Spinoza 1997).

3.7. References

Abbott, R.D. and Kaplan, D.L. (2015). Strategies for improving the physiological relevance of human engineered tissues. *Trends in Biotechnology*, 33, 401–407.

Abelmalek, A.A. (2004). Edgar Morin, sociologue et théoricien de la complexité : des cultures nationales à la civilisation européenne. *Sociétés*, 86, 99–117 [Online]. Available at: https://www.cairn.info/revue-societes-2004-4-page-99.htm.

Alberch, P. (1991). From genes to phenotype dynamical systems and evolvability. *Genetica*, 84, 5–11.

Anderson, D.L. (2002). Occam's razor: Simplicity, complexity, and global geodynamics. *Proceedings of the American Philosophical Society*, 146, 56–76.

André, J.C. (2017a). *From Additive Manufacturing to 3D/4D Printing 1: From Concepts to Achievements*. ISTE Ltd, London, and John Wiley & Sons, New York.

André, J.C. (2017b). *From Additive Manufacturing to 3D/4D Printing 2: Current Techniques, Improvements and their Limitations*. ISTE Ltd, London, and John Wiley & Sons, New York.

André, J.C. (2017c). *From Additive Manufacturing to 3D/4D Printing 3: Breakthrough Innovations: Programmable Material, 4D Printing and Bio-printing*. ISTE Ltd, London, and John Wiley & Sons, New York.

André, J.C. (2021). Bio-printing – de l'organe à la médecine personnalisée, entre résultats et promesses. Techniques de l'Ingénieur, RE268 v2.

Arendt, H. (1972). *La crise de la culture*. Gallimard, Paris.

Assis, L.C.S. (2015). Homology assessment in parsimony and model-based analyses: Two sides of the same coin. *Cladistics*, 31, 315–320.

Bachelard, G. (1970). *La formation de l'esprit scientifique*. Vrin, Paris.

Bachelard-Jobard, C. (2001). *L'eugénisme, la science et le droit*. PUF, Paris.

Ben Dhia, H. (1998). Multiscale mechanical problems: The Arlequin method. *Comptes Rendus de l'Académie des Sciences, Série IIB*, 326, 899–904.

Bergson, H. (2019). *L'évolution créatrice*. PUF, Paris.

Binet-Sanglé, C. (1918). *Le haras humain*. Albin Michel, Paris.

Blanchoin, L., Boujemaa-Paterski, R., Sykes, C., Plastino, J. (2014). Actin dynamics, architecture, and mechanics in cell motility. *Physiological Reviews*, 94, 235–263.

Bodaghi, M., Damanpack, A.R., Liao, W.H. (2016). Self-expanding/shrinking structures by 4D printing. *Smart Materials Structures*, 25, 105034.

Boehner, P. (1957). *Ockham: Philosophical Writings*. Thomas Nelson and Sons, London.

Bongard, J. and Lipson, H. (2007). Automated reverse engineering of nonlinear dynamical systems. *Proceedings of the National Academy of Science of the USA*, 104, 9943–9948.

Boutot, A. (1986). La dimension philosophique de la théorie des catastrophes. *Revue de synthèse*, 4, 385–410.

Carrozzino, M. and Bergamasco, M. (2010). Beyond virtual museums: Experiencing immersive virtual reality in real museums. *Journal of Cultural Heritage*, 11, 452–458.

Chaline, J. (2012). La relativité d'échelle dans la morphogenèse du vivant : fractal, déterminisme et hasard. *Annales de chirurgie plastique esthétique*, 5, 517–526.

Champion, K., Zheng, P., Aravkin, A.Y., Brunton, S.L., Kutz, J.N. (2020). A unified sparse optimization framework to learn parsimonious physics-informed models from data. *IEEE Access*, 8, 169260.

Chang, C.C., Boland, E.D., Williams, S.K., Hoying, B. (2011). Direct-write bio-printing three-dimensional bio-hybrid systems for future regenerative therapies. *Journal of Biomedical Material Research B: Application in Biomaterials*, 98, 160–170.

Charles, T. (2016). La Propriété intellectuelle et l'impression 3D : rencontre du 3ème type [Online]. Available at: http://www.finyear.com/La-Propriete-intellectuelle-et-l-Impression-3D-rencontre-du-3eme-type_a37216.html; and Livre Blanc sur la propriété industrielle de l'impression 3D dans la plasturgie et les enjeux réglementaires [Online]. Available at: http://www.laplasturgie.fr/livre-blanc-plasturgie-impression-3d/.

Chung, S., Song, S.E., Cho, Y.T. (2017). Effective software solutions for 4D printing: A review and proposal. *International Journal of Precision Engineering and Manufacturing Green Technology*, 4, 359–371.

Churchill, W. (2017). Le dictionnaire des citations [Online]. Available at: http://dicocitations.lemonde.fr/citations/citation-44664.php.

Cinquin, O. and Demongeot, J. (2002). Positive and negative feedback: Striking a balance between necessary antagonists. *Journal of Theoretical Biology*, 216, 229–241.

Cinquin, O. and Demongeot, J. (2005). High-dimensional switches and the modelling of cellular differentiation. *Journal of Theoretical Biology*, 233, 391–411.

Clavelin, M. (1996). *La philosophie naturelle de Galilée*. Albin Michel, Paris.

Comte, A. (2012). *Cours de philosophie positive*. Hachette/BnF, Paris.

Cranston, P.G., Veldhuis, J.H., Narasimhan, S., Brodland, G.W. (2010). Cinemechanometry (CMM): A method to determine the forces that drive morphogenetic movements from time-lapse images. *Annals of Biomedical Engineering*, 38, 2937–2947

Cresta, P., Allix, O., Rey, C., Guinard, S. (2007). Non-linear localization strategies for domain decomposition methods: Application to post-buckling analyses. *Computer Methods in Applied Mechanics and Engineering*, 196, 1436–1446.

D'Arcy-Thompson, D.W. (1917). *On Growth and Form*. Cambridge University, Cambridge.

Darwin, C. (2020). *Le corail de la vie – Carnet B*. Rivages Poche, Paris.

Delignières, D. (2015). Systèmes dynamiques et morphogénèse [Online]. Available at: http://didier.delignieres.perso.sfr.fr/Supports-doc/systemes_dynamiques_et_morphogenese.pdf.

Demoly, F. and André, J.C. (2021). Is order creation through disorder in additive manufacturing possible? *Cogent Engineering*, 8(1), 1889110.

Deng, D., Yang, Y., Chen, Y., Lan, X., Tice, J. (2017). Accurately controlled sequential self-folding structures by polystyrene film. *Smart Materials and Structures*, 26, 085040.

Denis, F. (2007). *Préface au livre de P. Berche. Une histoire des microbes*. John Libbey, Paris.

Dewey, J. (1997). *Democracy and Education*. The Free Press, New York.

Ding, Z., Weeger, O., Qi, H.J., Dunn, M.L. (2018). 4D rods: 3D structures via programmable 1D composite rods. *Materials and Design*, 137, 256–265.

Doursat, R., Sayama, H., Michel, O. (2012). *Morpho-genetic Engineering Toward Programmable Complex Systems*. Springer, Heidelberg.

Dubois, F. (2002). Modélisation mathématique de l'évolution : du continu classique au discret quantique. Afscet speech at Moulin d'Andé, 8–9 June [Online]. Available at: https://www.afscet.asso.fr/halfsetkafe/textes-2002/dubois.ande.html.

Durand, X. (2020). *Oser le risque*. Hermann, Paris.

Erdös, P. (2000). *Erdös, l'homme qui n'aimait que les nombres*. Belin, Paris.

Farhat, C. and Roux, F.X. (1991). A method of finite element tearing and interconnecting and its parallel solution algorithm. *International Journal for Numerical Methods in Engineering*, 32, 1205–1227.

Ferrari, J. (2002). *Aleph zero*. Babel, Paris.

Galton, F. (1869). *Hereditary Genius. An Enquiry on its Laws and Consequences*. MacMillan, London.

Gibbs, J.W. (1902). *Elementary Principles in Statistical Mechanics*. C. Scribner's Sons, New York .

Gladman, A.S., Matsumoto, E.A., Nuzzo, R.G., Mahadevan, L., Lewis, J.A. (2016). Biomimetic 4D printing. *Natural Materials*, 15, 413–441.

Goodwin, B.C. (1982). Development and evolution. *Journal of Theoretical Biology*, 97, 43–55.

Goodwin, B.C. (1999). D'Arcy Thompson and the problem of biological form. In *On Growth and Form: Spatio-temporal Pattern Formation*, Chaplain, M.A.J., Singh, G.D., McLachlan, J.C. (eds). John Wiley & Sons, London.

Goodwin, B.C., Kauffman, S., Murray, J.D. (1993). Is morphogenesis an intrinsically robust process? *Journal of Theoretical Biology*, 163, 135–144.

Gould, S.J. (1998). *La vie est belle*. Le Seuil, Paris.

Griesemer, J. (2019). Towards a theory of extended development. In *Perspectives on Evolutionary and Developmental Biology*, Fusco, G. (ed.). Padova UP, Pavia.

Guguin, G. (2014). Stratégie non-intrusive de couplage plaque/3D pour la simulation des assemblages de plaques composites stratifiées. Thesis, ENS Cachan [Online]. Available at: https://tel.archives-ouvertes.fr/tel-01215882/document.

Hamel, C.M., Roach, D.J., Long, K.N., Demoly, F., Dunn, M.L., Qi, H.J. (2019). Machine-learning based design of active composite structures for 4D printing. *Smart Material Structures*, 28, 065005.

Harel, D. and Pnueli, A. (1985). On the development of reactive systems. In *Logics and Models of Concurrent Systems*, Apt, K.R. (ed.). NATO ASI Series F-13 Springer-Verlag, New York.

Hayek, F. (2013). *Droit, législation et liberté*. PUF, Paris.

Heard, E. (2013). *Epigénétique et mémoire cellulaire*. Fayard, Paris.

Hello, E. (2020). De l'anthropogenèse à l'anthropo-génétique, l'œuvre de Charles Binet-Sanglé (1868–1941). *Bulletin d'histoire et d'épistémologie des sciences de la vie*, 27, 135–156.

Hénon, M. (1976). A two-dimensional mapping with a strange attractor. *Communication in Mathematical Physics*, 50, 69–77.

Hone, N. (2018). Tangible animation – Designing organic performance with multi-material 3D/4D printing [Online]. Available at: https://researcharchive.vuw.ac.nz/xmlui/bitstream/handle/10063/8063/thesis_access.pdf?sequence=1 [Accessed 25 March 2022].

Hunding, A. and Engelhardt, R. (1994). Early biological morphogenesis and nonlinear dynamics. *Journal of Theoretical Biology*, 173, 401–413.

Husserl, E. (1950). *Idées directrices pour une phénoménologie*. Gallimard, Paris.

Husserl, E. (1982). *Idées directrices pour une phénoménologie II : recherches phénoménologiques pour la constitution*. Presses Universitaires de France, Paris.

Jaeger, J. (2018). Shift happens: The developmental and evolutionary dynamics of the gap gene system. *Current Opinion in Systems Biology*, 11, 65–73.

Jaeger, J. (2019). Dynamic structures in Evo-devo: From morphogenetic fields to evolving organisms. In *Perspectives on Evolutionary and Developmental Biology*, Fusco, G. (ed.). Padova University Press, Pavia [Online]. Available at: https://www.researchgate.net/profile/Steffen_Harzsch/publication/330703268_Ecological_Developmental_Biology_and_global_ocean_change_brachyuran_crustacean_larvae_as_models/links/5c502aa8a6fdccd6b5d1ad79/Ecological-Developmental-Biology-and-global-ocean-change-brachyuran-crustacean-larvae-as-models.pdf#page=336 [Accessed 25 March 2022].

Jean, A. (2019). *De l'autre côté de la machine*. Editions de l'Observatoire, Paris.

Julien, V. (2020). *Ce que peuvent les sciences. Une enquête*. Éditions Matériologiques, Paris.

Jung, T., Dieck, M.C., Lee, H., Chung, N. (2016). Effects of virtual reality and augmented reality on visitor experiences in museum. In *Information and Communication Technologies in Tourism*, Inversini, A. and Schegg, R. (eds). Springer International Publishing, Cham.

Kalnad, C. (2016). Review on animatronics. *Imperial Journal of Interdisciplinary Research*, 2, 1168–1170.

Kant, E. (2000). *Critique de la faculté de juger*. Le livre de poche, Paris.

Kauffman, S.A. (1993). *The Origins of Order: Self-organization and Selection in Evolution*. Oxford University Press, Oxford.

Kauffman, S.A. (1996). *At Home in the Universe: The Search for the Laws of Self-organization and Complexity*. Oxford University Press, Oxford.

Keller, E.F. (2011). Towards a science of informed matter. *Studies in History and Philosophy of Biological and Biomedical Sciences*, 42, 174–179.

Kluge, A. (1994). Sophisticated falsification and research cycles: Consequences for differential character weighting in phylogenetic systematics. *Zoologica Scripta*, 26, 349–360.

Kwok, T.H. and Chen, Y. (2017). GDFE: Geometry-driven finite element for four-dimensional printing. *Journal of Manufacturing – Science and Engineering*, 139, 111006.

Ladevèze, P. and Dureisseix, D. (2000). A micro/macro approach for parallel computing of heterogeneous structures. *International Journal for Computational Civil and Structural Engineering*, 1, 18–28.

Lakatos, I. (1994). *Histoire et méthodologie des sciences*. PUF, Paris.

Laurent, M. (2013). *Systèmes biologiques à dynamique non linéaire*. Ellipses, Paris.

Le Biez, V. (2021). *Platon a rendez-vous avec Darwin*. Les Belles Lettres, Paris.

Le Guyader, H. (2008). Morphogenèse, structures physiques et évolution biologique. *Comptes rendus chimie*, 11, 186–191.

Lecointre, G. (2004). Avenir et pertinence des méthodes d'analyse en phylogénie moléculaire. *Biosystema*, 22, 7–14 [Online]. Available at: https://glecointre.mnhn.fr/docs/053_Bio22Lecointre05.pdf.

Lehn, J.M. (2004). Supramolecular chemistry: From molecular information towards self-organization and complex matter. *Reports on Progress in Physics*, 67, 249–265.

Lesne, A. (2002). Auto-organisation en biologie [Online]. Available at: https://www.lptmc.jussieu.fr/user/lesne/berder.pdf.

Letelier, J.C., Cárdenas, M.L., Cornish-Bowden, A. (2011). From l'Homme machine to metabolic closure: Steps towards understanding life. *Journal of Theoretical Biology*, 286, 100–113.

Li, W., Wang, F., Sigmund, O., Zhang, X.Z. (2021). Design of composite structures with programmable elastic responses under finite deformations. *Journal of the Mechanics and Physics of Solids*, 151, 104356.

Longo, G., Montévil, M., Kauffman, S. (2012). No entailing laws, but enablement in the evolution of the biosphere. In *GECCO 2012: Proceedings of the 14th Annual Conference on Genetic and Evolutionary Computation*, Soule, T. (ed.). ACM, New York.

Lorenz, E. (1960). Maximum simplification of the dynamic equations. *Tellus*, 12, 243–254.

Lusch, B., Kutz, J.N., Brunton, S.L. (2018). Deep learning for universal linear embedding of nonlinear dynamics. *Nature Communications*, 9, 4950.

Macquart, J. (2020). Pérégrination scientifico-philosophique. *Hegel*, 2, 174–182.

Mandel, J. and Brezina, M. (1993). *Balancing Domain Decomposition: Theory and Performance in Two and Three Dimensions*. University of Colorado at Denver, Denver.

Moës, N. and Belytschko, T. (2002). X-fem, de nouvelles frontières pour les éléments finis. *Revue européenne des éléments finis*, 11, 305–318.

Montévil, M., Speroni, L., Sonnenschein, C., Soto, A.M. (2016). Modeling mammary organogenesis from biological first principles: Cells and their physical constraints. *Progress in Biophysics and Molecular Biology*, 122, 58–69.

Moreno, A. and Mossio, M. (2015). *Biological Autonomy: A Philosophical and Theoretical Enquiry*. Springer, Dordrecht.

Mossio, M., Montévil, M., Longo, G. (2016). Theoretical principles for biology: Organization. *Progress in Biophysics and Molecular Biology*, 122, 24–35.

Naficy, S., Gately, R., Gorkin III, R., Xin, H., Spinks, G.M. (2017). 4D printing of reversible shape morphing hydrogel structures. *Macromolecular Materials and Engineering*, 302, 1600212.

Nicolescu, B. (2002). Niveaux de réalité. In *Implications philosophiques de la science contemporaine. Volume 2*, D'Espagnat, B. (ed.). PUF, Paris.

Nottale, L. (2008). Scale relativity, non-differentiability and fractal time-space. In *Scaling, Fractals and Wavelets*, Abry, P., Gonçalvès, J., Levy-Vehel, M. (eds). ISTE Ltd, London, and John Wiley & Sons, New York.

Ohlsson, T. (2012). Cutting with Occam's razor. *Physical Review*, D86, 097301.

Osborn, H.F. (1888). On the structure and classification of the mesozoic mammalia. *Journal of the Academy of Natural Sciences of Philadelphia*, 9, 186–265.

Oster, G. and Alberch, P. (1982). Evolution and bifurcation of developmental programs. *Evolution*, 36, 444–459.

Pathak, J., Hunt, B., Girvan, M., Lu, Z., Ott, E. (2018). Model-free prediction of large spatiotemporally chaotic systems from data: A reservoir computing approach. *Physical Review Letters*, 120, 024102.

Pavé, A. (2006). By way of introduction: Modelling living systems, their diversity and their complexity: Some methodological and theoretical problems. *C.R. Biologie*, 329, 3–12.

Perrier, P. (2003). Modélisation et complexité [Online]. Available at: http://www.asmp.fr/travaux/gpw/philosc/rapport3/4perrier.pdf.

Petitot, J. (1983). Structuralisme et phénoménologie : la théorie des catastrophes et la part maudite de la raison. *Séminaire de philosophie et mathématiques*, 6, 1–42 in Théorie des catastrophes et phénoménologie [Online]. Available at: http://www.numdam.org/item?id=SPHM_1983___6_A1_0.

Petitot, J. (2015). Les premiers textes de René Thom sur la morphogenèse et la linguistique : 1966–1970 [Online]. Available at: https://hal.archives-ouvertes.fr/hal-01265180/document.

Pichot, A. (1995). *L'eugénisme ou les généticiens saisis par la philanthropie*. Hatier, Paris.

Pidd, M. (1996). Five simple principles of modeling. In *Proceedings of the Winter Simulation Conference*, IEEE Computer Society, Washington.

Powell, S.G. (1995). The teacher's forum: Six key modeling heuristics. *Interfaces*, 25, 114–125.

Prochiantz, A. (2012). *Qu'est-ce que le vivant ?* Le Seuil, Paris.

Quine, W. (1966). Simple theories of a complex world. In *Ways of Paradox and Other Essays*, Quine, W. (ed.). Random House, New York.

Raimbault, J. (2018). *Co-evolution and Morphogenetic Systems*. arXiv:1803.11457v1.

Ramasubramanian, A. and Taber, L.A. (2008). Computational modeling of morphogenesis regulated by mechanical feedback. *Biomechanical Modeling in Mechano-biology*, 7, 77–91.

Ricard, J. (2003). Complexité, émergence, information et causalité dans les systèmes biologiques. In *Implications philosophiques de la science contemporaine, Volume 3*, D'Espagnat, B (ed.). PUF, Paris.

Rose, J. (2001). *Profession quasi-chercheur*. L'Harmattan, Paris.

Rosset, C. (1994). *Schopenhauer, philosophe de l'absurde*. PUF, Paris.

Rudnicki, L. (2014). Occam's razor cuts away the maximum entropy principle [Online]. Available at: https://arxiv.org/abs/1407.3738v2 [Accessed 25 March 2022].

Ruelle, D. (1988). Is our mathematics natural? The case of equilibrium statistical mechanics. *Bulletin of the American Mathematical Society*, 19, 259–268.

Rus, D. and Tolley, M.T. (2015). Design, fabrication and control of soft robots. *Nature*, 521, 467–475.

Russell, B. (1951). *Mysticism and Logic*. Penguin Books, Baltimore.

de Saint-Exupéry, A. (1953). *Lettre au Général X*. Editions universitaires, Paris/Brussels.

de Saint-Exupéry, A. (2007). *Le petit prince*. Folio Junior, Paris.

Schiano, S. (2012). De Schopenhauer à Darwin : le théâtre de l'évolution. *Romantisme*, 157, 117–125.

Schmidt, M. and Lipson, H. (2009). Distilling free-form natural laws from experimental data. *Science*, 324, 81–85.

Sève, L. (2005). De quelle culture logico-philosophique la pensée du non-linéaire a-t-elle besoin ? In *Emergence, complexité et dialectique*, Guespin-Michel, J. (ed.). Odile Jacob, Paris.

Sheldrake, R. (2003). *Une nouvelle science de la vie*. Editions du Rocher, Monaco.

Sheldrake, R. (2009). *Morphic Resonance – The Nature of Formative Causation*. Park Street Press, Rochester.

Shen, B., Erol, O., Fang, L., Kang, S.H. (2020). Programming the time into 3D printing: Current advances and future directions in 4D printing. *Multi-functional Materials*, 3, 012001.

Shepherd, R.F., Ilievski, F., Choi, W., Morin, S.A., Stokes, A.A., Mazzeo, A.D., Whitesides, G.M. (2011). Multigait soft robot. *Proceedings of the National Academy of Sciences*, 108, 20400–20403.

Sherman, Z.M., Howard, M.P., Lindquist, B.A., Jadrich, R.B., Truskett, T.M. (2020). Inverse methods for design of soft materials. *Journal of Chemical Physics*, 152, 140902.

Simmel, G. (2003). Luxe éternel, luxe émotionnel. In *Le luxe éternel ; de l'âge du sacré au temps des marques*, Lipovetsky, G. and Roux, E. (eds). Gallimard, Paris.

Sloterdijk, P. (2006). *Le palais de cristal : à l'intérieur du capitalisme planétaire*. Moren Sell, Paris.

Sober, E. (1981). The principle of parsimony. *British Journal of Philosophical Science*, 32, 145–156.

Sober, E. (1993). *Philosophy of Biology*. Oxford UP, Oxford.

Soman, S.S. and Vijayavenkataraman, S. (2020). Perspectives on 3D bio-printing of peripheral nerve conduits. *International Journal of Molecular Sciences*, 21, 5792.

Sommer, P. (2016). Les tissus humains, entre biologie et mécanique, propos d'un biologiste. *MECAMAT, CNRS*, Aussois, 18–22 January.

Sossou, G., Demoly, F., Montavon, G., Gomes, S. (2018). Design for 4D printing: Rapidly exploring the design space around smart materials. *Procedia CIRP*, 70, 120–125.

Soto, A.M., Longo, G., Miquel, P.-A., Montévil, M., Mossio, M., Perret, N., Pocheville, A., Sonnenschein, C. (2016a). Toward a theory of organisms: Three founding principles in search of a useful integration. *Progress in Biophysics and Molecular Biology*, 122, 77–82.

Soto, A.M., Longo, G., Montévil, M., Sonnenschein, C. (2016b). The biological default state of cell proliferation with variation and motility, a fundamental principle for a theory of organisms. *Progress in Biophysics and Molecular Biology*, 122, 16–23.

Spinoza, B. (1997). *Traité théologico-politique*. Garnier-Flammarion, Paris.

Stengers, I. (2020). *Réactiver le sens commun – lecture de Whitehead en temps de débâcle*. Les empêcheurs de penser en rond, Paris.

Stiegler, B. (2019). *Il faut s'adapter*. Gallimard NRF Essais, Paris.

Su, J.W., Tao, X., Deng, H., Zhang, C., Jiang, S., Lin, Y., Lin, J. (2018). 4D printing of a self-morphing polymer driven by a swellable guest medium. *Soft Matter*, 14, 765–772.

Su, J.W., Li, D., Xie, Y., Zhou, T., Gao, W., Deng, H., Xin, M., Lin, J. (2021). A machine learning workflow for 4D printing: Understand and predict morphing behaviors of printed active structures. *Smart Materials and Structures*, 30, 015028.

Thom, R. (1960). *Modèles mathématiques de la morphogénèse*. Bourgois, Paris.

Thom, R. (1968). *Biology and Structuralism Towards a Theoretical Biology*. Waddington C.H., University of Edinburgh Press, Edinburgh.

Thom, R. (1976). *Structural Stability and Morphogenesis*. W.A. Benjamin, Reading.

Thom, R. (1984). *Stabilité structurelle et morphogenèse*. Dunod, Paris.

Tibbits, S. (2013). The emergence of 4D printing [Online]. Available at: https://www.ted.com/talks/skylar_tibbits_the_emergence_of_4d_printing [Accessed 25 March 2022].

Tibbits, S. (2016). *Self-Assembly Lab: Experiments in Programming Matter*. Routledge, New York.

Turing, A. (1952). The chemical basis of morphogenesis. *Philosophical Transactions of the Royal Society of London. Series B, Biological Sciences*, 237, 37–72.

Vandeginste, P. (2012). Modéliser le vivant [Online]. Available at: http://www.larecherche.fr/savoirs/dossier/4-modeliser-vivant-01-05-2012-91189.

Varela, F.J. (1979). *Principles of Biological Autonomy*. North Holland/Elsevier, New York.

Varela, F.J., Maturana, H.R., Uribe, R. (1974). Auto-poiesis: The organization of living systems, its characterization and a model. *Bio-Systems*, 5, 187–196.

Varela, F.J., Thompson, E., Rosch, E. (1991). *The Embodied Mind: Cognitive Science and Human Experience*. MIT Press, Cambridge, MA.

Verd, B., Monk, N., Jaeger, J. (2018a). Modularity, criticality and evolvability of a developmental gene regulatory network [Online]. Available at: https://www.biorxiv.org/content/early/2018/09/10/413211 [Accessed 25 March 2022].

Verd, B., Clark, E., Wotton, K.R., Janssens, H., Jiménez-Guri, E., Crombach, A., Jaeger, J. (2018b). A damped oscillator imposes temporal order on posterior gap gene expression in Drosophila. *PLoS Biology*, 16, e2003174.

Waddington, C.H. (2017). *Organization Stability and Process: Volume 3: Toward a Theoretical Biology*. Routledge, New York.

Walsh, D.M. (2015). *Organisms, Agency, and Evolution*. Cambridge University Press, Cambridge.

Walsh, D.M. (2018). Objectcy and agency: Towards a methodological vitalism. In *Everything Flows: Towards a Processual Philosophy of Biology*, Nicholson, D.J. and Dupré, J. (eds). Oxford University Press, Oxford.

Webster, G. and Goodwin, B. (1996). *Form and Transformation: Generative and Relational Principles in Biology*. Cambridge University Press, Cambridge.

Weeger, O., Yeung, S.K., Dunn, M.L. (2018). Fully isogeometric modeling and analysis of nonlinear 3D beams with spatially varying geometric and material parameters. *Computer Methods in Applied Mechanics and Enginering*, 342, 95–115.

Weeger, O., Narayanan, B., Dunn, M.L. (2019). Isogeometric shape optimization of nonlinear, curved 3D beams and beam structures. *Computer Methods in Applied Mechanics and Engineering*, 345, 26–51.

Wojciechowski, R., Walczak, K., White, M., Cellary, W. (2004). Building virtual and augmented reality museum exhibitions. *Proceedings of the Ninth International Conference on 3D Web Technology*. New York.

Young, J. (2005). *Schopenhauer and Darwin*. Routledge, New York.

4

Toward a Possibly Programmable Self-organization?

> The fact remains that the construction of a tissue or the differentiation of an organ, macroscopic phenomena, must be considered as the integrated result of multiple microscopic interactions caused by proteins and based on their stereospecific recognition properties. (Monod 2014)

> Engineers hate complexity. I hate emergent properties. I don't want the plane I'm going to fly on tomorrow to have any emergent property in flight. (Drew 2010)

4.1. Introduction

Everyday objects (i.e. mechanical, electrical, computer-based and civil) have long been (and this is still true, even if it is no longer possible to take apart your old alarm clock) composed of heterogeneous elements assembled in a very precise and complicated way. The designers developed them to function as deterministically and predictably as possible according to a precise set of specifications. Traditional technological systems can be analyzed as simple objects subject to the linear causalities of traditional science. But, for some time now, these objects have been made non-dismantlable and it is becoming increasingly difficult to explore their mode of construction (see the now obsolete syndrome of the old alarm clock, of which there was always a gear left after reassembly) and, since 1984, they can be made by additive manufacturing (as opposed to the classical subtractive manufacturing by elimination of material of the elements of the objects).

In this case, according to a deterministic construction, the basic building blocks are generally of identical size and the number of materials composing them is quite small. In contrast, self-organizing natural systems use agents with

> [...] simple rules and random ensemble dynamics. Certainly, many non-trivial patterns and behaviors can emerge from these agents, which interact with each other mechanically, chemically or visually, and this appearance of 'spontaneous order' is often cited as a hallmark of complex systems. (Doursat et al. 2014)

They can be roughly described by "order parameters" or "phase transitions" but do not follow a set path from disorder to an ordered structure (Demoly and André 2021).

Physical chemists are well aware of these phenomena and, for example, in the case of photo-polymerization in stereolithography, can we not speak of a certain organization from an initial random distribution of monomers? Undoubtedly so, but this does not get us very far (though it does allow us to enter the subject). On the other hand, when we leave 3D printing to go to 4D printing, there is a disturbance of the local order inside one or several materials induced by an energetic disturbance. With the idea of movement or change in functionality, it is necessary to engage in a systemic analysis (Le Moigne 1977, 1990) of the technology through an ontological and functional approach to the technical system. In particular, the structural models constituting a 4D object, as well as the interrelationships that generate movement, must be taken into account in this systemic analysis. This is what the authors will try to explore in this chapter to examine whether order (a form of information in matter) can be the source of 4D effects or whether, on the contrary, through fatigue phenomena, this effect could be reduced or even disappear.

> It is signs that ensure the passage between our inner world and the outer world. The sign is an intermediary, an interface between the sensible and the immaterial. A conventional interface that reveals and veils at the same time. Secrets stick to words, gestures, expressions, symbols, they stick to our shown/hidden identity, to our strategies, to our interactions. (Petitat 1998)

For science, it is the question of knowing what is held to be real, in the sense of the unconventional, of that not produced by the theoretical and practical activity of the scientist. The ontological question, for epistemology, is first of all that of the reality of the theoretical entities of which science speaks.

Box 4.1. *Ontology*

4.2. A look at the technology

> The hybrid nature of developing systems, by their multiscale nature at all phases of their evolution, and by the ever-increasing imprint of their history, precludes the possibility of a unitary theory or set of laws of the same type as those that apply to single-scale physical systems (e.g. classical or quantum mechanics, or thermodynamics). (Newman 2014)

H. Simon (1969) put it this way:

> An artifact may be regarded as a meeting point, an interface [...] between an internal environment, the substance and organization of the artifact itself, and an external environment, the surroundings in which it is implemented. If the internal environment is adapted to the external environment, the artifact will serve its intended purpose.

This is what is expected of a 4D structure, that is to achieve the objective defined by the designers in a given environment, under a given stimulus. One of the axes of the sciences of the artificial is to build the relationship between these goals, the environment to which the 4D artifact must adapt with its internal structure (Aït-El-Hadj 2015). The latter reminds us:

> The characterization of technology within the field of the artificial thus requires the use of an additional criterion. The first is the enslavement of physical means and natural forces, which, at that moment and by that very fact, escape the sphere of the natural since they are integrated, manipulated and organized by Man according to a prior purpose.

This context is obviously well suited to 4D printing.

Any technological system is always an ordered aggregation of the use of pre-existing technologies, with a hierarchical, multi-level system. Existing know-how is exploited by the designer to constitute technical sets of increasing complexity, but which possess a functional unity and an architectural order (algorithm). Thus, the technological system can be seen by the addition of affirmed and validated knowledge as a system of systems with a hierarchical organization of knowledge and applications.

It is then a matter of exploiting through “prior knowledge of the art” additive methods which, through trial and error, have led by successive increments to the current technological packages that successfully “mix” scientific aspects with robust practical routines. However, in the spirit of 4D printing, all of these aspects are merged into the "object-material-stimulation" interaction set. The trial-and-error methods that made the success of manufacturing may no longer be exploitable for this emerging technology because it is necessary to explore the black box constituted by this complex system. One of the advantages of 4D printing relative to other technologies is that it can free itself from the inertia and habits linked to artifacts, but it will only be able to develop if it can prove, within the framework of substitution, an increase in performance relative to "classic" technologies. Or it will have to explore new niche applications or simply deal with obvious situations.

If the current artifact is the product of skillful combinations, resulting from the ordered confluence of several technical processes, the 4D object must integrate in its global logic, during its manufacture, all the sought-after functionalities. Only voxels, present in large numbers, make up the operational technical system… (with, however, the aspect of energy stimulation which must not be forgotten). "The technical object progresses by internal redistribution of functions into compatible units…" (Simondon 1969). Now, for this author, the artifact respects the following rules:

– Each function is performed by one organ and one organ only.

– Each organ has only one function.

– The main function is thus only a loose aggregation of particular functions (see also Cantot and Luzeaux (2011)).

In the case of 4D objects, is this rule respected? With simple objects, for example with origami, only one function exists and therefore the rules are followed; but for complex objects integrating several functionalities, these rules may no longer be valid, especially with probable interdependencies. There is therefore another conceptual break to be considered, that of the design, which must move from addition to complete (and complex) integration in the evolving object. Any act of convergence should then take place upstream.

4.3. Natural (spontaneous) self-organization

> Happy are the writers who take away the weight of thought and who weave with a light finger a luminous disguise of the complexity of things. (Valery 1957)

In his "Essay on the Immediate Data of Consciousness" (2013), Bergson wrote: "Duration seems to act in the manner of a cause... Under these conditions, a presumption may be invoked in favor of a conscious force or free will which, subject to the action of time and duration […] would thus escape the law of conservation of energy". This is indeed what happens in the field of life, where the order of the system increases by wasting energy (as a reminder, the global entropy of the universe continues to increase) and by giving the impression of determinism.

The morphogenesis of living organisms is a privileged field where the impotence of mechanistic materialism is revealed because of a "proximate" causality. As Vecchi and Hernandez (2014) remind us, the concept of "morphogenetic field" aims to explore the coordination phenomena of cellular behaviors essential to morphogenesis. The idea of sensitive information means the whole morphogenetic theory is based on the capacity of cells to "interpret" information issued by their "consorts", regardless of their physical support. There are thus feedbacks (downward causality) associated with regulation by networks and physico-chemical supports. Morphogenesis, according to Newman (2014), involves a multitude of generic physical mechanisms (viscosity, electrical excitability, rigidity, stickiness...), each of which is found (but not all together) in non-organic realities. Perhaps then the secret of evolutions is the subtraction of information in the interaction?

Thermodynamics can tell us what can happen, but a distinction must be made between thermodynamics and the kinetics of transformations. The latter allows us to distinguish, among a given number of thermodynamically possible events, the smallest number that actually take place at measurable speeds. For Boy (2007), "the properties of living matter manifest themselves as the maintenance, the self-maintenance of certain topological situations much more than as pure energetic and functional conditions. This is why living individuation must be thought of according to topological patterns".

But, to say that matter can, in certain situations, be ordered means that the form can, at first approximation, be described geometrically in a simple way, using for ease (?) the idea of symmetry (in the case of a crystal for example), even if there are more or less complicated substructures, which may even take on a fractal aspect, but the general form remains simple. We can think that the structures, more or less self-organized, must be associated with certain types of physical or physico-chemical conditions (Chossat 1996, 2015).

The term "self-organization" refers to a spontaneous/stimulated formation of dynamic order leading to forms of organization from a set of interacting homogeneous or non-homogeneous units. After placing energy in the right place at the right time in a given environment, the initial random (or rather non-random) system organizes itself to reach a form, if possible the one sought by the experimenter (Moreno 2004). From a thermodynamic point of view, this "shaping" implies that the transport of matter and energy in the system that is created must be under functional constraint in the absence of external energy input. As the object is constructed, there is an appearance of causality (universal physical laws), which can be represented by a kind of selection between several possible alternatives (Campbell 1974). "Thus, its role is to construct higher level entities, formed from the (re)structuring of units intrinsically constituted by material causality" (Moreno 2004).

4.3.1. *Nonlinearities*

> Far from equilibrium, irreversible processes are a source of coherence. The appearance of this coherent activity of matter – of "dissipative structures" – imposes on us a new way of looking at, a new way of situating ourselves in relation to the system that we define and manipulate. Whereas at equilibrium and near equilibrium, the behavior of the system is, for sufficiently long periods of time, entirely determined by the boundary conditions, from now on we will have to recognize a certain autonomy that allows us to speak of structures far from equilibrium as phenomena of "self-organization". (Prigogine and Stengers 2009)

Figure 4.1. *Geometric nonlinearity. For a color version of this figure, see www.iste.co.uk/demoly/4Dprinting1.zip*

Figure 4.1 shows, as an example, the nonlinear effects experienced by a beam. Dissipative structures are processes in which a fluctuation stabilizes to form a dynamic macroscopic configuration permitted by the energy flow through the system under construction. These macroscopic structures require a continuous supply of energy of a magnitude appropriate to its coherence in order to grow. "The term 'self-organization' implies a spontaneous creation of order or organization rather than an agential identity capable of determining or constituting itself in a way that implies functional plasticity" (Alvaro 2004). This is at heart respectful of what Bruter (1976) wrote: "Events shape the object over time, it is above all to have pierced the secrets of its history, of the lineage of which it is, at the same time, the outcome and a projection ...".

A classic example is that of living systems that display autonomous behaviors and capacities: growth, differentiation, self-repair (at least partial for some species), etc. They are able to modify their behavior and structure according to external parameters (e.g. nutrients, temperature and gravity).

Quoted by Thuillier (1980), Stéphane Leduc wrote: "The elementary act of life is diffusion and osmosis." It was this professor of medical physics in Nantes who produced forms and structures "similar to those of living beings" (very early 20th century). In 1910, he published a work entitled: "Physicochemical theory of life and spontaneous generations". He naturally came into conflict with the Academy of Sciences. Figure 4.2 represents one of Leduc's osmotic productions (Thuillier 1980). He wrote in 1912:

> When one has come to know the physical mechanism of the production of an object or a phenomenon, [...] it becomes possible [...] to reproduce the object or the phenomenon, science has become synthetic. Biology is a science like any other [...] it must be successively descriptive, analytical and synthetic. (Leduc 1912)

Figure 4.2. *Leduc osmotic productions (copy of Leduc (1910); see also Thuillier (1980) and Eastes et al. (2009))*

This causal theory and the methods that follow from it to be applied to these constructs aim at obtaining finished and optimized products after "starting up", especially with regard to the consumption of material and energy, the notion of voxel which can then reach the size of a molecule, or even a cation or an atom. In the idea, as for all (or most) engineering sciences, these methods are still often thought of as deterministic: phenomena are controlled by laws, which must be discovered and applied as faithfully as possible. There would thus be (too) little room for randomness.

If it were so, starting from energy that is spatially positioned in a known way, from cooperative elements that can assemble, it would be enough to wait for a "sufficient time" to reach an equilibrium leading to an object with a given shape. Repeating the same experiment, all other things being equal, should provide the same object (same shape, same size, same properties). Thus, under this causal hypothesis, it is not possible to predict the evolution of shapes by deducing them only from the application of simple laws that could have been identified. This result illustrates an evolution that develops in a non-deterministic way through the interaction of natural systems with each other and with their environment.

However, as an example, Prigogine (1994) recalls this experiment: if sodium chlorate is cooled at rest, the crystals produced are optically inactive. If, at the same time, the solution is shaken, either levorotatory or dextrorotatory crystals are obtained. This is a bifurcation; the first one gives either a left or a right form and the initial nuclei which will participate in the crystallization are either left or right. The symmetry of the reactive system is thus broken. This example serves to show that irreversible phenomena do not necessarily reduce to an increase in disorder, but that there can be constructive roles for this irreversibility of time, as is the case with life. Hawking (1989) reminds us: "A strong thermodynamic arrow is necessary for intelligent life to act ...." These aspects can be considered in another framework, that of bio-printing (4D printing of the living) (see André (2017c)).

Under these conditions, Turing (1952) showed from other examples, how reactions between chemical substances, which he called morphogens, coupled with a diffusion process of these substances through living tissues, could give rise to specific structures or regularities (spots on a fur coat, shape of a crystal or a shell, etc.). In his seminal paper, he introduced a mathematical model called "reaction-diffusion equations" and the study of this model in some relatively simple cases.

The dynamics of this basic model he makes explicit results from the coupling between chemical reactions and diffusion of reactants, in a continuously fed system. Turing structures correspond to spatially periodic variations in the concentrations of chemical species (Joshi and Davidson 2012; Lesne 2013; Zenil 2013; Pujo-Menjouet 2014; Shaygo 2015) that compete with molecular transport mechanisms. Therefore, to observe spatial variations in concentrations, other mechanisms must come into play. He introduces a scenario involving two species A and B, "morphogens" and the conjunction of the following properties:

- Species A is able to induce its own production.
- Species A can produce the second species B.
- Species B inhibits the production of A.
- B diffuses faster than A.

Thus, a small local fluctuation at a point in space induces a slight excess of A, so that the production of A and B is higher than the average at that point. B, diffusing faster than the excess of A, creates an inhibiting corona around the initial point, isolating the A peak by a zone richer in B. These random evolutions can occur at other locations in space with other peaks of species A.

> These activator peaks result from the selective amplification of the minute but innumerable fluctuations that spontaneously affect the homogeneous mixing state. The most dynamically stable distribution of these peaks forms a periodic pattern, like the spots that can be observed on a cheetah. (Chossat 2015)

For example, El Mellah et al. (2012) present examples of complex structures realized numerically from simple rules. For example, for skin staining, one can consider the coexistence of two types of secretions for cells, a short-range activating effect and a long-range inhibitory effect, as shown in Figure 4.3.

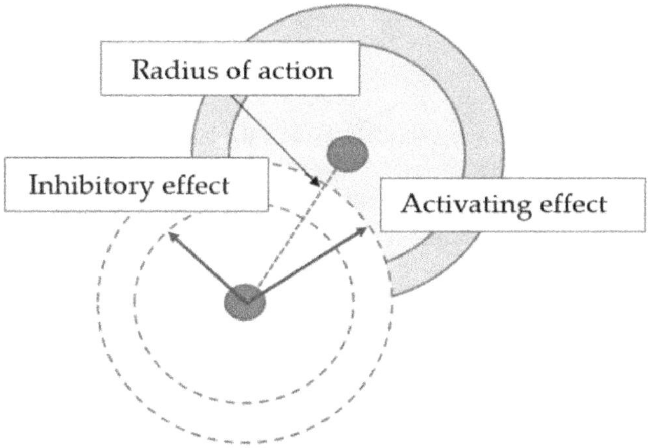

Figure 4.3. *Short- and long-range interactions. For a color version of this figure, see www.iste.co.uk/demoly/4Dprinting1.zip*

Mathematical modeling (which is outside the scope of this book) can be done. For example, the preceding authors show that it is possible to represent the spots of the royal cheetah through numerical modeling (see Figure 4.4).

Figure 4.4. *Illustrated (left) and calculated (right) spots from the Turing morphogenetic model. For a color version of this figure, see www.iste.co.uk/demoly/4Dprinting1.zip*

There is thus localization of a chemical entity in privileged zones. However, if there is localization, the spots are not located in the same way from one animal to another, from one side of the animal to another, in the same place or in a symmetrical way. Indeed, there are bifurcations that condition the distribution of the colored entities.

James Clerk Maxwell wrote in 1876 (quoted by Thom (1983)):

> When the state of things is such that an infinitesimally small variation in the present state will alter the future state by only an infinitesimally small quantity, the state of the system, whether at rest or in motion, is said to be stable: but when an infinitesimally small variation in the present state can cause a finite difference in a finite time, the condition of the system is said to be unstable. It is obvious that the existence of unstable conditions makes it impossible to predict future events, if our knowledge of the present state is only approximate and not exact …

The question of mastering the "right" main parameters influencing stability or instability is therefore essential, but it may not be sufficient.

For example, Yin et al. (2021), through a linear stability analysis show that bifurcations can trigger pattern formation (see above). They develop a numerical model that tracks the dynamic evolution of the pattern beyond the steady state and apply it to starfish oocytes, allowing them to decipher the synchronous protein dynamics and surface contraction pattern that emerges in the anaphase of meiosis, a model that takes into account global oscillation and traveling waves. This complex study highlights the crucial role of bio-chemical-mechanical feedback in the evolution of the morphogenetic model. It is clear that this study is difficult to extend in order to reach real genericity.

4.3.2. *Achieving the desired shape?*

> One might even argue that high criticism, being the purest form of personal impression, is in its own way more creative than any creation, since having less relation to any external object, it finds in itself its own raison d'être and as the Greeks would have said, is an end in itself and for itself. (Bayard 2007)

Once science begins to understand how nature has selected certain forms and not others, both in the physical and biological world, it can seek to apply or transpose the corresponding mechanisms to solve engineering problems – that is, of interest to us, the making of artifacts, tools or final objects. The wide variety of applications given to a generic form such as the fabrication of a functional object is not necessarily inspired by what nature does, but it is interesting to consider whether knowledge of "morphogenesis" can be used to achieve the desired form through self-organization (Baquiast and Jacquemin 2004). Taken in a broad sense, self-organization corresponds to a set of phenomena (natural and artificial, real and virtual) of spontaneous formation of dynamic order. It is a process of spontaneous creation of organized forms (e.g. a global pattern) from a set of interacting entities. The concept of self-organization originates from a scientific movement called "Second Cybernetics" (see Yovits and Cameron (1960), Von Foerster and Zopf (1962) and Lévy (1985)) and from research in the thermodynamics of irreversible processes (Glansdorff and Prigogine 1971; Stengers 1985; McMullin and Varela 1997).

For Whitehead (1906):

> What is required at this stage is a simple hypothesis concerning the motion of real objective entities and correlating this with the motion of electric points and electrons. From such a hypothesis it would be possible to deduce in the simplest way the whole of the electromagnetic and gravitational laws. This concept taken in its fullness implies the hypothesis that the universe is formed by only one class of entities. Properties of "space" and physical phenomena "in space" simply become the properties of this single class of entities. To

> simplify the above axioms [...] the ideal to aim for should be to deduce them in part or in whole from more general axioms that would also embrace the laws of physics. Thus, these laws would not presuppose geometry: they would create it.

The idea of this author is to examine how mathematics can allow us to conceive the nature of our material world, in particular "matter" in space.

For our purposes, it is an inverse problem, applicable to the direct manufacture of objects (and not by additive manufacturing, voxel after voxel), to create (by order) 3D structures of variable sizes from a pseudo-homogeneous medium (a classical problem that is "normally" contrary to entropy, which means that we must know what energy and in what form it is needed, because the process is anti-dissipative; in fact, there is nothing in the demonstration of the second principle that allows us to assert that there is a causal relationship between entropy variation and evolution toward a chaotic state). But, as soon as several entities interact, where the effects are not proportional to the causes, there may be the possibility of nonlinear characteristics with specific effects (e.g. snowflakes in non-equilibrium systems with continuous material input or "photochemical muscles" which introduce tensions in the manner of mammalian muscles). A multi-scale approach could be an asset to dissipate an excess of energy. Under these conditions of recursion, the structure is in a way a cause and a consequence of this dissipation (self-similarity; see Figure 4.5). Self-organization thus gives the impression that the structure emerges from the combination of interactions between constituent entities (including energy).

Thom (1983) analyses the problems associated with the control of morphogenesis by relying on the notion of a "black box" model with its inputs and outputs (some of which become inputs again by recursion). When one chooses inputs Ui, it is possible to measure outputs Sj as long as their existence has been identified, and consequently to define a relation between Ui and Sj (as proposed by Turing in 1952). The central question, well known in process engineering, is, from the experiments carried out, to restore the intimate mechanism occurring inside the black box. In complex systems, there may be "exceptional" situations for which the relationships {Ui, Sj} are outside of what is "normally" expected. The search for these "divergent" but real situations can lead to new and necessary information to know more precisely what is going on inside the "box". So, if we do not know where to look for (and what to find in) these situations from an experimental point of view, it is up to modeling to help us look for singularities at different scales (see Figure 4.5).

Figure 4.5. *Self-similarity as illustrated by the fractal pyramid realized in the 1990s (Dionnet et al. 1992/1993). For a color version of this figure, see www.iste.co.uk/demoly/4Dprinting1.zip*

According to Laurent Larger (2015), nonlinear dynamical systems apply to any practical situation where movements occur in space and/or time and for which there is a deterministic description of the evolution, in the form of a differential equation (temporal dynamics only) or a partial differential equation (spatio-temporal dynamics). The world around us is of course 100% dynamic and nonlinear, but the current technological mastery of its functioning is very often limited to situations of linear approximations. The example of morphogenesis illustrates very well not only the relevance of the nonlinear approach (a simple, deterministic, local dynamic model can explain how nature builds stable, reproducible and robust forms at the macroscopic level), but also the existing margin for scientific progress in the understanding, and then in the control and exploitation, of the associated mechanisms for concrete applications, with enormous technological potential.

Finally, it is necessary to underline here the importance of the notions of local and global in nonlinear dynamic phenomena, and to understand that a very complex solution at the global level can indeed emerge in a deterministic way from a property that is only local (the deterministic dynamic equation and the local nonlinear couplings, which applies to the whole of the medium studied) and relatively simple in its formulation.

> The doctrine of finality, in its extreme form, as we find it in Leibniz for example, implies that things and beings merely carry out a program once laid down. But if there is nothing unforeseen, no invention or creation in the universe, time becomes unnecessary. As in the mechanistic hypothesis, it is still assumed here that everything is given. (Bergson 1907)

From a modeling point of view, we can consider a system that evolves in time from a given form to reach another form (dynamic approach) or to impose on a stable form the constraints of different forms (chemical, energetic, etc.) that will lead to deformation (thermodynamic approach). In the first case, it may be useful to introduce temporal elements that allow the evolution to be defined at a given time; in the second case, thermodynamic and chemical potential elements are introduced. This being the case, starting from a given set Ui, it is possible in principle to define Sj, provided that all the influencing factors are known. The inverse problem aims to define the sets Sj that determine a setpoint and to seek the Ui that meet the objective. This is a classical problem which is always difficult in the case of linear systems which are generally ill-conditioned (see de-convolution, if only because of noise and measurement errors; see Mugnier (2008)); it will be even more difficult in the case of nonlinear approaches with noisy signals. This being the case, the stakes are high because the control of the inverse problem can constitute an integral means of realizing an object from an easily realizable matrix whose deformation as a function of time or energy supplied could allow the realization of a complex object.

Chauvet (2006) considers that when two units A and B are subject to certain conditions, they cannot do anything other than associate (see the work of Turing mentioned in the introduction). For example, the dynamics of two biochemical pathways with transport of a reactant to each other would, according to this principle, be more stable than the dynamics of each separate entity (stabilizing self-association principle according to Chauvet (2006), Staetzler et al. (2011) and Nykter et al. (2008)). The same can be true with enzymes (see Kekenes-Huskey et al. (2015)) with separations between species that require molecular diffusion-limited transport (solving the Von Schmoluchowski equation in a potential field (Kekenes-Huskey et al. 2012)). Self-organizing systems can be very stable relative to the magnitude of many of the parameters, but they can diverge suddenly with small changes in one or more of the influencing variables to reach a new stable state (Nykter et al. 2008). Turing-like evolutions with fractal approaches can also be proposed (see Bizzarri et al. (2011)).

Thus, between the illusion (?) of an observable self-organization and the opposite problem, constituting an important challenge, allowing the realization of the organized structure desired by the designer, there is a considerable gap to overcome. Perhaps it would be necessary to reflect on nonlinear, non-equilibrium and multi-scale phenomena? It would also be necessary to look at the biological side of things with original experimental methods that go beyond traditional physical chemistry to create 3D structures (going beyond the nanometric sizes of the DNA origami already mentioned). It would undoubtedly be necessary to move away from classical reductionist methods to engage in holistic thinking to explore this small part of the complexity. This reflection is engaged hereafter from a work carried out for 3D printing (Demoly and André 2021), while knowing that it corresponds to an immense construction site which, at least for the authors, still needs to be dug.

To put it simply, what the engineer is looking for is to place matter "in the right place" and "at the right time" so that the organization leads, at a chosen moment, to the desired object (and not something close to it). Since we are dealing with the insufficient control of systems, they have to look for situations where they regain control over the complexity of the whole by informing the matter. An example can be found in Afshar and Di Leo (2021) who propose to exploit reaction-diffusion principles in solids whose functionality can evolve through ion transport in metals. This transport generates mechanical stresses that deform the solid structure (but the inverse problem is not studied).

Efforts in this area are presented below.

> If our cells are free, their organization must obey the principles of economy or ecology that structure societies or ecosystems. In an ecosystem such as the forest as a whole, there are general regulatory phenomena that control temperature or soil composition. However, there is not a forest genome that, programming each of its inhabitants, ensures the realization of global phenomena. Homeostasis, i.e. the constancy of the forest's internal environment, is not written anywhere. There is no such thing as a nature police force. (Sonigo 2000)

4.4. Self-organization and 3D/4D printing

> Propaganda does not aim to arouse inspiration; it closes, it condemns all the orifices through which an inspiration might pass. (Weil 2021)

4.4.1. *General considerations*

We have already recalled Blanqui's illusory idea of the existence of objects that meet selected criteria in "Eternity by the Stars" (2012). The world of technology has undergone a major transformation in recent decades. Elements that previously existed in a purely mechanical or electrical (i.e. physical) form, and in particular elements describing logic, control and decision making, are increasingly taking the form of embedded systems and software (i.e. cyber elements). There is no longer any manufacturing without the digital. "Thus appears the image of a moment when everything would be transferred, transposed, gesture, force, thought, into a perfect double" (Guièze 1983). The engineer, in charge of the efficient, economical and robust production of objects, is the intermediary in achieving this objective.

An intentional artifact (related to the work of the engineer and/or designer, see Figure 4.6) can be seen as a means of linking an "internal" environment, the substance, functioning and organization of the object itself, and an external environment, the surroundings in which it is implemented. If the two environments are compatible, the artifact meets the specifications. As H.A. Simon (2004) pointed out in another framework, knowledge of an artifact such as a machine "has an advantage over knowledge of nature, because it is based on valid prior foundations, the purposes of which will be deviated from with a certain amount of new intentionality to give the projects intelligibility and openings to society".

Figure 4.6. *Examples of artifacts (man-made according to Mathis (2011)). For a color version of this figure, see www.iste.co.uk/demoly/4Dprinting1.zip*

> The engineer's job is to mobilize the technical knowledge at his disposal to imagine innovations that represent a compromise acceptable to all stakeholders. Technical perfection is the poison of the economy! One must be prepared to tinker to satisfy the greatest number of people. The resources of technology are such that we cannot believe an engineer who assures you without batting an eyelid that there is only one efficient and profitable technical solution and that we must choose it. (Callon and Lacoste 2011)

Thus, in the cultural approach in engineering sciences, the search for an understanding of reality relies largely on "strategies of detour, in other words to the construction of models or the production of virtual facts" (Granger 1995). But these possible detours exploit the knowledge of a robust possibility and, as far as possible, of the simplest possible by imposing majority causalities and the greatest possible determinism.

It is indeed on this basis that additive manufacturing emerged in 1984 (André et al. 1984). The idea is to place material precisely "in the right place" and "at the right time" so that the organization leads by addition to the desired object (and not something close to it). To achieve this goal, it is necessary to look for situations where it is possible to take control of the complexity of the overall by choosing specific processes: polymerization induced by light or electromagnetic waves, localized fusion of fusible wires, powders, etc. (see André (2017a)). The notion of voxel (transformed elementary volume) intervenes in this process: an object is thus defined by the addition of N voxels.

Thermodynamics describes the change of a system from an initial state to a final equilibrium state, without taking into account transitions. In fact, it is the path and "distance" of the system from its final equilibrium that generates and maintains the structural complexity found in living organisms and that we seek to reproduce with "non-living" matter (Monteiro 2018). To design a set of components that will self-assemble in solution to form a desired thermodynamically stable structure, the interactions between the components must be strong enough that an assembled structure has a lower free energy than its unassembled components. They must also be "specific" so that the desired shape is lower in free energy than other possible assembled structures. However, these strength and specificity requirements that may apply to a modest number of dilute elements (lack of shielding effect) may be inapplicable to a large number of systems and the "meeting" time between elements may be very long, preventing fabrication in reasonable time frames (Whitelam and Jack 2015).

Dissipative structures are processes in which a fluctuation stabilizes to form a dynamic macroscopic configuration permitted by the energy flow through the system under construction. These macroscopic structures require a continuous supply of energy of an amplitude appropriate to its coherence in order to grow. "The term 'self'-organization implies a spontaneous creation of order or organization rather than an agential identity capable of determining or constituting itself in a sense that implies functional plasticity" (Alvaro 2004). A classic example is that of living systems that display autonomous behaviors and capacities: growth, differentiation, self-repair (at least partial), etc. They are capable of modifying their behavior and their environment in a way that is not only autonomous but also self-representative. They are able to modify their behavior and structure according to external parameters (e.g. nutrients, temperature and gravity).

Heinz Von Foerster (1960) introduced the apparently paradoxical idea of the principle of creating "order from noise", formulated as early as 1960, which was taken up by Henri Atlan (2006, 2011) in his theory of self-organization. The thesis consists of arguing that the noise that is introduced into self-organizing systems generates, in the final analysis, less entropy than a reorganization of the system that is oriented toward new purposes. (The contemporary revival of this idea in the cognitive sciences appears in the form of "stochastic resonances" (Varela 1983)). It is therefore a direction envisaged in this article by imagining how, by means of a disorder (more or less organized), forms can be created that meet the expectations of a user.

Manufacturing strategies differ in the way they store and use assembly information. For Cademartiri and Bishop (2015), in top-down approaches, information is stored in a centralized location (a computer) and executed by a machine. The complexity of the assembled structure (i.e. the amount of information required to describe its features) derives from the capabilities of the machine rather than from the basic elements, which may be simple voxels. In contrast, in bottom-up approaches – that is relying on self-assembly principles – the information is distributed among the components (e.g. in their interactions), and is executed by spontaneous physical and/or chemical processes with little or no external guidance. Relative to the top-down approach, the massive parallelism inherent in self-assembly is one of its most attractive attributes for the idea of "autonomous" or even spontaneous construction (Bhalla et al. 2015).

4.4.2. *Creation of 3D artifacts*

> Half-knowledge triumphs more easily than full knowledge: it sees things as simpler than they are, and thereby gives a more comprehensible and convincing idea of them. (Nietzsche 2014)

Kheifetz (2018) considers that it is necessary to go through 3D printers to produce objects and that self-organization processes cannot meet an object realization goal. In Demoly and André (2021), we agreed with him because it seems difficult to precisely master a shape responding to a prior instruction. It is during the self-assembly of the elements from the beginning of the manufacturing process that the object acquires its final shape (Jakab et al. 2004) and the difficulty is to solve the inverse problem (as long as at least one solution exists and we know it). In this paragraph concerning self-assembly processes, we find the old dream of having a universal assembler that could make any object (or organism) from atoms/molecules/voxels taken from the environment. We are far from it, but this does not prevent us from trying to build less ambitious models of self-assemblers.

The morphology of a 3D object is designed in advance or it can be developed over time through the interactions between the different elements that will participate in the creation of the shape. A successful generation process must take into account global and local effects of the environment during morphogenesis (Zahadat et al. 2017). Here, the evolution of many systems is governed by the external environment. For synergetics, this translates mathematically into the variation of the order (or control) parameters of the system by the environment. This causes the system to have its stability disrupted, which can lead to other patterns, other regularities and subsequently, the achievement of desired goals (Firat et al. 2020). The creation of transformative novelty is therefore possible (but not very likely) because of this break in the logical sequence by bifurcation between a set of interacting elements (Schwarz 1977).

A morphogenetic approach to programmable matter can be organized around two broad classes of objects: active assemblers and passive components. Together they fulfill roles analogous to those of cells and their products in embryonic morphogenesis. Components include structural elements and physical carriers of signals (e.g. chemical morphogens). These components are passive to the assembly process but may be active in the assembled system. For example, components could include objects that will function as sensors, actuators or amplifiers in a robot, but are passively manipulated during assembly and reconfiguration (McLennan 2015).

If the possible revolution, into which the theme of self-organization could plunge us in order to realize in one operation a 3D object that meets a set of specifications, consists of applying knowledge to knowledge itself, it is probably appropriate for the moment to leave the final word to Jean Staune (2015) to conclude with this paradoxical and pessimistic sentence: "Why study problems that have no solution or whose solution is too complex to be found?"

4.4.3. *What about 4D printing? Stimulated self-organizing systems: bottom-up coupling*

> Oxymoron has gained popularity because we live in a world where ideologies having declined [...] one now moves only in contradictory situations. (Eco 2015)

To illustrate the point, Rajasekharan et al. (2017) showed that it is possible to fabricate bio-inspired composites based on combining bottom-up molecular self-assembly and top-down additive manufacturing. This combination allows for bottom-up tuning of material properties resulting in mechanical behaviors similar to those observed in natural mineralized composites. The general idea is to have acceptable self-organization conditions without affecting the spatial resolution, one of the fundamental constraints of 3D printing. There are several ways to achieve this goal, the main one being to achieve self-organization in a voxel of size close to the desired resolution. This compromise moves away from the principle of exploiting the massive parallelism inherent to self-assembly, which was a particularly attractive promise.

4.4.3.1. *Chemical robots*

> Lorenz thought that there must be a link between this refusal of weather to reproduce and the impotence of meteorologists to make predictions – a link between a-periodicity and unpredictability. (Gluck 1987)

Richard Feynman predicted as early as the 1950s the development of molecular nano-machines such as those produced by Jean-Pierre Sauvage as early as 1983 when he invented a chemical method to intertwine two molecules in the form of rings, thus forming a chain named catenane (Collin et al. 2001; CNRS 2016). Through electrical or light stimulation, it is possible to drive the rotation of a molecule in a controlled, motor-like manner. The cooperative behavior that is exploited is characterized by a situation where the different agents interact to solve a task or situation. Other sizes can also be considered as recalled by Pinho et al. (2020).

Maeda et al. (2011) showed that self-oscillating gel actuators can be realized for chemical robotics. The gel has a cyclic reaction network as a metabolic process. During the reaction, the gel has a small but significant volume change by chemical energy. The periodic self-oscillating motion of the gel is produced by the dissipation of chemical energy from the oscillatory Belousov–Zhabotinsky reaction. Other systems where chemistry is absent can be proposed (see Mathews et al. (2017)). The final question is to control not the displacement but its stopping (requirement to eliminate a chemical reactant).

Generally speaking, the collective behavior that could be exploited in additive manufacturing is not sufficient on its own. It would be necessary to find ways to increase the capacities of the voxels alone with the help of a pilot. This is the whole scientific debate opened in this chapter.

4.4.3.2. *Some results of stimulated or constrained self-organization*

> It can be said, in a graphic way, that what happens, in all these areas, is not evolution, but rather an efflorescence in which spread-out forms continually unfold and die, to be replaced by new forms that are different but similar enough to give the impression of an entity in perpetual change. (Böhm 1979)

In the field of additive manufacturing, one seeks to provide the form of an object that meets prior instructions; one also seeks to provide the object with other performances of functionality, evolution (4D printing, bio-printing), etc. (André 2017c). Table 4.1 gathers some (non-biological) results that are based on the different items of section 4.4.

Title	Comments	Reference(s)
Out-of-balance hardware system	Self-organization following the return to equilibrium	Li et al. 2017; Vatankhah-Varnosfaderani et al. 2018; Beloshenko et al. 2019; Estrin et al. 2019
Bi-stability	Reconfigurable structures	Mousanezhad et al. 2017; Chen et al. 2018; Chen and Shea 2018; Liu et al. 2019
Self-organization of environmentally sensitive nanoparticles	Inkjet 3D printing of the self-organization of aqueous colloidal suspensions containing mono-dispersed silica particles and/or polystyrene nanoparticles as a function of solvent evaporation	Sowade et al. 2015; Hsiao et al. 2017
Colloidal self-assembly	Direct-write colloidal assembly is a process that combines the bottom-up principle of colloidal self-assembly with the versatility of direct-write 3D printing	Roger 2013; Jia et al. 2015; Jian et al. 2016; Tan et al. 2018
Magnetic field deposition	Handling and non-contact deposition of particles from a colloidal solution (guided electrophoresis) followed (or not) by relaxation	Grzelczak et al. 2010; Abelmann et al. 2020; Pritchet et al. 2020
Electric field deposition	Construction of hierarchical structures inspired by mother-of-pearl with complex three-dimensional (3D) shapes	Grzelczak et al. 2010; Yadav et al. 2017; Yang et al. 2019
Orientation in a shear	Particle/filler alignment under shear stresses applied to concentrated inks during direct writing in additive manufacturing	Hausmann et al. 2018
Orientation of a phase	Use of an inhomogeneous reception surface	Palacin 2017
Directed self-assembly in additive manufacturing (stereolithography)	Ultrasonically directed self-assembly to organize a user-specified particle pattern to fabricate 3D macro-scale materials with a user-specified microstructure of particles of any material	Greenhall and Raeymaekers 2017; Niendorf and Raeymaekers 2020; Wadsworth et al. 2020
Self-organization of polymers in additive manufacturing	Self-assembly of photonic crystal polymers with color change	Patel et al. 2020
Pixelated voxels	Directed self-assembly of liquid crystal monomers into cross-linked polymer networks with stimulus-induced property changes	Kowalski et al. 2017

Polymer-ceramic composite	Macroscopic alignment of a self-assembled nanostructure of lyotropic liquid crystals polymerizable by 3D printing	Rajasekharan et al. 2017
Bio-bot	Machines using natural muscles	Cheney et al. 2014; Raman et al. 2017
Biofilms	Biological organization	Needleman and Dogic 2017; Hartmann et al. 2019
Thermal gradient pixelation	Method that optimizes synthesis with initial heating rates up to 100°C/minute generating a rich family of nanoscale building blocks with distinct morphologies	Ye et al. 2010
Modification of adhesion properties	pH control is used to increase adhesion during handling and cancel adhesion during release in a micro clamp	Dejeu et al. 2009
Modification of rheological properties in additive manufacturing	Application of self-assembly principles to modify the rheology of "non-printable" materials, while offering the possibility of having properties characteristic of supramolecular materials, sensitive to stimuli	Manning et al. 2019

Table 4.1. *Examples of coupling between top-down and bottom-up approaches*

It is, as we can see, essentially a matter of defining situations where materials are sensitive to an artificially produced environment which allows them to have their physical properties modified, such as shape, electrical conductivity, color or reflection, according to an external input. But these few examples represent only a modest contribution to the total number of publications on additive manufacturing (160,000 on the CNRS database), which is increasing by about 20% per year.

4.4.4. *Can we envisage a "learning" 4D system?*

> You don't know [...] that there is life only in differences: difference of temperature, difference of potential. And if the same heat or the same cold reigns everywhere in the universe, they must be shaken to give rise to fire, explosion, Gehenna. We will shake them. (Zamiatine 2020)

As can be seen from Table 4.1, it does not seem possible to produce simple methods leading to the exploitation of couplings from self-organizations to achieve 4D printing. However, one work has caught our attention, that of Montana Garcia's

(2017) thesis, complemented in part by those of Estrin et al. (2019) and Beloshenko et al. (2019).

4.4.4.1. *Information gathering*

First, Vatankhah-Varnosfaderani et al. (2018) synthesized an elastomer composed of a central block on which side chains extended by two terminal blocks are grafted (see Figure 4.7). This material follows approximately the same deformation curve as a biological tissue. Its color changes according to its state of deformation; indeed, atomic force microscopy observations and X-ray scattering experiments have shown that the terminal blocks of these structures gather in nanometric spheres, distributed in a matrix formed by the "pin-like" structures. Light interferes with this architecture according to the distance between the ends; a stretching of the material thus results in a change of color (chameleon effect). It is therefore possible to encode mechanical properties (elastomer) and optical properties in a single polymer, with modulation possibilities. By imposing a mechanical deformation on an object, it is therefore possible, in principle, to show where the mechanical stress zones are located.

Chinese researchers have developed ionic conductive materials (ionic organohydrogel). Dyes from the spiropyran family change from pale yellow to bluish purple under the effect of mechanical stress. Based on this idea, the team designed a unique membrane named "I-Skin" (see Qiu et al. (2021)). In 2017, Di Giacomo et al. had already developed an artificial skin, using a snake-like mechanism to sense heat. Even more recently, according to Kieczinski, in 2019, scientists from EPFL (Swiss Federal Institute of Technology in Lausanne, Switzerland), presented an artificial skin which provided a sense of touch via tiny air pockets.

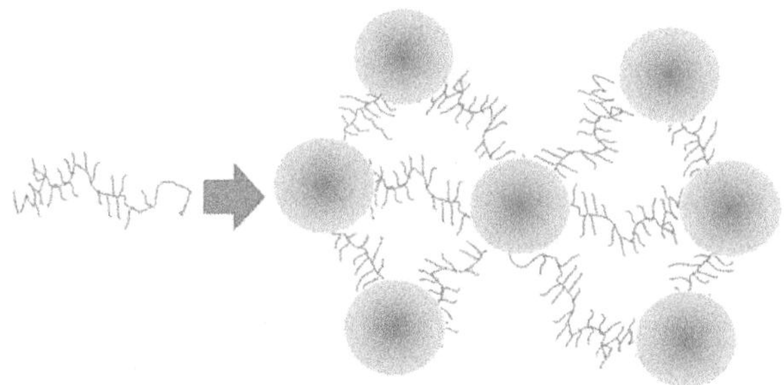

Figure 4.7. *Active "chameleon" structure. For a color version of this figure, see www.iste.co.uk/demoly/4Dprinting1.zip*

NOTE 4.1.– Jin et al. (2019) showed that information can be achieved through distributed decision making from a soft, stretchable silicone composite doped with thermo-chromic pigments and innervated with liquid metal. The ability to deform the liquid metal couples geometric changes to Joule heating, allowing tunable thermo-mechano-chromic sensing of touch and deformation. The principle is different, but the objective is essentially the same.

NOTE 4.2.– Martinez et al. (2020) used cholesteric liquid crystals that change color as an elastomeric material is stretched.

4.4.4.2. *The act of learning*

> If the question of the priority of the egg and the hen or of the hen over the egg embarrasses you, it is because you suppose that animals were originally what they are now. (Diderot 1994)

> Self-organization constitutes an energy-material process occurring in systems far from equilibrium where fluctuations can amplify and generate a dynamic configuration. This configuration is the non-trivial result of a long-scale correlation at the microscopic level, maintained recursively by the energy flow and boundary conditions, and exists only as a dynamic process. (Moreno 2004)

If now, staying in concepts, we have a homogeneous material sensitive to a mechanical deformation (validated for example by physical or physico-chemical systems such as those briefly presented above), if this material is transformed by mechanical manipulation, the areas subjected to this stimulus will have a specific mechanical behavior.

This can be achieved by a life-inspired approach, based on the interactions of a large number of simple agents with each other and with their environment, capable of generating structures or organizations that constitute solutions to given problems. This approach is known to generate robust and self-adapting solutions to variable and unknown conditions of the environment or system (Simonin 2010). But while the multi-agent systems envisioned by Simonin can be successfully studied and applied in the digital sciences, more material solutions are not abundant.

For Thys and Lehn (2015), self-organization can result from the molecular or supramolecular association of constituents in thermodynamic equilibrium, allowing adaptation, in response to stimuli (passive change from one equilibrium state to another). Systems with "active" self-organization, where non-equilibrium dissipative chemical or physical processes correspond to structural and functional properties in response to stimuli; they are adaptive. The emergence of new physical or chemical characteristics is associated with inter- and intramolecular forces of low amplitudes (to correspond to phenomena measurable under conditions of use where the force of the experimenter is sufficient to effect the change).

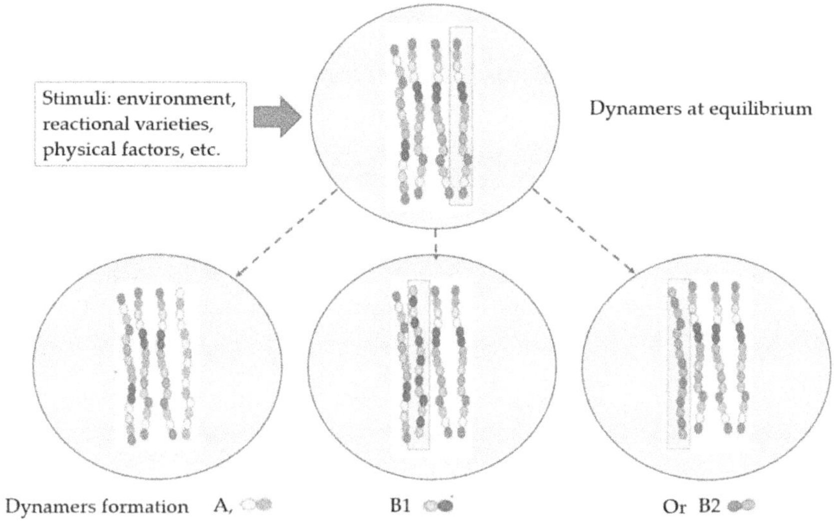

Figure 4.8. *Adaptation of a dynamers library (center) to the application of various stimuli (A, B, C) by generating a stimulus-specific dynamer. For a color version of this figure, see www.iste.co.uk/demoly/4Dprinting1.zip*

In the case of materials, the response to various stimuli or energetic effectors opens the way to adaptive functional materials, such as "dynamers", or dynamic polymers (see Figure 4.8). According to these two authors, the combination of various characteristics (information, programming, dynamics, reversibility, structural and constitutional diversity) opens up vast perspectives toward an adaptive and evolutionary chemistry, a chemistry or physical chemistry that could be exploited in 4D printing. Indeed, adaptations correspond to transformations of dynamic systems induced by physical or chemical agents, such as modifications of the environment, phase changes, physical (temperature, light, pressure etc.) or chemical (protons, ions) stimuli.

In composite materials, mixtures of immiscible polymers can lead to original properties (mechanical, optical, thermal, etc.). In such multiphase systems, all properties are strongly influenced by the presence of interfaces between polymers. In many cases, the development of surfaces with specific properties involves the use of "decorated interfaces" (Chennevière 2014). These interfaces are composed of a substrate on which polymer chains are more or less well-anchored. These chains couple mechanically to the surrounding material and control the transmission of stress. This coupling depends on the penetration of the surface chains into the matrix; the dynamics of these chains play an important role in the fabrication kinetics of these interfaces as well as in their mechanical properties.

Thus, thermoplastic polymers are normally not cross-linked and the polymer chains are linked together by secondary bonds: a small increase in temperature causes the breaking of these low-energy bonds, which can lead to the liquefaction of the polymer following the mobility of the chains between them. This behavior is reversible. The glassy state, obtained at temperatures below the glass transition zone, is a disordered non-equilibrium state that can evolve. Indeed, any dynamical system in a perturbed equilibrium state evolves in time to a new equilibrium state in accordance with the dynamical equations of motion. During a change in temperature, the system evolves from a viscous fluid state with moving chains and involved in very large-scale motions (creeping motions or Rouse modes (see Kalathi et al. (2015))), to a stable glassy state.

According to Feige (2019), elastomer reinforcement strategies play on the architecture of the polymer network, independently of its chemical nature. A dual elastomeric network is then a set of two interpenetrating polymer networks that have different architectures from each other (see Figure 4.9). One of the networks called “filler” is highly cross-linked and contains chains that are pre-stretched. The second network is called “matrix” and instead has a low density of crosslinking. Deformation of this macro-architecture causes the load network to break, which dissipates mechanical energy and redistributes its stress to the stretchable chains of the matrix component. In this structure, the stretching of the charge network is an important parameter that controls the ability of the chains to separate and dissipate energy. The question that will now be addressed corresponds to the effect of the adverse phenomenon where it is a mechanical stimulation that is considered to reorient the polymeric chains.

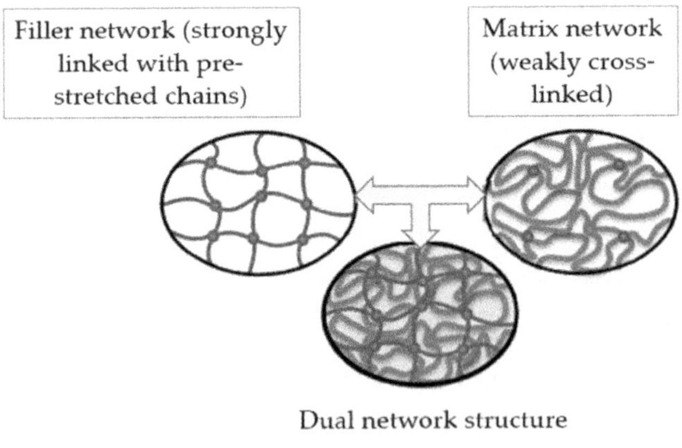

Figure 4.9. *Schematic representation of a “dual network structure” – the charge lattice is shown in blue, and the matrix lattice in red. The two networks are interpenetrated at the molecular level. For a color version of this figure, see www.iste.co.uk/demoly/4Dprinting1.zip*

Identity is a whole, a mixture of what moves and remains. (Robert 2019)

Something persists by changing, that is what it means to last. (Ricoeur 1991)

4.4.4.2.1. First example

Montana Garcia (2017) has done research work on polymer mixtures whose miscibility can be characterized by thermodynamics (the entropy variation is always favorable to the mixture and represents the disorder level of the system). Figure 4.10 represents the possibilities of mixing in the case of two simple liquids (regular solution), a polymer in a solvent (polymer solution) and two immiscible polymers.

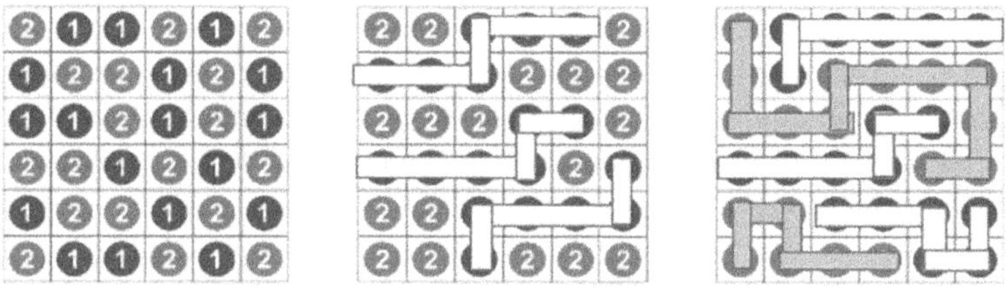

a) b) c)

Figure 4.10. *Lattice model of a binary mixture of: a) two simple liquids (regular solution), b) polymer in solvent (polymer solution) and c) immiscible polymers. For a color version of this figure, see www.iste.co.uk/demoly/4Dprinting1.zip*

The lines represent the polymer chains in the mixture. The lattice model assumes that the movement of the molecules is not entirely free and random, but limited to a specific area within a matrix. Following this principle, the Flory–Huggins theory (Flory 1941; Huggins 1941) describes the entropy of a binary mixture in a matrix arrangement of two species without change of volume, using a combinatorial expression of mixing that applies to molecules with flexible chains of different sizes.

The majority of polymer blends are often made up of immiscible polymers (generating two phases). Of different natures, they tend to repel each other. The Flory–Huggins interaction parameter (Aid 2017) is then positive and the attractions between chains of the same nature are greater than the steric order repulsions. Since most polymers are immiscible with each other, there will be phase separation during mixing. These different morphologies can appear during the mixing process, with a memory role in the final properties of the multiphase material. Thus, temperature changes will have a role in the distribution of the constituents within the global system.

Montana Garcia (2017) shows that during the mixing step of immiscible polymer melts, the final morphology depends on various factors such as the mixing temperature relative to the melting temperatures of the polymers, the rotational speed of the extruder rotors (the intensity of shear stress), the mixing time, the ratio of the viscosities of the polymers as well as the composition of the mixture. Mechanical effects can thus be observed. It is therefore possible to envisage a situation where the structure is modified locally by mechanical shear, giving it local mechanical properties. It is on this basis that we could describe a notion of learning applicable to 4D objects.

4.4.4.2.2. Second example

Shape memory polymers (SMPs) possess the ability to change shape in response to changing environmental conditions. Generally, according to Li (2016), these materials in a permanent shape can be manipulated and fixed into a temporary shape. This temporary deformation remains stable until a stimulus is applied to trigger the resumption of the permanent form, induced by the release of elastic energy stored in the temporary form. This ability to remember different shapes can be exploited provided that one is able to mechanically modify polymer bonds as illustrated in Figure 4.11.

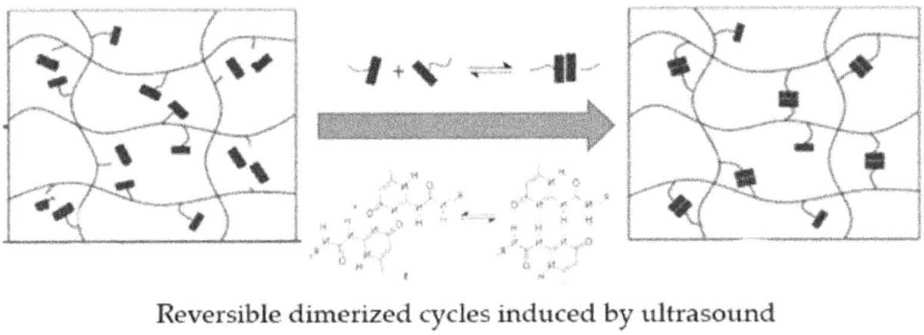

Figure 4.11. *Removal of loose links between polymers according to Li (2016). For a color version of this figure, see www.iste.co.uk/demoly/4Dprinting1.zip*

In his thesis work, Li (2016) used ultrasound to trigger the reshaping of polymers and showed that the process could be observed through freeze/thaw cycles. This area should be of interest for further development of the idea.

4.4.4.3. *Toward an operating manual*

> Science is by definition anti-dogmatic because it knows that it proceeds by trial and error, and because [...] its implicit principle is that of "fallibilism" according to which it is always on the alert to correct its own errors. (Eco 2016)

Figure 4.12 recalls the different 4D printing methods retained by the authors. If we can respect what has been presented above, it would become possible to:

– make an object by a homogeneous method;

– achieve mechanical bending of the object (with possible visualization of the bending regions) to reveal regions with specific elastic properties;

– achieve a 4D effect from a global stimulation having only a local effect (with possible gradients of effects);

– make the operation reversible by carrying out a "Return to Zero" for example by thermal means.

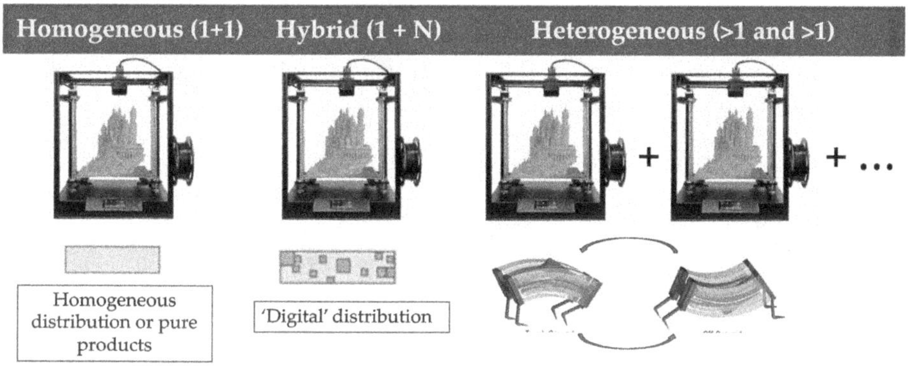

Figure 4.12. *Reminder on 4D printing methods (figure presented in other chapters). For a color version of this figure, see www.iste.co.uk/demoly/4Dprinting1.zip*

> With non-equilibrium environments, the movements of the stimulated areas can lead to a return to thermodynamic equilibrium; the 4D system is therefore likely to lose its "memory".

Box 4.2. *Memory loss*

4.4.5. *Removal of a blocking element*

In the work we have just presented, we involved a medium likely to evolve over time, a medium which participates in the 4D transition. On homogeneous manufacturing bases, one can, for example, consider a photoactive material embedded in an elastomeric matrix containing a dye degradable at a wavelength different from that used for actuation. Under these conditions, the areas that are to become active are previously irradiated to destroy the dye preventing the photoactive effect. Normally this effect is not reversible, but it is possible to work with reversible photochromes.

4.5. Conclusion

> It is necessary to preserve the possibility of acting in accordance with impulses transformed by socio-cultural achievements, constantly challenged by the imagination and creativity. (Laborit 1976)

In Demoly and André (2021), it has been shown that the principles of self-organization could not be successfully exploited to realize 3D objects (except in small areas that do not disturb the spatial resolution). In this chapter, on the other hand, the 3D-4D transition can take advantage of these principles by providing an intermediate realization mode between homogeneous 4D printing and hybrid 4D printing on the one hand, and a possibility of reversibility on the other hand.

It is a pity that the skills of the two authors in the fields of material transformation (polymer chemistry in particular) do not allow them to try to validate the concepts presented in this chapter. However, we intend to go a little further on this register, as soon as it is possible, but for that, we will need some time.

4.6. References

Abelmann, L., Hageman, T., Löthman, P., Mastrangeli, M., Elwenspoek, M. (2020). Three-dimensional self-assembly using dipolar interaction. *Science Advances*, 6, 2375–2548.

Afshar, A. and Di Leo, C.V. (2021). A thermodynamically consistent gradient theory for diffusion-reaction-deformation in solids: Application to conversion-type electrodes. *Journal of the Mechanics and the Physics of Solids*, 151, 104368.

Aid, S.M. (2017). Étude de la miscibilité des polymères par la méthode de coalescence des grains en vue du recyclage des DEEE par rotomoulage. PhD Thesis, Paris-Tech, Paris [Online]. Available at: https://pastel.archives-ouvertes.fr/tel-01769980/document.

Aït-El-Hadj, S. (2015). Éléments de modélisation systémique de la dynamique technologique. *Marché et organisations*, 23, 99–121.

Alvaro, M. (2004). Auto-organisation, autonomie et identité. *Revue internationale de philosophie*, 228, 135–150.

André, J.C. (2017a). *From Additive Manufacturing to 3D/4D Printing 1: From Concepts to Achievements*. ISTE Ltd, London, and John Wiley & Sons, New York.

André, J.C. (2017b). *From Additive Manufacturing to 3D/4D Printing 2: Current Techniques, Improvements and their Limitations*. ISTE Ltd, London, and John Wiley & Sons, New York.

André, J.C. (2017c). *From Additive Manufacturing to 3D/4D Printing 3: Breakthrough Innovations: Programmable Material, 4D Printing and Bio-printing*. ISTE Ltd, London, and John Wiley & Sons, New York.

André, J.C., Le Méhauté, A., De Witte, O. (1984). Dispositif pour réaliser un modèle de pièce industrielle. French Patent n° 84 11 241, 16 July.

Atlan, H. (2006). *L'organisation biologique et la théorie de l'information*. Le Seuil, Paris.

Atlan, H. (2011). *Le vivant post-génomique ou qu'est-ce que l'auto-organisation ?* Odile Jacob, Paris.

Baquiast, J.P. and Jacquemin, C. (2004). La morphogénèse [Online]. Available at: http://www.automatesintelligents.com/echanges/2004/jan/morphogenese.html.

Bayard, P. (2007). *Comment parler des livres qu'on n'a pas lus ?* Editions de Minuit, Paris.

Beloshenko, V., Vozniak, I., Beygelzimer, Y., Estrin, Y., Kulagin, R. (2019). Severe plastic deformation of polymers. *Materials Transactions*, 60, 1192–1202.

Bergson, H. (1907). *L'évolution créatrice*. PUF, Paris.

Bergson, H. (2013). *Essai sur les données immédiates de la conscience*. Garnier-Flammarion, Paris.

Bhalla, N., Ipparthi, D., Klemp, E., Dorigo, M. (2015). A geometrical approach to the incompatible substructure problem in parallel self-assembly. *International Conference on Parallel Problem Solving from Nature PPSN 2014: Parallel Problem Solving from Nature – PPSN XIII* [Online]. Available at: https://iridia.ulb.ac.be/~mdorigo/Published_papers/All_Dorigo_papers/BhaIppKleDor2014ppsn.pdf.

Bizzarri, M., Giulani, A., Cucina, A., D'Anselmi, F., Soto, A.M., Sonnenschein, C. (2011). Fractal analysis in a systems biology approach to cancer. *Seminars in Cancer Biology*, 21, 175–182.

Blanqui, A. (2012). *L'éternité par les astres*. Impressions nouvelles, Paris [cited in Verdo, Y. (2018). *Le violon d'Einstein*. Odile Jacob, Paris].

Böhm, D. (1979). Unicité de l'"ordre enveloppé". *Le Monde* [Online]. Available at: https://www.lemonde.fr/archives/article/1979/10/24/david-bohm-unicite-de-l-ordre-enveloppe_2792439_1819218.html.

Bruter, C.P. (1976). *La théorie du système général ; théorie de la modélisation*. PUF, Paris [Online]. Available at: http://www.mcxapc.org/inserts/ouvrages/0609tsgtm.pdf.

Cademartiri, L. and Bishop, K.J.M. (2015). Programmable self-assembly. *Nature Materials*, 14, 2–9.

Callon, M. and Lacoste, A. (2011). Défendre l'innovation responsable. *Debating Innovation*, 1, 5–18.

Campbell, D.T. (1974). Downward causation in hierarchically organized biological systems. In *Studies in the Philosophy of Biology*, Ayala, F. and Dobzhansky, T. (eds). University of California Press, Berkeley.

Cantot, P. and Luzeaux, D. (2011). *Simulation and Modeling of Systems of Systems*. ISTE, London, and John Wiley & Sons, New York.

Chauvet, G. (2006). *Comprendre l'organisation du vivant et son évolution vers la conscience*. Vuibert, Paris.

Chen, T. and Shea, K. (2018). An autonomous programmable actuator and shape reconfigurable structures using bi-stability and shape memory polymers. *3D Printing and Additive Manufacturing*, 5, 91–101

Chen, T., Bilal, O.R., Shea, K., Daraio, C. (2018). Harnessing bi-stability for directional propulsion of soft, untethered robots. *Proceedings of the National Academy of Sciences of the USA*, 115, 5698–702.

Cheney, N., Clune, J., Lipson, H. (2014). Evolved electrophysiological soft robots. *ALIFE 14: The Fourteenth Conference on the Synthesis and Simulation of Living Systems*, 14, 222–229.

Chennevière, A. (2014). Dynamique de chaînes de polymère greffées et glissement aux interfaces. Thesis, Université de Paris-Sud, Orsay [Online]. Available at: https://tel.archives-ouvertes.fr/tel-01174918/file/VD2_CHENNEVIERE_ALEXIS_12122014.pdf.

Chossat, P. (1996). *Les symétries brisées*. Sciences d'avenir, Pour la Science/Belin, Paris.

Chossat, P. (2015). Les mathématiques de la morphogénèse : l'autre grande contribution d'Alan Turing (I) [Online]. Available at: http://images.math.cnrs.fr/Les-mathematiques-de-la.html.

CNRS (2016). Jean-Pierre Sauvage, un Nobel pour les machines moléculaires. *Le journal CNRS*. [Online]. Available at: https://lejournal.cnrs.fr/articles/jean-pierre-sauvage-un-nobel-pour-les-machines-moleculaires.

Collin, J.P., Dietrich-Buchecker, R., Sauvage, J.P. (2001). Vers des machines et des moteurs moléculaires. *Actualité chimique*, 27–32.

Dejeu, J., Gauthier, M., Rougeot, P., Boireau, W. (2009). Adhesion forces controlled by chemical self-assembly and pH – Application to robotic micro-handling. *Applied Material & Interfaces*, 1, 1966–1977 [Online]. Available at: https://hal.archives-ouvertes.fr/hal-00413803/document [Accessed 25 March 2022].

Demoly, F. and André, J.C. (2021). Is order creation through disorder in additive manufacturing possible? *Cogent Engineering*, 1889110 [Online]. Available at: https://www.tandfonline.com/doi/full/10.1080/23311916.2021.1889110 [Accessed 25 March 2022].

Di Giacomo, R., Bonanomi, L., Costanza, V., Maresca, B., Daraio, C. (2017). Biomimetic temperature sensing layer for artificial skins [Online]. Available at: https://authors.library.caltech.edu/73994/4/1512.01161.pdf [Accessed 25 March 2022].

Diderot, D. (1994). *Le rêve de d'Alembert*. Robert Laffont, Paris.

Dionnet, V., Héliodore, F., Le Méhauté A., André, J.C. (1992). Procédé de fabrication d'un objet fractal par stéréolithographie et objet fractal obtenu par un tel procédé, Brevet français, 9206652 du 02.06.92, Brevet européen, 93401389.7 du 28.07.1993, brevet américain US5355318 A, demande US 08/070,172 du 11 octobre 1994, US Patent 5,355,318.

Doursat, R., Sayama, H., Michel, O. (2014). L'ingénierie morphogénétique : modèles de processus dynamiques pour la morphogenèse. In *Modéliser & simuler 2 : épistémologies et pratiques de la modélisation et de la simulation*, Varenne, F., Silberstein, M., Dutreuil, S., Huneman P. (eds). Éditions Matériologiques, Paris.

Drew, E. (2010). Symposium on opportunities and challenges in the emerging field of synthetic biology. Synthesis report, OECD and The Royal Society [Online]. Available at: https://www.oecd.org/sti/biotech/45144066.pdf [Accessed 25 March 2022].

Eco, U. (2015). *Chroniques d'une société liquide*. Le Livre de Poche, Paris.

El Mellah, I., Loh, A., Pottier, A. (2012). Méthode de Turing et morphogénèse : illustration des équations de réaction – diffusion [Online]. Available at: https://media4.obspm.fr/public/M2R/appliquettes/Turing/AutomateTuring_website.html.

Estrin, Y., Beygelzimer, Y., Kulagin, R. (2019). Design of architectured materials based on mechanically driven structural and compositional patterning. *Advanced Engineering Materials*, 21, 1900487.

Feige, L. (2019). Introduction of anisotropy in double network elastomers by orientation and two-step curing. PhD ESPCI, Paris [Online]. Available at: https://pastel.archives-ouvertes.fr/tel-02887289/document [Accessed 25 March 2022].

Firat, Z., Ferrante, E., Gillet, Y., Tuci, E. (2020). On self-organized aggregation dynamics in swarms of robots with informed robots. *Neural Computing and Applications* [Online]. Available at: https://doi.org/10.1007/s00521-020-04791-0.

Flory, P.J. (1941). Thermodynamics of high polymer solutions. *The Journal of Chemical Physics*, 9, 660–660.

Glansdorff, P. and Prigogine, I. (1971). *Structure, stabilité et fluctuations*. Masson, Paris.

Gluck, J. (1987). *Chaos*. The Viking Press, New York.

Granger, G.G. (1995). *Le probable, le possible et le virtuel*. Odile Jacob, Paris.

Grzelczak, M., Vermant, J., Furst, E.M., Liz-Marza, L.M. (2010). Directed self-assembly of nanoparticles. *ACS Nano*, 4, 3591–3605.

Guièze, G. (1983). Les énoncés de Mind and machines comme image de l'Objet technique. In *Pensée et machine*, Anderson, A.R. (ed.). Champ Vallon, Paris.

Hartmann, R., Singh, P.K., Pearce, P., Mok, R., Song, B., Diaz-Pascual, F., Dunkel, J., Drescher, K. (2019). Emergence of three-dimensional order and structure in growing biofilms. *Nature Physics*, 15, 251–256.

Hausmann, M.K., Rühs, P.A., Siqueira, G., Läuger, J., Libanori, R., Zimmermann, T., Studart, A.R. (2018). Dynamics of cellulose nano-crystal alignment during 3D printing. *ACS Nano*, 12, 6926–6937.

Hawking, S. (1989). *A Brief History of Time*. Bantam Book, New York.

Hsiao, L.C., Md Badruddoza, A.Z., Cheng, L.C., Doyle, P.S. (2017). 3D printing of self-assembling thermos-responsive nano-emulsions into hierarchical mesostructured hydrogels. *Soft Matter*, 13, 921.

Huggins, M.L. (1941). Solutions of long chain compounds. *The Journal of Chemical Physics*, 9, 440–450.

Jakab, K., Neagu, A., Mironov, V., Markwald, R.R., Forgacs, G. (2004). Engineering biological structures of prescribed shape using self-assembling multicellular systems. *Proceedings of the National Academy of Science of the USA*, 101, 2864–2869.

Jia, K., Cheng, Y., Liu, X., Li, X., Dong, J. (2015). Thermal, light and pH triple stimulated changes in self-assembly of a novel small molecular weight amphiphilic binary system [Online]. Available at: https://pubs.rsc.org/en/content/getauthorversionpdf/c4ra13038c [Accessed 25 March 2022].

Jiang, F., Chen, S., Cao, Z., Wang, G. (2016). A photo, temperature, and pH responsive spiropyran-functionalized polymer: Synthesis, self-assembly and controlled release. *Polymer*, 83, 85–91.

Jin, Y., Lin, Y., Kiani, A., Joshipura, I.D., Ge, M., Dickey, M.D. (2019). Materials tactile logic via innervated soft thermos-chromic elastomers. *Nature Communications*, 10, 4187.

Joshi, S.D. and Davidson, L.A. (2012). Epithelial machines of morphogenesis and their potential application in organ assembly and tissue engineering. *Biomechanics Modeling in Mechano-biology*, 11, 1109–1121.

Kalathi, J.T., Kumar, S.K., Rubinstein, M., Grest, G.S. (2015). Rouse mode analysis of chain relaxation in polymer nanocomposites. *Soft Matter*, 11, 4123–4132.

Kekenes-Huskey, P.M., Gillette, A., Hake, J., McCammon, J.A. (2012). Finite-element estimation of protein–ligand association rates with post-encounter effects: Applications to calcium binding in troponin C and SERCA. *Computer Science Discovery*, 5, 014015.

Kheifetz, M.L. (2018). From information and additive technologies to self-reproduction of machines and organisms [Online]. Available at: https://cyberleninka.ru/article/n/from-information-and-additive-technologies-to-self-reproduction-of-machines-and-organisms [Accessed 25 March 2022].

Kieczinski, N. (2019). Réalité virtuelle : cette peau artificielle stimule le sens du toucher pour "ressentir la VR". *Neozone*. [Online]. Available at: https://www.neozone.org/jeux-video/realite-virtuelle-cette-peau-artificielle-stimule-le-sens-du-toucher-pour-ressentir-la-vr/.

Kowalski, B.A., Guin, T.C., Auguste, A.D., Godman, N.P., White, T.J. (2017). Pixelated polymers: Directed self-assembly of liquid crystalline polymer networks. *ACS Macro Letters*, 6, 436–441

Laborit, H. (1976). *Eloge de la fuite*. Robert Laffont, Paris.

Larger, L. (2015). *Groupe de veille INSIS – contribution*. INSIS, CNRS, Paris.

Le Moigne, J.L. (1977). *La théorie du système général*. PUF, Paris.

Le Moigne, J.L. (1990). *La modélisation des systèmes complexes*. Dunod, Paris

Leduc, S. (1910). *Théorie physico-chimique de la vie et générations spontanées*. Poinat, Paris.

Leduc, S. (1912). *La biologie synthétique : étude de biophysique*. Poinat, Paris.

Lesne, A. (2013). Une conjonction de réactions chimiques et de diffusion : Alan Turing, les motifs et les structures du vivant [Online]. Available at: https://interstices.info/jcms/int_71868/alan-turing-les-motifs-et-les-structures-du-vivant.

Lévy, P. (1985). "Analyse de contenu des travaux du Biological Computer Laboratory" and "Le théâtre des opérations. Au sujet des travaux du BCL". *Cahiers du Centre de recherche, épistémologie et autonomie*, 8. Ecole polytechnique, Paris.

Li, G. (2016). Polymères et Hydrogels à mémoire de forme ultrason-répondants. PhD, Université de Sherbrooke, Quebec [Online]. Available at: https://savoirs.usherbrooke.ca/bitstream/handle/11143/8359/Li_Guo_PhD_2016.pdf?sequence=9&isAllowed=y.

Liu, L., Qiao, C., An, H., Pasini, D. (2019). Encoding kirigami bi-materials to morph on target in response to temperature. *Scientific Reports*, 9, 19499.

Maeda, S., Hara, Y., Nakamaru, S., Nakagawa, H., Hashimoto, S. (2011). Chemical robots [Online]. Available at: https://www.intechopen.com/books/on-biomimetics/chemical-robots [Accessed 25 March 2022].

Manning, K.B., Wyatt, N., Hughes, L., Cook, A., Giron, N.H., Martinez, E., Campbell, C.G., Celina, M.C. (2019). *Self-Assembly Assisted Additive Manufacturing: Direct Ink Write 3D Printing of Epoxy-Amine Thermosets*. Sandia National Laboratories [Online]. Available at: https://www.osti.gov/pages/servlets/purl/1487420 [Accessed 25 March 2022].

Martinez, A.M., McBride, M.K., White, T.J., Bowman, C.N. (2020). Reconfigurable and spatially programmable chameleon skin-like material utilizing light responsive covalent adaptable cholesteric liquid crystal elastomers. *Advanced Functional Materials*, 30, 2003150.

Mathews, N., Christensen, A.L., O'Grady, R., Mondada, F., Dorigo, M. (2017). Mergeable nervous systems for robots. *Nature Communications*, 8, 439 [Online]. Available at: https://www.nature.com/articles/s41467-017-00109-2 [Accessed 25 March 2022].

Mathis, R. (2011). Bronze artéfacts [Online]. Available at: https://fr.wikipedia.org/wiki/Fichier:Bronze_artéfacts_Montjustin.jpg.

McLennan, D.J. (2015). The morphogenetic path to programmable matter. *Proc. IEEE*, 103, 1226–1232 [Online]. Available at: http://citeseerx.ist.psu.edu/viewdoc/download?doi=10.1.1.705.775&rep=rep1&type=pdf [Accessed 25 March 2022].

McMullin, B. and Varela, F. (1997). Rediscovering computational autopoiesis. In *Fourth European Conference on Artificial Life*, Husbands, P. and Harvey, I. (eds). MIT Press, Cambridge, MA.

Monod, J. (2014). *Le hasard et la nécessité : essai sur la philosophie naturelle de la biologie moderne*. Points Essais, Paris.

Montana Garcia, J.S. (2017). Architecture multi-échelle de matériaux polymères : de l'auto-assemblage à l'assemblage forcé. Thesis, ENSAM, Paris [Online]. Available at: https://pastel.archives-ouvertes.fr/tel-01762645/document.

Monteiro, M.J. (2018). Order from disorder through dissipation of free energy. *Nature Nanotechnology*, 13, 771–772.

Moreno, A. (2004). Auto-organisation, autonomie et identité. *Revue internationale de philosophie*, 228, 135–150 [Online]. Available at: https://www.cairn.info/revue-internationale-de-philosophie-2004-2-page-135.htm?contenu=article.

Mousanezhad, D., Kamrava, S., Vaziri, A. (2017). Origami-based building blocks for modular construction of foldable structures. *Scientific Reports*, 7, 14792.

Mugnier, L. (2008). Introduction aux problèmes inverses : application à l'imagerie optique à haute résolution en astronomie, C9. In *Des données à l'objet : le problème inverse*, Léna, P. (ed.). EDP Sciences/CNRS, Paris.

Needleman, D. and Dogic, Z. (2017). Active matter at the interface between materials science and cell biology. *Nature Reviews Materials*, 2, 17048 [Online]. Available at: https://www.nature.com/articles/natrevmats201748.

Newman, S.A. (2014). Physico-genetics of morphogenesis: The hybrid nature of developmental mechanisms. In *Towards a Theory of Development*, Minelli, A. and Pradeu, T. (eds). Oxford University Press, Oxford.

Nietzsche, F. (2014). *Œuvres complètes*. Arvensa, Paris.

Nykter, M., Price, N.D., Aldana, M., Ramsey, S.A., Kauffman, S.A., Hood, L.E. (2008). Gene expression dynamics in the macrophage exhibit criticality. *Proceedings of the National Academy of Sciences of the USA*, 105, 1897–1900.

Palacin, S. (2017). Approche bottom-up : le principe de l'auto-assemblage [Online]. Available at: https://www.fun-mooc.fr/c4x/UPSUD/42003/asset/Cours_2-3b.pdf.

Patel, B.B., Walsh, D.J., Kim, D.H., Kwok, J., Lee, B., Guironnet, D., Diao, Y. (2020). Tunable structural color of bottlebrush block copolymers through direct-write 3D printing from solution. *Science Advances*, 6, [Online]. Available at: https://advances.sciencemag.org/content/6/24/eaaz7202.

Petitat, A. (1998). *Secret et formes sociales*. PUF, Paris.

Pinho, A.C., Buga, C.S., Piedade, A.P. (2020). The chemistry behind 4D printing. *Applied Materials Today*, 19, 100611.

Prigogine, I. (1994). *Les lois du chaos*. Flammarion, Paris.

Prigogine, I. and Stengers, I. (2009). *Le temps et l'éternité*. Champs Sciences, Paris.

Pritchet, D., Ehmann, K., Cao, J., Huang, J. (2020). Manipulation and localized deposition of particle groups with modulated electric fields. *Micro-machines*, 11, 226 [Online]. Available at: https://www.ncbi.nlm.nih.gov/pmc/articles/PMC7074757/.

Pujo-Menjouet, L. (2014). Théorie sur l'apparition de structures de Turing pour les biologistes, ou éclaircissements sur deux intuitions ingénieuses. In *Modéliser & simuler 2 : épistémologies et pratiques de la modélisation et de la simulation*, Varenne, F., Silberstein, M., Dutreuil, S., Huneman, P. (eds). Éditions matériologiques, Paris.

Qiu, W., Zhang, C., Chen, G., Zhu, H., Zhang, Q., Zhu, S. (2021). Colorimetric ionic organohydrogels mimicking human skin for mechanical stimuli sensing and injury visualization. *ACS Applied Materials Interfaces*, 13, 26490–26497.

Rajasekharan, A.K., Bordes, R., Sandström, C., Ekh, M., Andersson, M. (2017). Hierarchical and heterogeneous bioinspired composites – Merging molecular self-assembly with additive manufacturing. *Small*, 13, 1700550.

Raman, R., Cvetkovic, C., Bashir, R. (2017). A modular approach to the design, fabrication, and characterization of muscle-powered biological machines. *Nature Protocols*, 12, 519–533.

Ricœur, P. (1991). *Œuvres – Temps et Récit*. Livre de poche, Paris.

Robert, M. (2019). *Descartes pour les jours de doute*. J'ai Lu, Paris.

Roger, K. (2013). Nano-émulsions auto-assemblées par des méthodes physico-chimiques. Thesis, Université Pierre et Marie Curie, Paris.

Shalygo, Y. (2015). The kinetic basis of morphogenesis, ECAL 2015. *13th European Conference on Artificial Life*, July 20–24, York, ASME/MIT (ed.), Boston, 122–129.

Simon, H.A. (1969). *Science des systèmes, science de l'artificiel*. Dunod, Paris.

Simon, H.A. (2004). *Les sciences de l'artificiel*. Folio-Essais, Paris.

Simondon, G. (1969). *Du mode d'existence des objets techniques*. Aubier-Montaigne, Paris.

Simonin, O. (2010). *Contribution à la résolution collective de problème – modèles d'auto-organisation par interactions directes et indirectes dans les SMA réactifs et robotiques*. HDR, Nancy [Online]. Available at: https://tel.archives-ouvertes.fr/tel-00546300/document.

Sonigo, P. (2000). Cellules en liberté. In *Ni Dieu, ni gène ; pour une autre théorie de l'hérédité*, Kupiec, J.J. and Sonigo, P. (eds). Points Sciences, Paris.

Sowade, E., Blaudeck, T., Baumann, R.R. (2015). Inkjet printing of colloidal nano-spheres: Engineering the evaporation-driven self-assembly process to form defined layer morphologies. *Nanoscale Research Letters*, 10, 362.

Staetzler, K., Sonnenschein, C., Soto, A.M. (2011). Systems biology beyond networks: Generating order from disorder through self-organization. *Seminars in Cancer Biology*, 21, 165–174.

Staune, J. (2015). *Les clés du futur ; réinventer ensemble la société, l'économie et la science*. Plon, Paris.

Stengers, I. (1985). Les généalogies de l'auto-organisation. In *Cahiers du Centre de reserche en épistémologie et autonomie*, 8. Ecole polytechnique, Paris.

Tan, A.T.L., Beroz, J., Kolle, M., Hart, A.J. (2018). Direct-write freeform colloidal assembly. *Advanced Materials*, 30, 1803620.

Thom, R. (1983). *Paraboles et catastrophes*. Flammarion, Paris.

Thuillier, P. (1980). *Le petit savant illustré*. Le Seuil, Paris.

Thys, H. and Lehn, J.M. (2015). L'auto-organisation : vers une chimie de la matière complexe. *Actualité chimique*, 399, 9–15 [Online]. Available at: https://www.lactualitechimique.org/IMG/pdf/2015-399-aout-sept.-p9-this_hd.pdf?5588/d48cd0c0dec95bc4a74312bfe78f09e303ebc542.

Turing, A. (1952). The chemical basis of morphogenesis. *Philosophical Transactions of the Royal Society of London*, B237, 37–72.

Valéry, P. (1957). *Œuvres*. Gallimard, Paris.

Varela, F.J. (1983). *L'auto-organisation : de la physique au politique*. Le Seuil, Paris.

Vatankhah-Varnosfaderani, M., Keith, A.N., Cong, Y., Liang, Y., Rosenthal, M., Sztucki, M., Clair, C., Magonov, S., Ivanov, D.A., Dobrynin, A.V. et al. (2018). Chameleon-like elastomers with molecularly encoded strain-adaptive stiffening and coloration. *Science*, 359, 1509–1513.

Vecchi, D. and Hernandez, I. (2014). The epistemological resilience of the concept of morphogenetical field. In *Towards a Theory of Development*, Minelli, A. and Pradeu, T. (eds). Oxford University Press, Oxford.

Von Foerster, H. (1960). On self-organizing systems and their environments. In *Self-Organizing Systems*, Yovits, M.C. and Cameron, S. (eds.). Pergamon, New York.

Von Foerster, H. and Zopf, G. (1962). *Principles of Self-Organization*. Pergamon Press, New York.

Weil, S. (2021). *L'enracinement – prélude à une déclaration des devoirs envers l'être humain*. Payot, Paris.

Whitehead, A.N. (1906). On mathematical concepts of the material world. *Philosophical Transactions of the Royal Society of London, Series A, Containing Papers of a Mathematical or Physical Character*, 205, 465–525.

Whitelam, S. and Jack, R.L. (2015). The statistical mechanics of dynamic pathways to self-assembly. *Annual Review of Physical Chemistry*, 66, 143–163.

Yadav, R., Naebe, M., Wang, X., Kandasubramanian, B. (2017). Review on 3D prototyping of damage tolerant interdigitating brick arrays of nacre. *Industrial & Engineering Chemistry Research*, 56, 10516–10525.

Yang, Y., Li, X., Chu, M., Sun, H., Jin, J., Yu, K., Wang, Q., Zhou, Q., Chen, Y. (2019). Electrically assisted 3D printing of nacre-inspired structures with self-sensing capability. *Science Advances*, 5 [Online]. Available at: https://advances.sciencemag.org/content/5/4/eaau9490?utm_source=TrendMD&utm_medium=cpc&utm_campaign=TrendMD_1 [Accessed 25 March 2022].

Ye, X., Collins, J.E., Kang, Y., Chen, J., Chen, D.T.N., Yodh, A.G., Murray, C.B. (2010). Morphologically controlled synthesis of colloidal up-conversion nanophosphors and their shape-directed self-assembly. *Proceedings of the National Academy of Sciences*, 107, 22430–22435 [Online]. Available at: https://www.pnas.org/content/pnas/107/52/22430.full.pdf [Accessed 25 March 2022].

Yin, S., Li, B., Feng, X.Q. (2021). Bio-chemo-mechanical theory of active shells. *Journal of the Mechanics and Physics of Solids*, 152, 104419.

Youssef, M., Hueckel, T., Yi, G.R., Sacanna, S. (2016). Shape-shifting colloids via stimulated dewetting. *Nature Communications*, 7, 12216.

Yovitz, M. and Cameron, S. (1960). *Self-organizing Systems*. Pergamon Press, Oxford.

Zahadat, P., Hofstadler, D.N., Schmickl, T. (2017). Vascular morphogenesis controller: A generative model for developing morphology of artificial structures. *GECCO 17: Proceedings of the Genetic and Evolutionary Computation Conference*, 163–170 [Online]. Available at: http://zool33.uni-graz.at/artlife/sites/default/files/paper_2.pdf

Zamiatine, E. (2020). *Nous autres*. Gallimard, Paris.

Zenil, H. (2013). Turing patterns with Turing machines: Emergence and low-level structure formation. *Natural Computing*, 12, 291–303.

Index

L, M

N, O, P

R, S

T

Other titles from

ISTE

in

Systems and Industrial Engineering – Robotics

2022

BOURRIÈRES Jean-Paul, PINÈDE Nathalie, TRAORÉ Mamadou Kaba, ZACHAREWICZ Grégory
From Logistic Networks to Social Networks: Similarities, Specificities, Modeling, Evaluation

HAJJI Rafika, JARAR OULIDI Hassane
Building Information Modeling for a Smart and Sustainable Urban Space

KROB Daniel
Model-based Systems Architecting: Using CESAM to Architect Complex Systems (Volume 3 - Systemes of Systems Complexity Set)

LOUIS Gilles
Dynamics of Aircraft Flight

2020

BRON Jean-Yves
System Requirements Engineering

KRYSINSKI TOMASZ, MALBURET FRANÇOIS
Energy and Motorization in the Automotive and Aeronautics Industries

PRINTZ Jacques
System Architecture and Complexity: Contribution of Systems of Systems to Systems Thinking

2019

ANDRÉ Jean-Claude
Industry 4.0: Paradoxes and Conflicts

BENSALAH Mounir, ELOUADI Abdelmajid, MHARZI Hassan
Railway Information Modeling RIM: The Track to Rail Modernization

BLUA Philippe, YALAOU Farouk, AMODEO Lionel, DE BLOCK Michaël, LAPLANCHE David
Hospital Logistics and e-Management: Digital Transition and Revolution

BRIFFAUT Jean-Pierre
From Complexity in the Natural Sciences to Complexity in Operations Management Systems
(Systems of Systems Complexity Set – Volume 1)

BUDINGER Marc, HAZYUK Ion, COÏC Clément
Multi-Physics Modeling of Technological Systems

FLAUS Jean-Marie
Cybersecurity of Industrial Systems

JAULIN Luc
Mobile Robotics – Second Edition Revised and Updated

KUMAR Kaushik, DAVIM Paulo J.
Optimization for Engineering Problems

TRIGEASSOU Jean-Claude, MAAMRI Nezha
Analysis, Modeling and Stability of Fractional Order Differential Systems 1: The Infinite State Approach
Analysis, Modeling and Stability of Fractional Order Differential Systems 2: The Infinite State Approach

VANDERHAEGEN Frédéric, MAAOUI Choubeila, SALLAK Mohamed, BERDJAG Denis
Automation Challenges of Socio-technical Systems

2018

BERRAH Lamia, CLIVILLÉ Vincent, FOULLOY Laurent
Industrial Objectives and Industrial Performance: Concepts and Fuzzy Handling

GONZALEZ-FELIU Jesus
Sustainable Urban Logistics: Planning and Evaluation

GROUS Ammar
Applied Mechanical Design

LEROY Alain
Production Availability and Reliability: Use in the Oil and Gas Industry

MARÉ Jean-Charles
Aerospace Actuators 3: European Commercial Aircraft and Tiltrotor Aircraft

MAXA Jean-Aimé, BEN MAHMOUD Mohamed Slim, LARRIEU Nicolas
Model-driven Development for Embedded Software: Application to Communications for Drone Swarm

MBIHI Jean
Analog Automation and Digital Feedback Control Techniques
Advanced Techniques and Technology of Computer-Aided Feedback Control

MORANA Joëlle
Logistics

SIMON Christophe, WEBER Philippe, SALLAK Mohamed
Data Uncertainty and Important Measures
(Systems Dependability Assessment Set – Volume 3)

TANIGUCHI Eiichi, THOMPSON Russell G.
City Logistics 1: New Opportunities and Challenges
City Logistics 2: Modeling and Planning Initiatives
City Logistics 3: Towards Sustainable and Liveable Cities

ZELM Martin, JAEKEL Frank-Walter, DOUMEINGTS Guy, WOLLSCHLAEGER Martin
Enterprise Interoperability: Smart Services and Business Impact of Enterprise Interoperability

2017

ANDRÉ Jean-Claude
From Additive Manufacturing to 3D/4D Printing 1: From Concepts to Achievements
From Additive Manufacturing to 3D/4D Printing 2: Current Techniques, Improvements and their Limitations
From Additive Manufacturing to 3D/4D Printing 3: Breakthrough Innovations: Programmable Material, 4D Printing and Bio-printing

ARCHIMÈDE Bernard, VALLESPIR Bruno
Enterprise Interoperability: INTEROP-PGSO Vision

CAMMAN Christelle, FIORE Claude, LIVOLSI Laurent, QUERRO Pascal
Supply Chain Management and Business Performance: The VASC Model

FEYEL Philippe
Robust Control, Optimization with Metaheuristics

MARÉ Jean-Charles
Aerospace Actuators 2: Signal-by-Wire and Power-by-Wire

POPESCU Dumitru, AMIRA Gharbi, STEFANOIU Dan, BORNE Pierre
Process Control Design for Industrial Applications

RÉVEILLAC Jean-Michel
Modeling and Simulation of Logistics Flows 1: Theory and Fundamentals
Modeling and Simulation of Logistics Flows 2: Dashboards, Traffic Planning and Management
Modeling and Simulation of Logistics Flows 3: Discrete and Continuous Flows in 2D/3D

2016

ANDRÉ Michel, SAMARAS Zissis
Energy and Environment
(Research for Innovative Transports Set - Volume 1)

AUBRY Jean-François, BRINZEI Nicolae, MAZOUNI Mohammed-Habib
Systems Dependability Assessment: Benefits of Petri Net Models (Systems Dependability Assessment Set - Volume 1)

BLANQUART Corinne, CLAUSEN Uwe, JACOB Bernard
Towards Innovative Freight and Logistics (Research for Innovative Transports Set - Volume 2)

COHEN Simon, YANNIS George
Traffic Management (Research for Innovative Transports Set - Volume 3)

MARÉ Jean-Charles
Aerospace Actuators 1: Needs, Reliability and Hydraulic Power Solutions

REZG Nidhal, HAJEJ Zied, BOSCHIAN-CAMPANER Valerio
Production and Maintenance Optimization Problems: Logistic Constraints and Leasing Warranty Services

TORRENTI Jean-Michel, LA TORRE Francesca
Materials and Infrastructures 1 (Research for Innovative Transports Set - Volume 5A)
Materials and Infrastructures 2 (Research for Innovative Transports Set - Volume 5B)

WEBER Philippe, SIMON Christophe
Benefits of Bayesian Network Models
(Systems Dependability Assessment Set – Volume 2)

YANNIS George, COHEN Simon
Traffic Safety (Research for Innovative Transports Set - Volume 4)

2015

AUBRY Jean-François, BRINZEI Nicolae
Systems Dependability Assessment: Modeling with Graphs and Finite State Automata

BOULANGER Jean-Louis
CENELEC 50128 and IEC 62279 Standards

BRIFFAUT Jean-Pierre
E-Enabled Operations Management

MISSIKOFF Michele, CANDUCCI Massimo, MAIDEN Neil
Enterprise Innovation

2014

CHETTO Maryline
Real-time Systems Scheduling
Volume 1 – Fundamentals
Volume 2 – Focuses

DAVIM J. Paulo
Machinability of Advanced Materials

ESTAMPE Dominique
Supply Chain Performance and Evaluation Models

FAVRE Bernard
Introduction to Sustainable Transports

GAUTHIER Michaël, ANDREFF Nicolas, DOMBRE Etienne
Intracorporeal Robotics: From Milliscale to Nanoscale

MICOUIN Patrice
Model Based Systems Engineering: Fundamentals and Methods

MILLOT Patrick
Designing Human–Machine Cooperation Systems

NI Zhenjiang, PACORET Céline, BENOSMAN Ryad, RÉGNIER Stéphane
Haptic Feedback Teleoperation of Optical Tweezers

OUSTALOUP Alain
Diversity and Non-integer Differentiation for System Dynamics

REZG Nidhal, DELLAGI Sofien, KHATAD Abdelhakim
Joint Optimization of Maintenance and Production Policies

STEFANOIU Dan, BORNE Pierre, POPESCU Dumitru, FILIP Florin Gh., EL KAMEL Abdelkader
Optimization in Engineering Sciences: Metaheuristics, Stochastic Methods and Decision Support

2013

ALAZARD Daniel
Reverse Engineering in Control Design

ARIOUI Hichem, NEHAOUA Lamri
Driving Simulation

CHADLI Mohammed, COPPIER Hervé
Command-control for Real-time Systems

DAAFOUZ Jamal, TARBOURIECH Sophie, SIGALOTTI Mario
Hybrid Systems with Constraints

FEYEL Philippe
Loop-shaping Robust Control

FLAUS Jean-Marie
Risk Analysis: Socio-technical and Industrial Systems

FRIBOURG Laurent, SOULAT Romain
Control of Switching Systems by Invariance Analysis: Application to Power Electronics

GROSSARD Mathieu, REGNIER Stéphane, CHAILLET Nicolas
Flexible Robotics: Applications to Multiscale Manipulations

GRUNN Emmanuel, PHAM Anh Tuan
Modeling of Complex Systems: Application to Aeronautical Dynamics

HABIB Maki K., DAVIM J. Paulo
Interdisciplinary Mechatronics: Engineering Science and Research Development

HAMMADI Slim, KSOURI Mekki
Multimodal Transport Systems

JARBOUI Bassem, SIARRY Patrick, TEGHEM Jacques
Metaheuristics for Production Scheduling

KIRILLOV Oleg N., PELINOVSKY Dmitry E.
Nonlinear Physical Systems

LE Vu Tuan Hieu, STOICA Cristina, ALAMO Teodoro, CAMACHO Eduardo F., DUMUR Didier
Zonotopes: From Guaranteed State-estimation to Control

MACHADO Carolina, DAVIM J. Paulo
Management and Engineering Innovation

MORANA Joëlle
Sustainable Supply Chain Management

SANDOU Guillaume
Metaheuristic Optimization for the Design of Automatic Control Laws

STOICAN Florin, OLARU Sorin
Set-theoretic Fault Detection in Multisensor Systems

2012

AÏT-KADI Daoud, CHOUINARD Marc, MARCOTTE Suzanne, RIOPEL Diane
Sustainable Reverse Logistics Network: Engineering and Management

BORNE Pierre, POPESCU Dumitru, FILIP Florin G., STEFANOIU Dan
Optimization in Engineering Sciences: Exact Methods

CHADLI Mohammed, BORNE Pierre
Multiple Models Approach in Automation: Takagi-Sugeno Fuzzy Systems

DAVIM J. Paulo
Lasers in Manufacturing

DECLERCK Philippe
Discrete Event Systems in Dioid Algebra and Conventional Algebra

DOUMIATI Moustapha, CHARARA Ali, VICTORINO Alessandro, LECHNER Daniel
Vehicle Dynamics Estimation using Kalman Filtering: Experimental Validation

GUERRERO José A, LOZANO Rogelio
Flight Formation Control

HAMMADI Slim, KSOURI Mekki
Advanced Mobility and Transport Engineering

MAILLARD Pierre
Competitive Quality Strategies

MATTA Nada, VANDENBOOMGAERDE Yves, ARLAT Jean
Supervision and Safety of Complex Systems

POLER Raul *et al.*
Intelligent Non-hierarchical Manufacturing Networks

TROCCAZ Jocelyne
Medical Robotics

YALAOUI Alice, CHEHADE Hicham, YALAOUI Farouk, AMODEO Lionel
Optimization of Logistics

ZELM Martin *et al.*
Enterprise Interoperability –I-EASA12 Proceedings

2011

CANTOT Pascal, LUZEAUX Dominique
Simulation and Modeling of Systems of Systems

DAVIM J. Paulo
Mechatronics

DAVIM J. Paulo
Wood Machining

GROUS Ammar
Applied Metrology for Manufacturing Engineering

KOLSKI Christophe
Human–Computer Interactions in Transport

LUZEAUX Dominique, RUAULT Jean-René, WIPPLER Jean-Luc
Complex Systems and Systems of Systems Engineering

ZELM Martin, *et al.*
Enterprise Interoperability: IWEI2011 Proceedings

2010

BOTTA-GENOULAZ Valérie, CAMPAGNE Jean-Pierre, LLERENA Daniel, PELLEGRIN Claude
Supply Chain Performance / Collaboration, Alignement and Coordination

BOURLÈS Henri, GODFREY K.C. Kwan
Linear Systems

BOURRIÈRES Jean-Paul
Proceedings of CEISIE'09

CHAILLET Nicolas, REGNIER Stéphane
Microrobotics for Micromanipulation

DAVIM J. Paulo
Sustainable Manufacturing

GIORDANO Max, MATHIEU Luc, VILLENEUVE François
Product Life-Cycle Management / Geometric Variations

LOZANO Rogelio
Unmanned Aerial Vehicles / Embedded Control

LUZEAUX Dominique, RUAULT Jean-René
Systems of Systems

VILLENEUVE François, MATHIEU Luc
Geometric Tolerancing of Products

2009

DIAZ Michel
Petri Nets / Fundamental Models, Verification and Applications

OZEL Tugrul, DAVIM J. Paulo
Intelligent Machining

PITRAT Jacques
Artificial Beings

2008

ARTIGUES Christian, DEMASSEY Sophie, NÉRON Emmanuel
Resources–Constrained Project Scheduling

BILLAUT Jean-Charles, MOUKRIM Aziz, SANLAVILLE Eric
Flexibility and Robustness in Scheduling

DOCHAIN Denis
Bioprocess Control

LOPEZ Pierre, ROUBELLAT François
Production Scheduling

THIERRY Caroline, THOMAS André, BEL Gérard
Supply Chain Simulation and Management

2007

DE LARMINAT Philippe
Analysis and Control of Linear Systems

DOMBRE Etienne, KHALIL Wisama
Robot Manipulators

LAMNABHI Françoise *et al.*
Taming Heterogeneity and Complexity of Embedded Control

LIMNIOS Nikolaos
Fault Trees

2006

FRENCH COLLEGE OF METROLOGY
Metrology in Industry

NAJIM Kaddour
Control of Continuous Linear Systems

CPSIA information can be obtained
at www.ICGtesting.com
Printed in the USA
BVHW022110190323
R14754500001B/R147545PG660041BVX00007B/1

9 781786 307316